Bruce Robins
2130 Grand Ave
St. Paul MN 55105
651 690 3424

AF869058

THE

Joseph Bulova

SCHOOL *of* WATCHMAKING

TRAINING MANUAL

BULOVA SCHOOL *of* WATCHMAKING

Bruce A. Robins
2130 Grand Ave.
Saint Paul, MN 55105

JOSEPH BULOVA SCHOOL OF WATCHMAKING

TABLE OF CONTENTS

Thirteenth Edition
2003

ISBN 0-918845-21-1

The AWI-ELM Charitable Trust acknowledges that
The Joseph Bulova School of Watchmaking is the principal author and developer
of *The Joseph Bulova School of Watchmaking Training Manual*.

Published by:
The American Watchmakers-Clockmakers Institute
701 Harrison Avenue
Harrison, Ohio 45030
www.awi-net.org

JOSEPH BULOVA SCHOOL OF WATCHMAKING

FOUNDED IN 1945

The Joseph Bulova School of Watchmaking was founded in 1945 by the late Arde Bulova, Chairman of the Board of the Bulova Watch Company, who wished to repay, in some small measure, the sacrifice and service of returning disabled veterans after the Second World War.

To have known Arde Bulova would be to understand why he established this School and named it in honor of his beloved father and teacher, Joseph Bulova. His father not only taught his son the art of expert craftsmanship but also instilled in Arde Bulova an awareness of the importance of training men for a trade so they might lead purposeful and meaningful lives.

Mr. Arde Bulova was one of the rare men of vision who wanted to combine his love and insistance for perfection in craftsmanship with his desire to aid men in finding a useful vocation.

After the Second World War, Arde Bulova was able to turn his dream into a working reality. He set up a tuition-free school to teach disabled servicemen the skilled art of watchmaking under the most expert supervision and with an all-inclusive curriculum in a pleasant environment where similar interests and problems developed a close-knit, affable group of men working toward common goals.

So, in 1945, without fan-fare or headlines heralding his contribution to society, the cornerstone of the Joseph Bulova School of Watchmaking was laid, stating simply, "To Serve Those Who Served Us."

The School provided a dramatic answer to a real need . . . to train the disabled veterans in a trade for which there was an increasing demand.

To insure gainful employment for these disabled men for the future, the School's operation was based on the principle that from the finest curriculum, teaching methods and most experienced staff would emerge the most skilled craftsmen. This theory has been proven and remains as the basis for continued operation to the present time. Students of the Bulova School enjoy the culmination of years of watchmaking achievements and progress which has been carefully compiled and organized into their comprehensive training program.

By 1950, as the demand for the skilled watchmaker continued to increase, the School began to accept disabled civilians for training. Many were and still are referred by the State Divisions of Vocational Rehabilitation and Rehabilitation Centers.

Striving to be forward thinking in its approach to advances in education, 1958 marked the commencement of a new course of instruction conceived with the intention of meeting the growing demands of industry for precision workers. The new Precision Technician Course provides the student with the basic fundamentals for employment in areas such as instrument makers, instrument repairs, mechanical-electrical assembler, bench machinist and other comparable precision objectives. This course takes an average time of eight months for completion.

The Watchmaking Course, considered one of the finest of its kind in the world, allows an average time of 18 months for completion depending on the individual's progress according to his ability and effort. Upon completion of the course, the graduate is able to repair all kinds of watches including the more complicated types such as automatics, calendars, chronographs, electric and the new Accutron timepiece.

Unique in its training procedure, the School is also outstanding for the other facilities it provides.

These include such conveniences as a medical department on the School premises which consists of an Infirmary Dispensary and an Exercise Room. The department has the services of a Physician, a Physical Therapist, a full time Registered Nurse and a visiting Physical Medicine Consultant. Medical service and commonly used medications are free to the student.

Additionally, the School has a library with a wealth of material on the science of horology; many rare volumes are available for student use. Also included in the library are books for reference and relaxation.

The School has recreation space provided in the main building for music, cards and games.

There are complete kitchen facilities which provide well-balanced and nutritive meals.

A two-story air-conditioned dormitory is available for those wishing to reside at the School. The lower floor of dormitory is especially designed to accommodate wheelchair students. Each room is equipped with adequate closet and chest space and a television set.

A new building, opened in 1965, is a dual-purpose air-conditioned structure adjacent to the main building and existing dormitory. Entering the new addition from the main building via a ramped tunnel, the lower floor of the structure contains a magnificent 50' x 30' heat-controlled swimming pool expertly designed for the disabled with the paraplegic given prime consideration. The pool area is complete with locker room facilities, lavatory and shower space all planned to meet the specific requirements of wheelchair students. The second floor of the building, accessible by elevator or ramp, houses additional classrooms, machine shop, T.V. lounge room and supplementary shower and lavatory facilities.

Adjoining the new building is a gymnasium with a complete basketball court, archery targets, table tennis equipment, snooker table and other sports provisions which implement a full wheelchair sports program. The gym is used for Bulova Watchmaker Team League and demonstration games, basketball practice, archery meets, school assemblies, parties and meetings.

All of the buildings of the School are linked together by a complete tunnel ramp system for optimum use by wheelchair students and other disabled.

One of the most unique services offered by the Bulova School is its Job Placement Bureau. Presently, and for many years past, there are more job opportunities than qualified graduates to fill the positions. Although the student may not be able to work in a locale of his own choosing, he will, if properly completing his training, be placed in a good position if one is desired.

Requirements for admission to the School are based on those factors necessary for success in watchmaking and precision work, namely: Interest, Technical Comprehension, Mechanical Aptitude and Finger Dexterity. Applicants are required to take a battery of tests which can be administered by the Rehabilitation Counselor at the School or by any approved psychological service.

Graduates of the School have come from forty of the fifty states and foreign countries as distant as Israel, Lebanon, Hong Kong, Taiwan and Surinam in Dutch Guiana, plus Italy and Puerto Rico.

All of the facilities coupled with the excellent teaching staff provide depth and quality of training plus comprehensive rehabilitation for all of the students of the Joseph Bulova School of Watchmaking.

Preliminary
Training
Unit

★

MIND TO CONCEIVE

EYE TO PERCEIVE

HANDS TO ACHIEVE

Coordination in Finger Dexterity

PRELIMINARY TRAINING UNIT

The success of almost any pursuit in life depends, to a certain extent, upon a system of coordination between the eye, mind and hands. The ability to coordinate the activities of these three facilities is developed from infancy throughout life.

The successful pursuit of the arts, such as sculpture, music and painting, requires an extremely high degree of coordination between a person's eyes, mind and hands.

It is the high degree to which this coordinative ability must be developed in the watchmaker's trade that makes it so necessary for the student to apply himself to the following Training Unit conscientiously.

This unit is designed to develop finger dexterity and reasonable proficiency in the use of the basic tools of the trade. It is during the performance of this unit that the student forms the correct habits relative to his working position at the bench and its associated equipment.

Practical work . . .	40 hours
Related theory . . .	8 hours

PRELIMINARY TRAINING UNIT

★

Subject:

Finger Dexterity Exercises.

Objectives:

1. To develop finger dexterity.
2. To develop proficiency in the use of an eye glass, tweezer, and screwdriver.
3. To accustom the student to the bench and the proper working position.

Tools Required:

1 eye loupe (A)	1 balance block (D)
1 screwdriver (B)	1 balance screwdriver (E)
1 pr. tweezers (C)	1 holder (F)

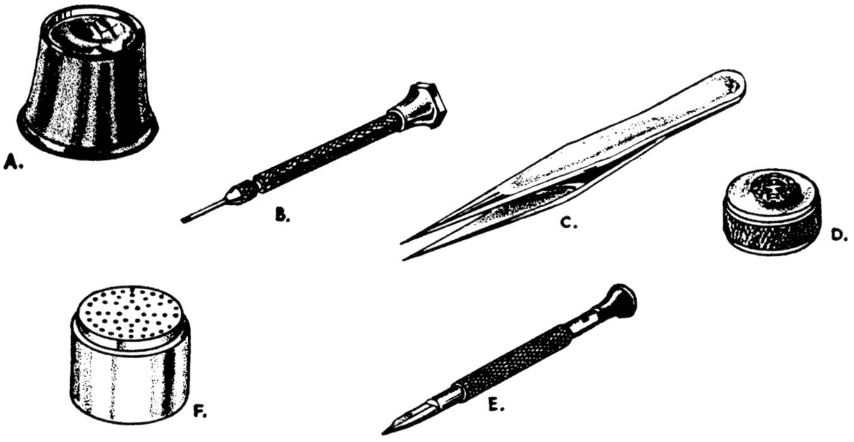

Fig. 1

NOTE: A clean, orderly bench is a prerequisite to neat, accurate work.

Proficient use of tools is of prime importance where good workmanship is desired. There is no substitute for practice in the many operations that are a part of the watchmaking trade. The conscientious application of a student to this training unit will immeasurably improve subsequent work.

OPERATION No. 1

In this operation actual plate screws are used. These screws are to be inserted in pillar plates each containing fifty tapped holes.

Care should be taken in this procedure to assure the formation of proper habits pertaining to the actual use of the tools.

Fig. 2 and 3 illustrate the proper use of tools needed for this operation.

Fig. 4 will serve as a guide in establishing proper habits relative to the working position and the location of the various components which go to make up the watchmaker's working unit.

This entire operation consists of the placement of the screws in the holes with a tweezer and the seating of the screws using a screwdriver of the proper size.

The completed plate will be graded according to:

1. The condition of the finished work, i.e., scratches, burrs, loose screws, etc.

2. The time consumed in the actual performance of the operation.

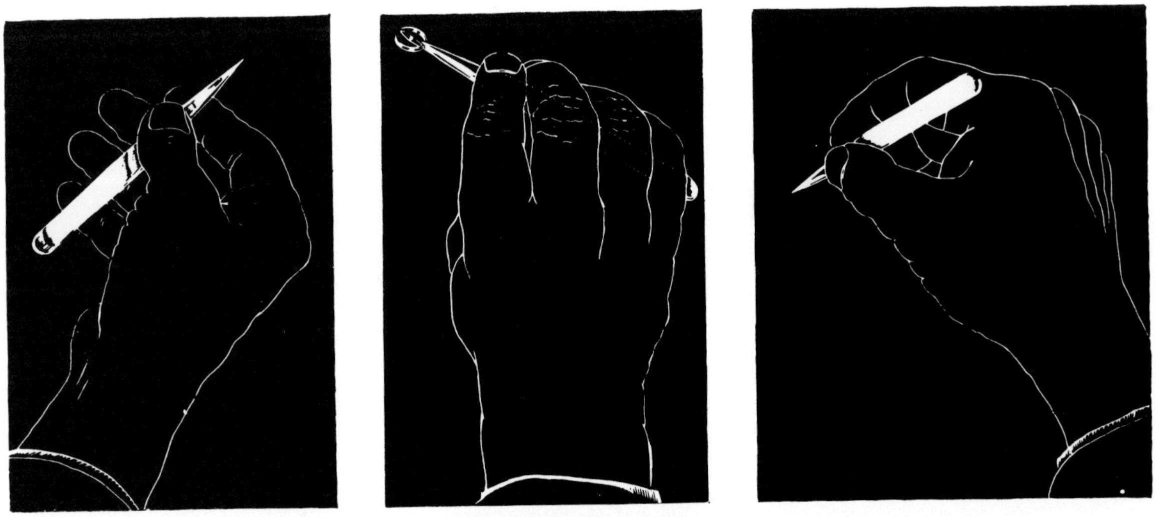

Fig. 2 (Use of tweezers.)

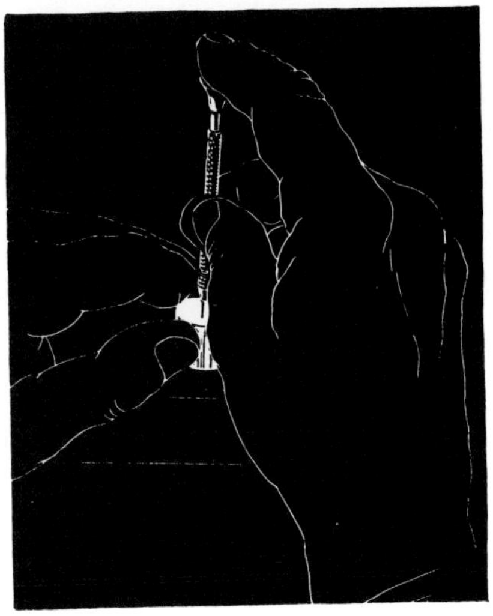

Fig. 3 (Use of Screwdriver.)

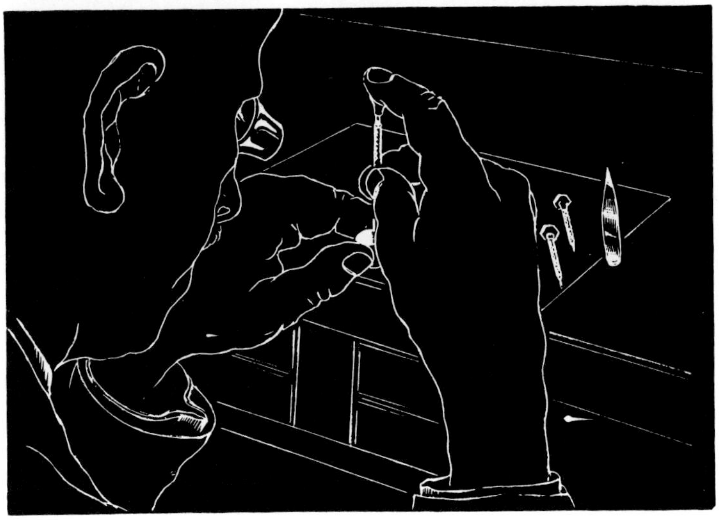

Fig. 4

OPERATION No. 2

This operation is performed in the same manner as Operation No. 1, but a smaller screw is used. These screws are actual pallet bridge screws as used in Bulova watches.

The grading of this operation is identical with that of Operation No. 1.

OPERATION No. 3

This operation is again similar to the preceding ones, but this time a still smaller screw is used. These screws are actual stud screws or jewel screws and will be graded on the same basis as Operations No. 1 and No. 2.

★

OPERATION No. 4

This operation deals with the insertion of balance screws into bi-metallic balance wheels. All balance wheels will fall into one of two general types. The older type, and a type not often used in newer watches is the bi-metallic or two metal type. As the name implies, this balance is usually constructed of two dissimiliar metals, brass and steel, and has an open or cut rim. These constructional details were used to minimize the effects of temperature change on the hairspring.

In Fig. 5 a typical bi-metallic balance wheel is shown, first without balance screws installed and second, with balance screws in their proper places.

The actual installation of the balance screws is an operation requiring some degree of skill on the part of a student. First, the balance screws are put into the slotted block and the block agitated until the screws are laying in the slots head up. Fig. 6A.

Place a balance wheel on the balance block with the balance arm down. Fig. 7A. Select a screw driver with a blade so ground that a slight wedging action will occur when the blade is pressed into the screw slot. Fig. 6B. The small amount of pressure thus created between the screw slot and blade is enough to support the screw without the aid of any additional mechanical device. Fig. 7B. Pick up a screw from the slotted block as shown in Fig. 6A and insert in the desired hole in the balance wheel. Fig. 7A and 7B.

In order to aid the student in the performance of this operation, the blade of the screw driver should be kept in near perfect condition. If, during the procedure the blade is broken, bent or deformed in any way it should be reground to the proper shape on a number 100 carborundum wheel before any attempt is made to use it further.

Throughout this operation care should be taken that the screw heads or the balance wheels are not marred in any way.

Grading of this operation will be based on the same standards as the previous operations in this unit.

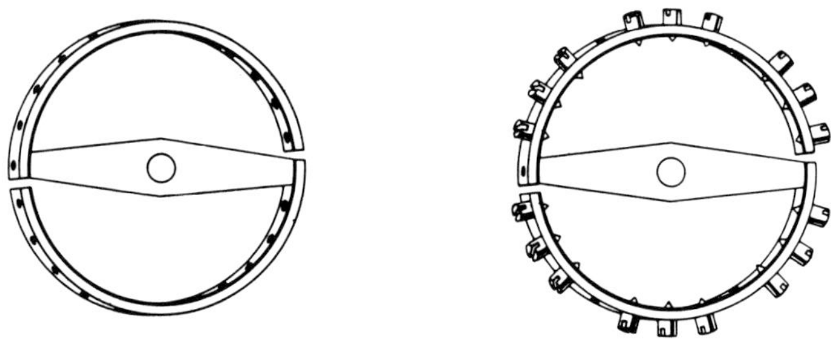

Fig. 5

Bi-metallic Balance Wheel without screws.

Bi-metallic Balance Wheel with balance screws inserted.

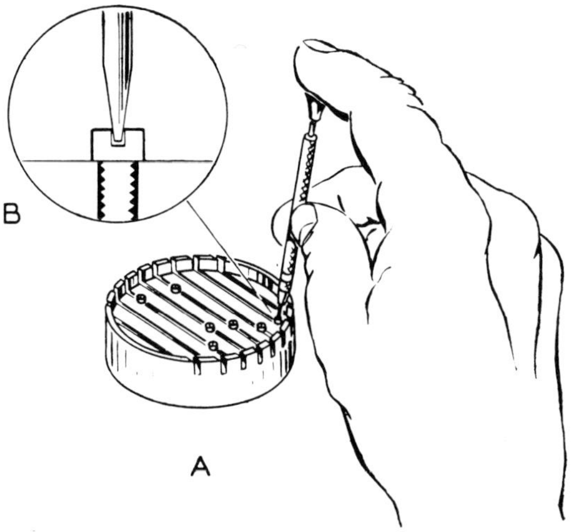

Fig. 6

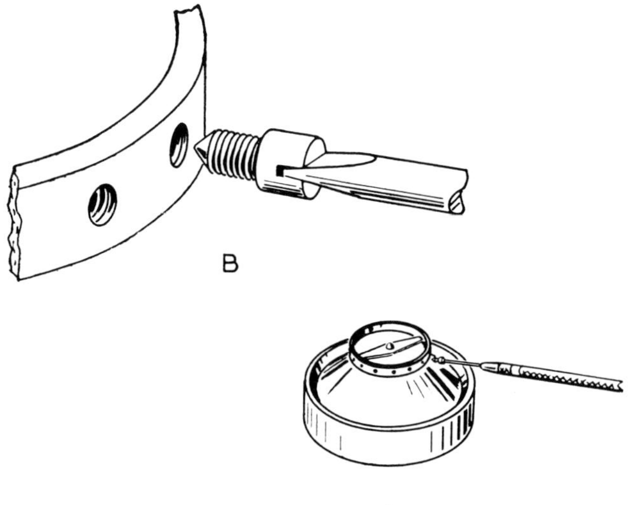

Fig. 7

OPERATION No. 5

This operation is identical with the preceding one, with the exception of the balance wheels used. In this case a Bulova 10-AK, mono-metallic balance wheel is used. This balance is illustrated in Fig. 8 and is shown with and without balance screws installed.

This balance differs from the one previously used in that this rim is solid and constructed of only one metal.

Temperature effects have been minimized through the use of a hairspring whose elasticity remains reasonably constant despite any normal temperature change, there-by eliminating the necessity for the weaker and more unstable construction of the bi-metallic balance wheel.

These balances should be handled in such a manner that warping and distortion is held at an absolute minimum.

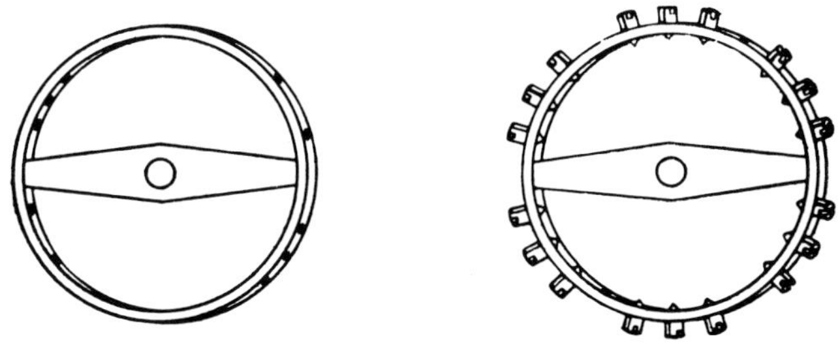

Fig. 8

Mono-metallic Balance Wheel —without balance screws.

Mono-metallic Balance Wheel —with screws inserted.

OPERATION No. 6

This operation deals with the cleaning of the balance wheels used in the preceding operations, preparatory to their use in succeeding training units.

When handling the various parts of a watch, there is always the possibility of these parts coming in contact with the hands of the operator. This will invariably leave a deposit of perspiration on the metal. If this moisture is allowed to dry and remain on the part, actual corrosion will take place with subsequent damage to the piece, thus rendering it useless for further service.

In addition to body moisture there is usually the problem of dust, grease, condensation and vapor reactions to mar the surface and shorten the life of the material.

With this in mind it is evident that all precautions must be taken to assure a clean, trouble-free job, and a clean bench will aid greatly in this respect.

Following is a brief cleaning procedure for watch parts in general and for balance wheels specifically. A detailed discussion of the complete cleaning service will be found in General Repair Information.

Cleaning Procedure

1. Dip Parts in Soap Solution
2. Dip Parts in No. 1 Rinsing Solution
3. Dip Parts in No. 2 Rinsing Solution
4. Dry Parts in Box-wood Sawdust

NOTE: When a cleaning machine is used, step No. 4 is eliminated and the parts are spun dry in the machine.

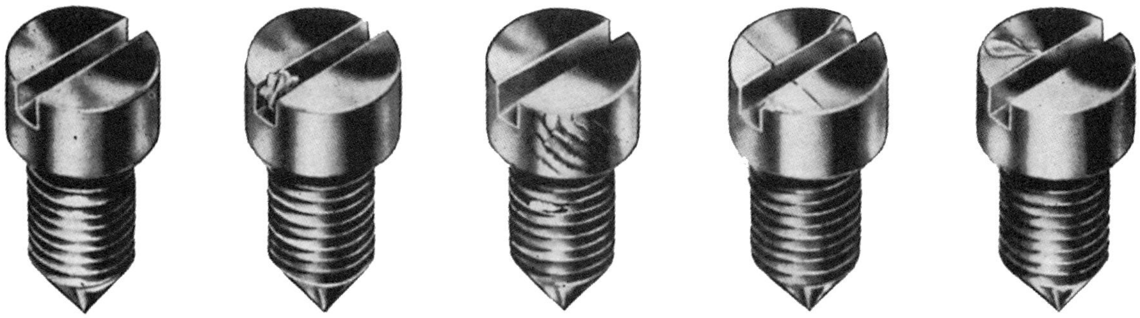

Faultless Screw — ***Saw Dust in Screw Slot*** — ***Finger Marks on Screw Head*** — ***Marred Slot and Scratched Head*** — ***Oil Spot on Screw Head***

Examples of dirt, grease, etc. on watch parts.

Training Unit
Number 1

★

STAKING BALANCE STAFFS

The balance wheel and hairspring unit is the time-keeping element of a watch. The shaft on which the balance wheel is mounted is known as the balance staff. This staff is supported in the plate and bridge by two pivots running in jeweled bearings. The balance wheel is fastened firmly to the balance staff by an operation known as "Staking".

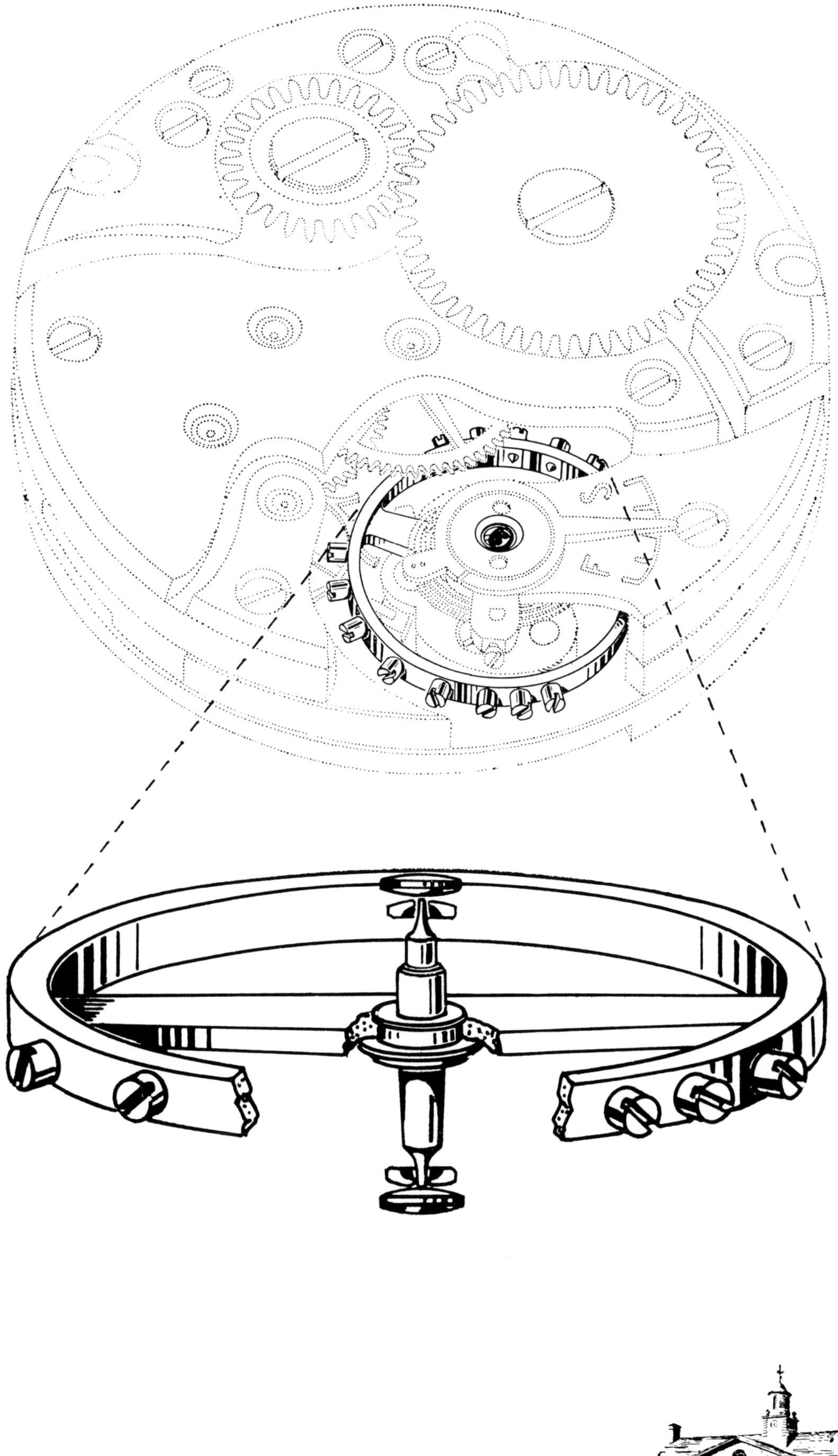

BULOVA SCHOOL *of* WATCHMAKING

TRAINING UNIT NUMBER I.

SUBJECT: Staking Balance Staffs

OBJECTIVES:

1. To develop reasonable skill in staking balance staffs.
2. To familiarize the student with various types of staffs and the staking methods used for each of these types.

Practical work — 20 hours

Related theory — 4 hours

TOOLS NEEDED:

1 eye loupe	3 punches
1 pair tweezers	1 stump
1 staking stand	1 brass hammer

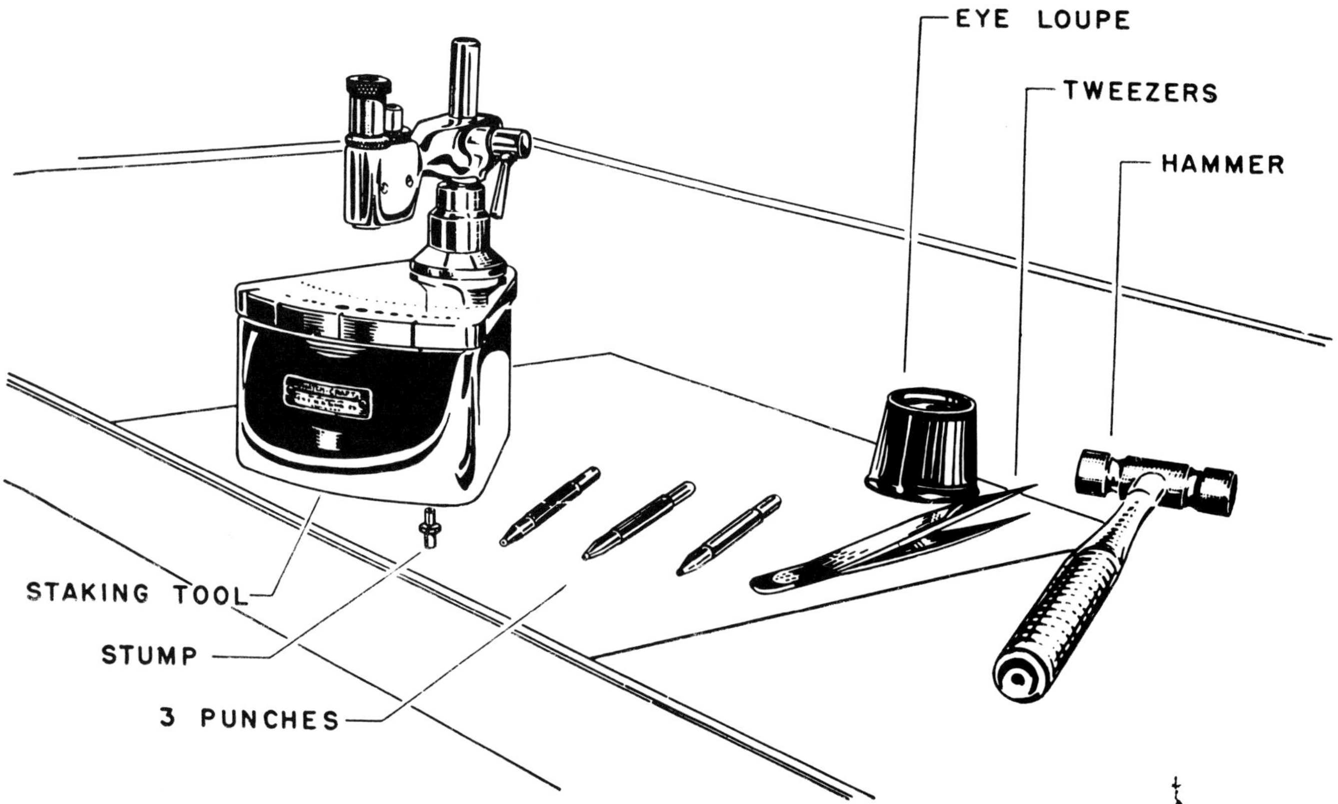

Fig. 1

A balance staff is the shaft on which the balance wheel is mounted. The rotating axis around which the balance wheel oscillates is formed by a line running through the pivots at each end of the staff. These pivots fit into jeweled bearings and are lubricated by minute oil reservoirs.

Fig. 2. Shows a typical balance staff with all parts named.

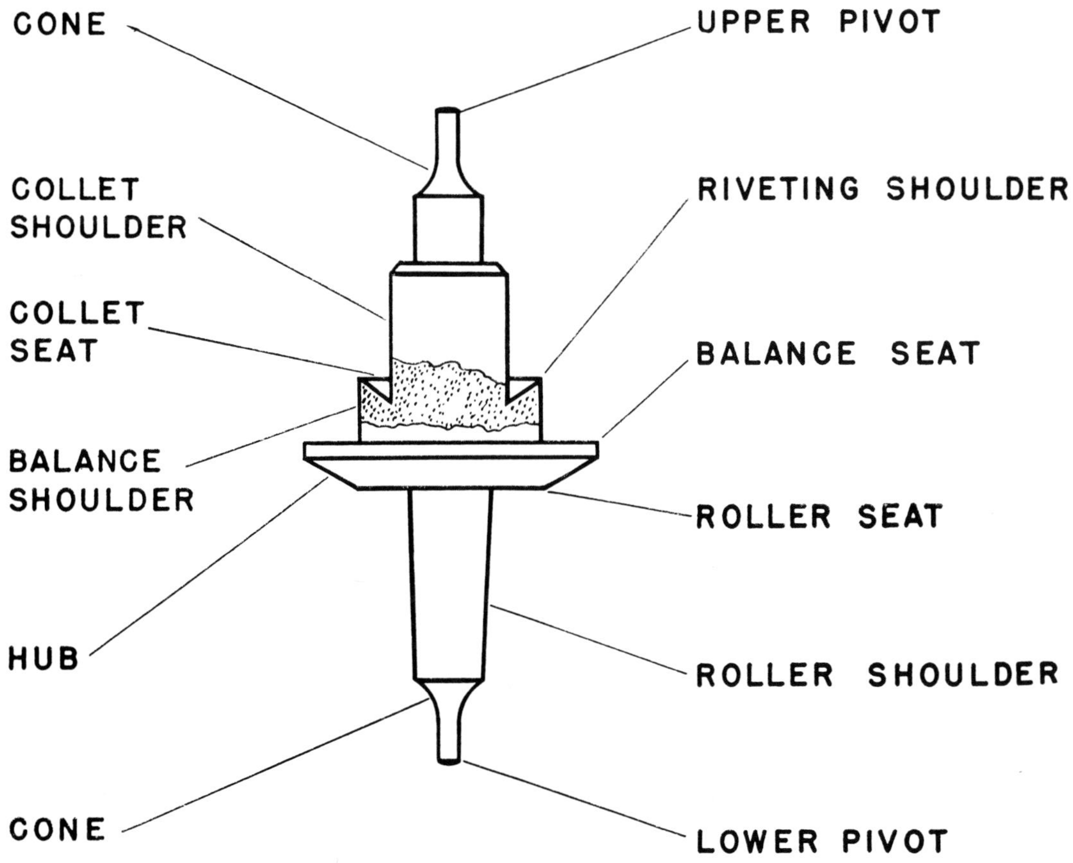

Fig. 2

In this entire course all reference to the staff will be made using the names shown here.

Two general staking methods are used in modern watches:

1. RIVETED STAFFS

In Fig. 3 several common forms of riveted staffs are shown.

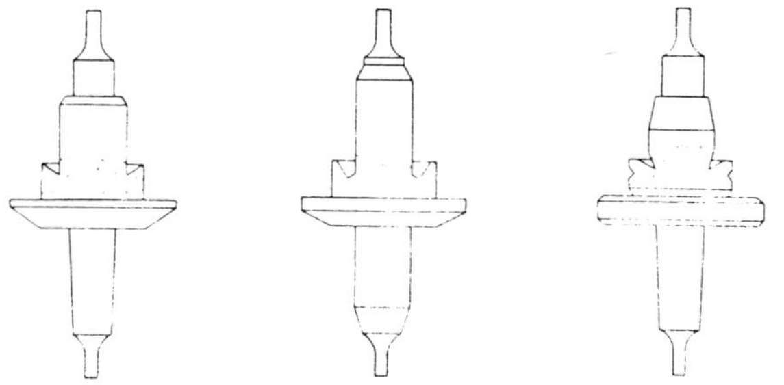

STANDARD RIVET TYPE — TOP GROOVE TYPE — SIDE GROOVE TYPE

Fig. 3

A. STANDARD RIVET TYPE

When staking a staff of this type the balance arm is brought firmly in contact with the balance seat with a staking punch. Fig. 4A.
The riveting shoulder is now spread with a spreading punch to facilitate the final riveting. Fig. 4B. The riveting shoulder is now headed over with a flat punch, Fig. 4C, completing the staking operation. A detailed job breakdown of this procedure is included at the end of the Training Unit.

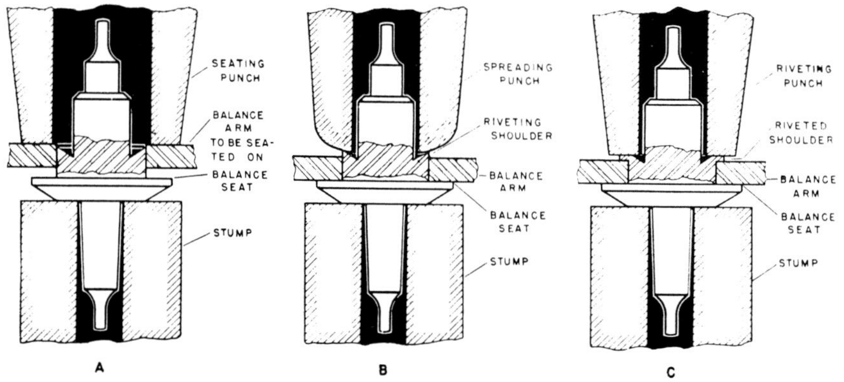

A — USE OF SEATING PUNCH
B — USE OF SPREADING PUNCH
C — USE OF RIVETING PUNCH

Fig. 4

B. TOP GROOVE TYPE

In Fig. 5 a Hamilton top grooved staff is shown. In staking this staff the spreading punch must *NOT* be used. A flat punch is used first to bring the balance arm firmly in contact with the balance seat. Fig. 5A. The final riveting is now accomplished with another flat punch as shown in Fig. 5B and C.

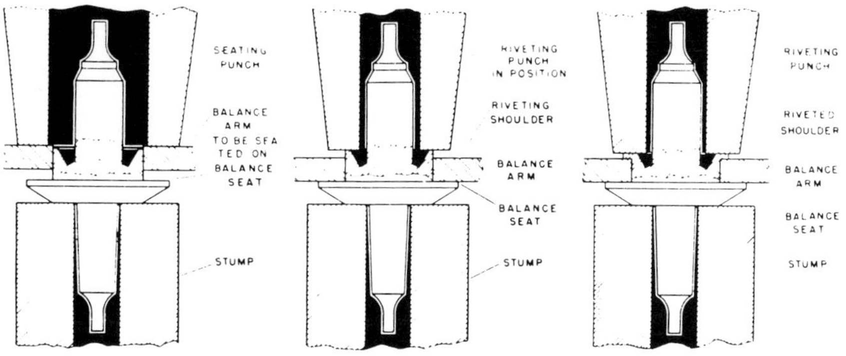

A — USE OF SEATING PUNCH | B — USE OF RIVETING PUNCH | C — RESULT OF RIVETING

Fig. 5

C. SIDE GROOVE TYPE

The Elgin side grooved staff shown in Fig. 6A is staked in exactly the same manner as the Hamilton top grooved staff (see 6B and 6C). The key point to observe here is that the spreading punch is *NOT* used.

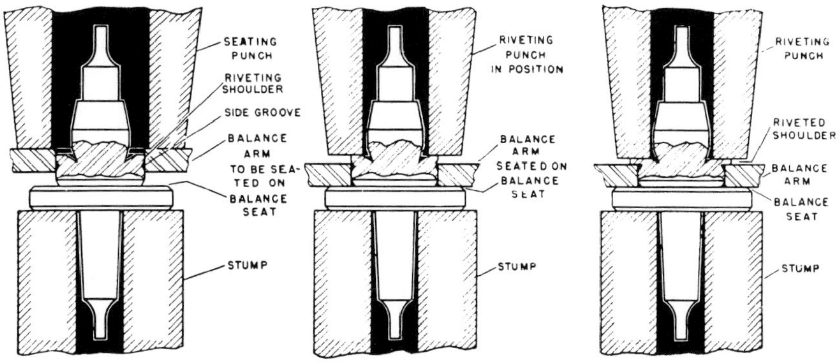

A — USE OF SEATING PUNCH | B — USE OF RIVETING PUNCH | C — RESULT OF RIVETING

Fig. 6

2. FRICTION FITTED STAFFS

Fig. 7A shows a Waltham friction fitted staff. The staking process is clearly indicated. A round nosed staking punch is brought in contact with the roller seat. This insures that the hub seat can be brought into positive seating position in the blue hub.

The Hamilton friction fitted staff is shown in Fig. 7B. The balance wheel with the hub attached is placed on the stump and the staff is put into position for staking. A flat faced punch is used to press the staff into its seated position with the hub seat on the staff in positive contact with the balance wheel hub.

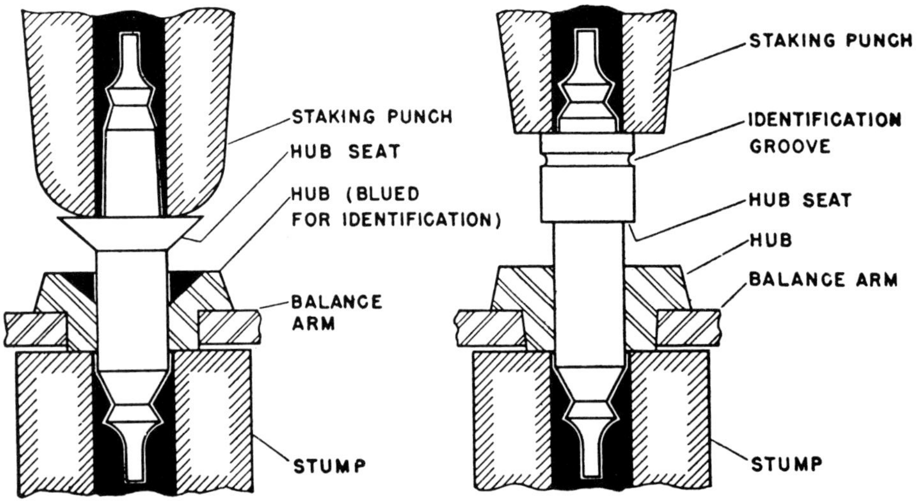

A B

Fig. 7

SUMMARY *of the* STAKING PROCEDURE

Step 1. Select the proper stump and necessary punches as follows:

a. The diameter of the stump hole (inverted flat nose punch) must be large enough to accommodate the roller shoulder and small enough to insure a firm support for the staff during the staking operation. (See Fig. 8B).

b. The *seating punch* (flat nose) must have a hole diameter very slightly larger than the balance wheel shoulder so it will clear this shoulder and rest only on the balance arm when the wheel is seated.

c. The round nosed *spreading punch* must have a hole diameter very slightly larger than the collet shoulder and small enough to contact the riveting shoulder well in from the edge.

d. The flat nosed *riveting punch* must have a hole the same diameter as the spreading punch.

NOTE: An inverted punch may be used as a stump in this operation.

Step 2. Place staking stand on the bench.

Step 3. Insert stump in the staking stand. Fig. 8A.

Step 4. Place balance staff in the stump with the balance seat up. Fig.8B.

Step 5. Place balance wheel on staff with the balance arm down. Fig. 8C.

Step 6. Insert Seating Punch in staking stand and lower it to the balance arm. Fig. 8D.

Step 7. Force the Seating Punch downward with the left hand until the balance arm is in firm contact with the balance seat.

Step 8. Remove the Seating Punch and insert the spreading punch in its place.

Step 9. Lower the spreading punch until it contacts the riveting shoulder and hold it firmly in place with the left hand. Fig. 8E.

Step 10. Spread the riveting shoulder by tapping the punch several times with the brass hammer. Rotate the balance in the stump with the little finger of the left hand while using the hammer.

Step 11. Remove the spreading punch and insert the Riveting Punch in its place.

Step 12. Lower Riveting Punch to the riveting shoulder and hold it firmly in place with the left hand.

Step 13. Head over the rivet by tapping the riveting punch several times with a brass hammer. Rotate the balance in the stump as in Step 10 while hammer is being used. Fig. 8F.

Step 14. Remove the Riveting Punch from the staking stand.

Step 15. Remove the staked balance from the stump.

Stake the remaining balances using the above precedures.

on the STANDARD RIVET TYPE STAFF

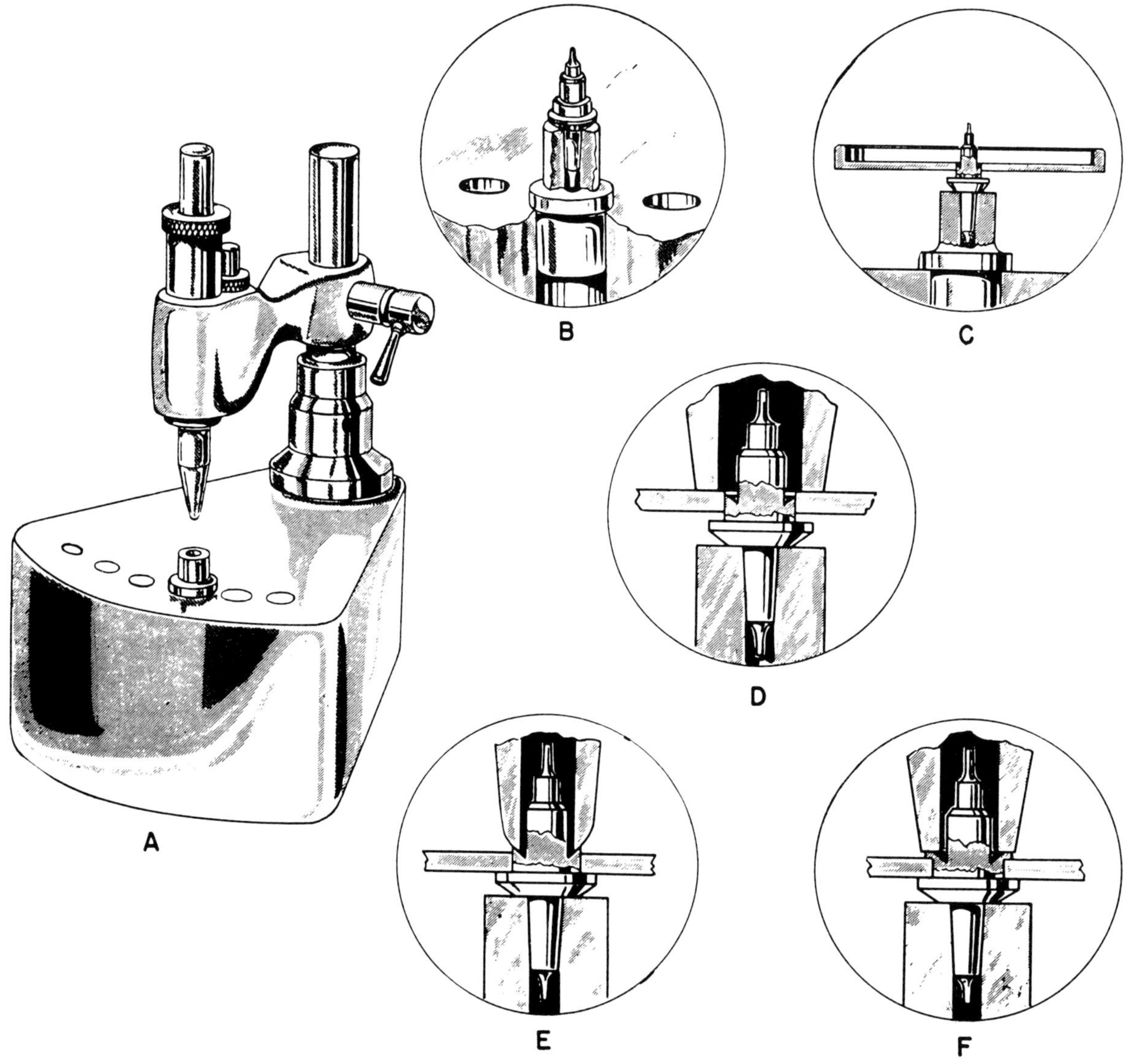

Fig. 8

SHOCKPROOF STAFFS

Fig. 9 shows a Bulova 11 AF shockproof staff which is staked exactly like the standard type staff. The only difference between the shockproof and the standard staffs is in the roller shoulder. The shockproof staff is reduced in diameter at the lower end of the roller shoulder to permit it to enter the shock jewel assembly.

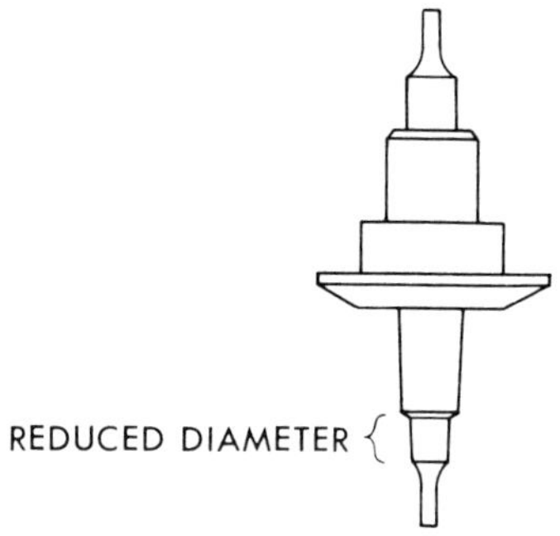

Fig. 9

When ordering a balance staff, be sure to specify if shockproof.

THE INCABLOC ROLLER

Figure 10 compares the Incabloc Roller with the standard type of roller.

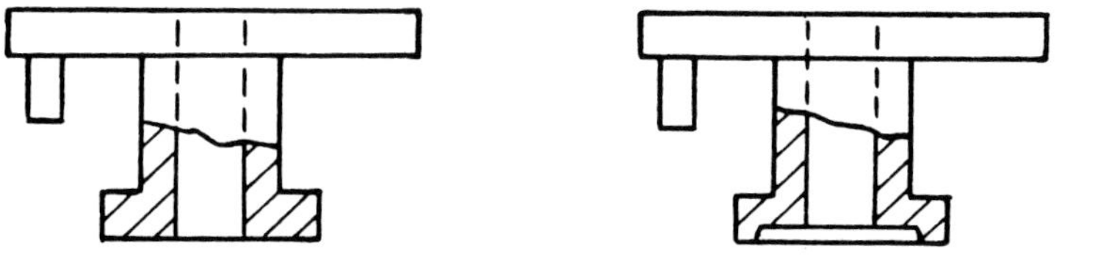

Fig. 10

METHODS OF REPLACING INCABLOC ROLLER

1. Support the roller in the V slot of the bench anvil and lightly tap the balance wheel and staff on to the roller.
2. Support the balance wheel and staff on a stump in the staking tool so that the staff is resting on its riveted shoulder. Now use a four pronged punch to lightly tap the roller (safety roller side up) onto the staff. Be sure the prongs provide safe clearance for the roller jewel.
3. Using the same procedure as in Method 2, select an Incabloc Punch in place of the four pronged punch to lightly tap the roller onto the staff.

THREE ARM BALANCE WHEEL

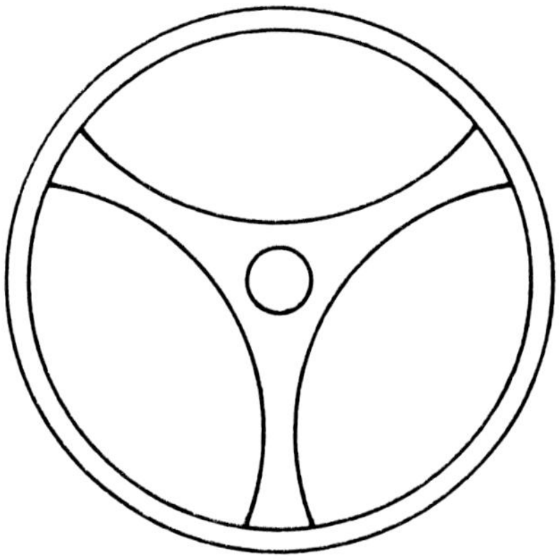

Fig. 11

When ordering staff for a three arm balance wheel, always specify that it is for a three arm balance wheel (See Fig. 11). Usually the standard type of roller remover is not suitable for removing the roller from a three arm balance wheel and staff. In such cases, the roller can be twisted off by mounting the balance wheel and staff in the lathe. Frequently, a suitable pin vise can also be used to twist off the roller as the wheel is held between the fingers.

Training Unit
Number 2

★

TRUING BALANCE WHEELS

The balance staff and balance wheel form an integral part of the timekeeping unit of the watch.

The function of the staff is to support the balance wheel and the function of the balance wheel is to serve as an accurately oscillating member of the timekeeping unit under the control of the hairspring.

If this function of the balance wheel is to be fulfilled it is necessary that it be mounted on the staff in such a manner that:

1. The entire rim lies in one plane.
2. The plane in which the rim lies must be perpendicular to the staff axis.
3. The entire rim must be concentric with the staff axis.

The operation designed to establish this condition is known as BALANCE TRUING.

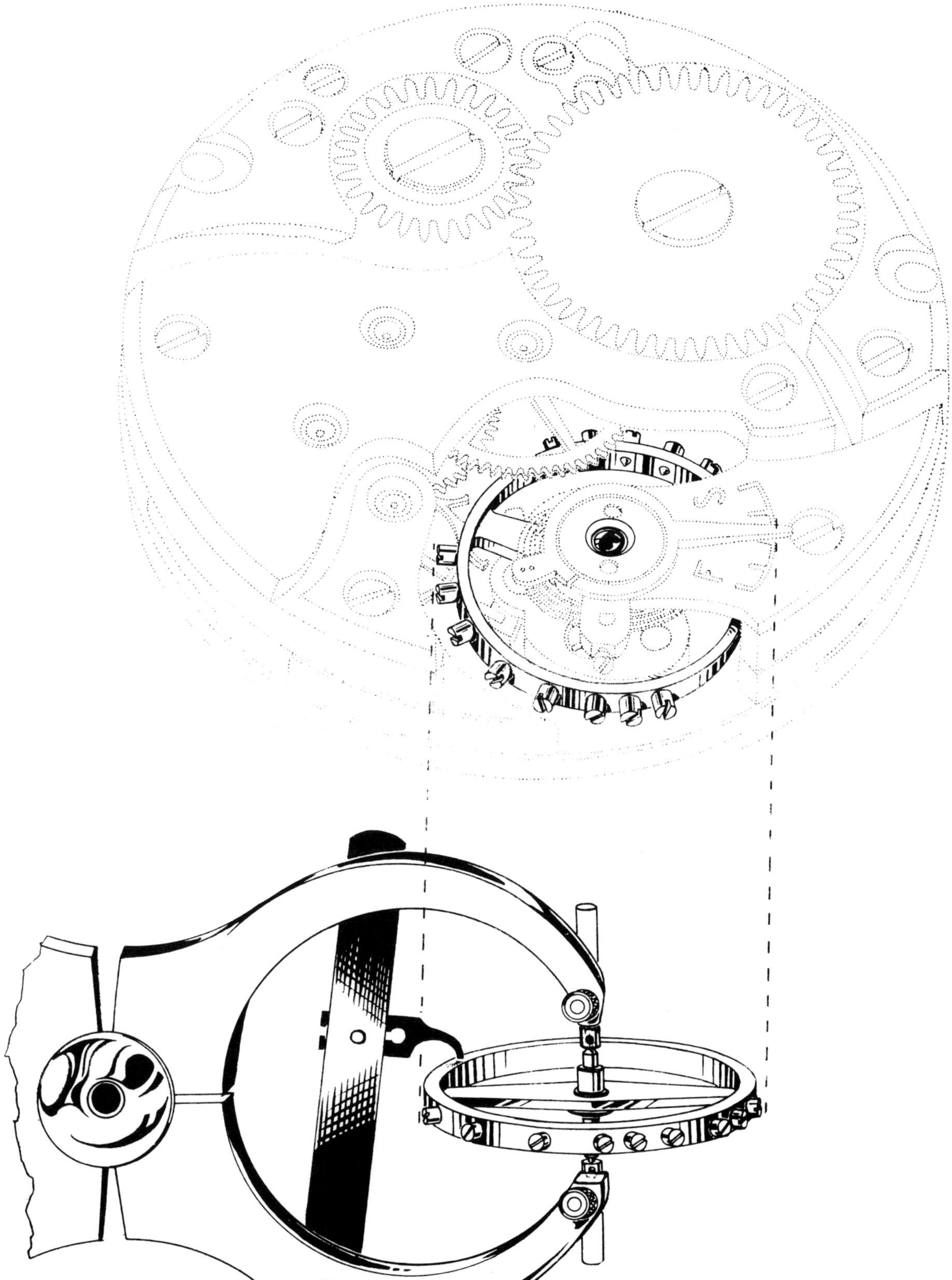

BULOVA SCHOOL *of* WATCHMAKING

TRAINING UNIT NUMBER 2

SUBJECT: Balance Truing

OBJECTIVES:

1. To enable the student to recognize the various conditions that may exist in untrued balances, and to determine the proper methods to be used in correcting these conditions.
2. To develop a reasonable degree of skill in the manipulative procedures involved in Balance Truing.

   Practical work — 40 hours
   Related theory — 8 hours

TOOLS REQUIRED:
Balance Truing Calipers
Balance Truing Wrench
Eye Loupe
Pair of Tweezers

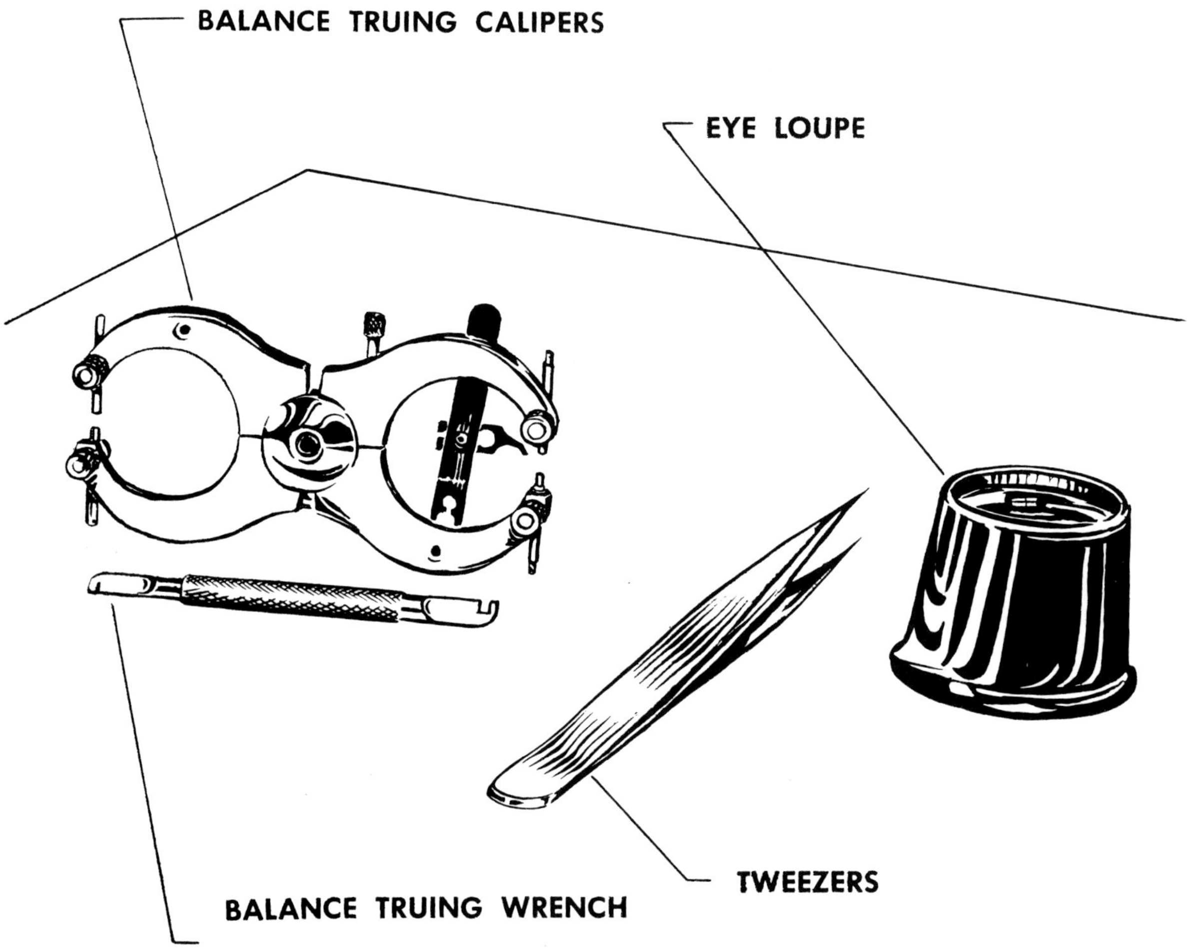

Fig. 1

PRACTICAL MANIPULATIVE OPERATIONS FOR BALANCE WHEEL TRUING

I. USE OF BALANCE TRUING CALIPERS

Throughout the truing operation the balance is held in the balance truing caliper. These calipers are so constructed that the balance is supported on the conical portion of the pivots. Fig. 2. Care must be taken that the pivot holes in the stumps are kept clean to prevent damaged or broken pivots.

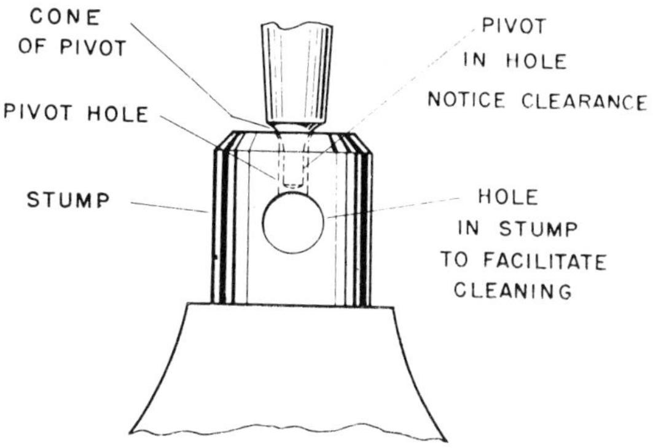

Fig. 2

HINGE TYPE CALIPERS

A further consideration when mounting the balance in the hinge type calipers is the alignment of the stumps with the staff.

In order to achieve this condition the stumps themselves must first be aligned. This alignment is present when, and only when the bows of the calipers are completely closed. Therefore, when mounting a balance for truing in these calipers, make certain that the bows are closed, then, by using the individual stump adjusting nuts, set the stumps so that the distance between them will be *slightly* less than the distance from cone to cone on the staff.

With these adjustments properly set the balance may be inserted and trued with no danger of damage to the pivots and with full confidence that indicated errors are due to the condition of the balance wheel and not to faulty mounting.

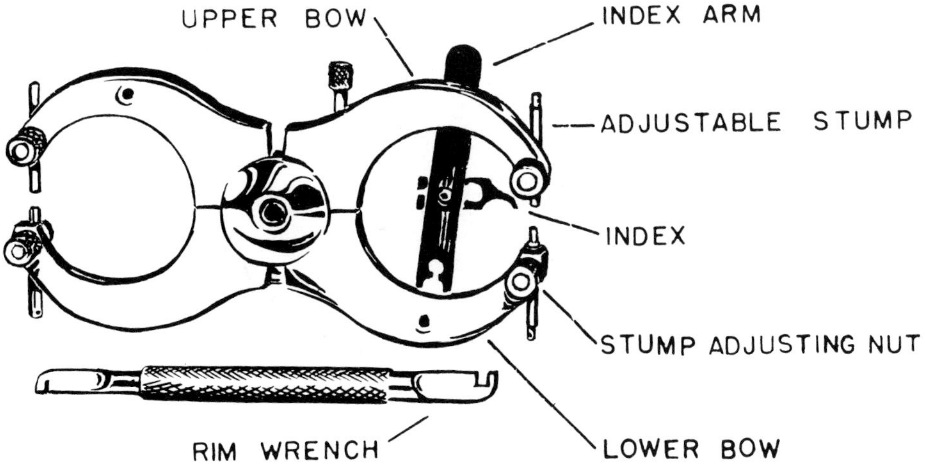

Fig. 3

SCREW-TYPE CALIPER

The screw type calipers are equipped with an adjustable index mounted on a sliding bar fastened to the frame. Fig. 4. The calipers are held in the left hand in a position similar to that of the hinge type. A thumb screw mounted between the two halves of the frame is used to open and close the calipers. This screw is turned by the thumb of the left hand, leaving the right hand free to hold and guide the balance.

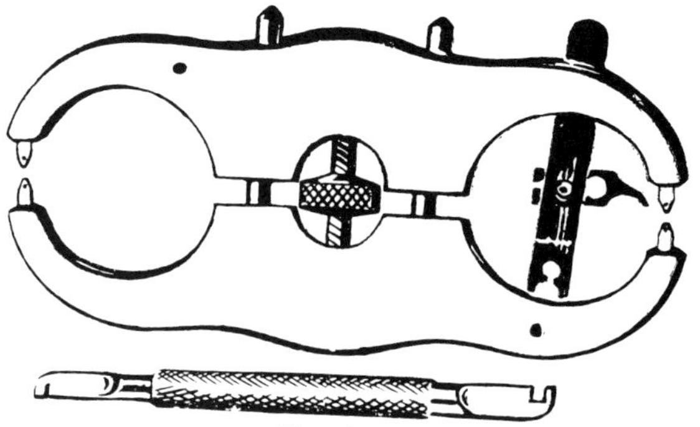

Fig. 4

★ ★ ★ ★ ★

2. HOW TO GRASP AND HOLD CALIPERS

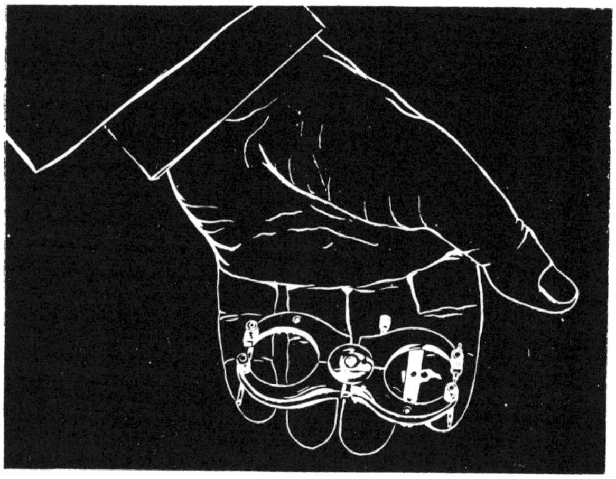

Fig. 5

Lay the calipers across the four fingers of the left hand as shown in Fig. 5.

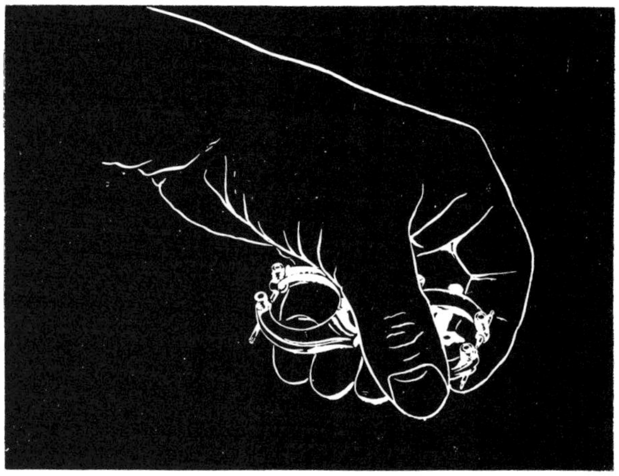

Fig. 6

Bring the thumb down over the calipers so as to rest on the lower bow. Fig. 6.

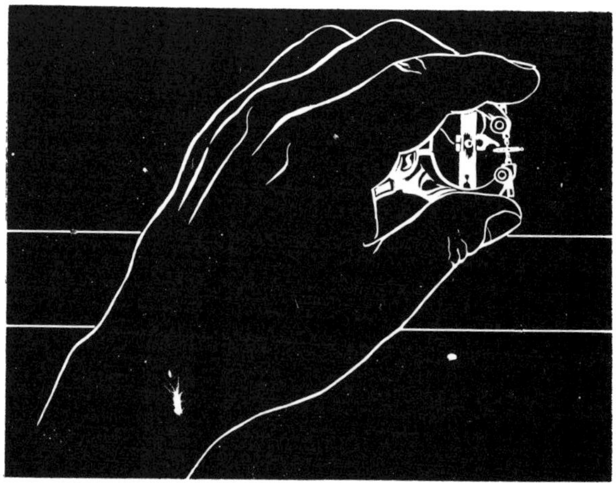

Fig. 7

Lift the caliper to the correct position at the bench. Fig. 7.

In the proper working position the forefinger is used to adjust the index arm and also provides leverage for the thumb when opening and closing the jaws. With practice, the moving of the index arm as well as the adjustment of the jaws can be performed with one hand.

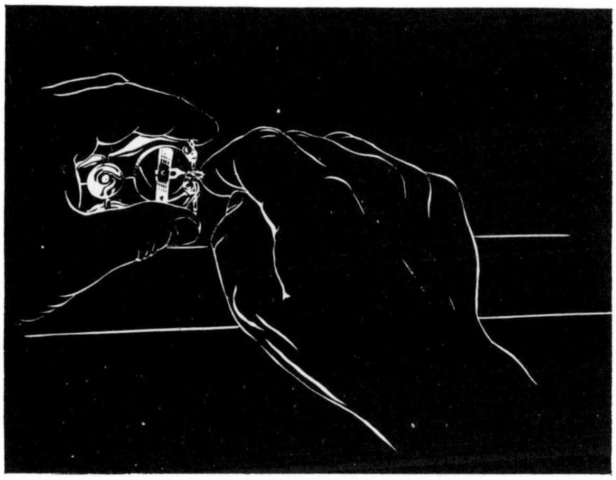

Fig. 8

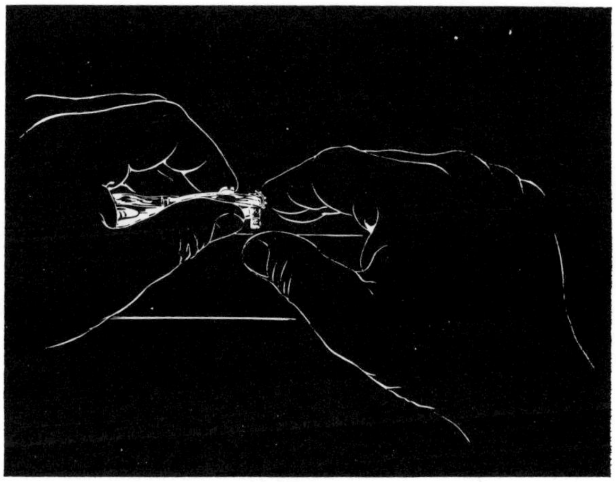

Fig. 9

Fig. 8 shows the proper position of the calipers for truing in the flat.

When truing a wheel in the round the calipers are tipped forward as shown in Fig. 9.

★ ★ ★ ★ ★

SPECIFICATIONS FOR A TRUE BALANCE WHEEL

Balance Truing is an important part of the watchmaker's trade. The operation involves the bending of the balance wheel rim until it conforms to the following requirements of a true balance.

1. All parts of the rim must lie in the same plane. Fig. 10A.
2. The plane in which the rim lies must be perpendicular to the staff axis. Fig. 10A.
3. The entire rim must be concentric with the staff axis. Fig. 10B.

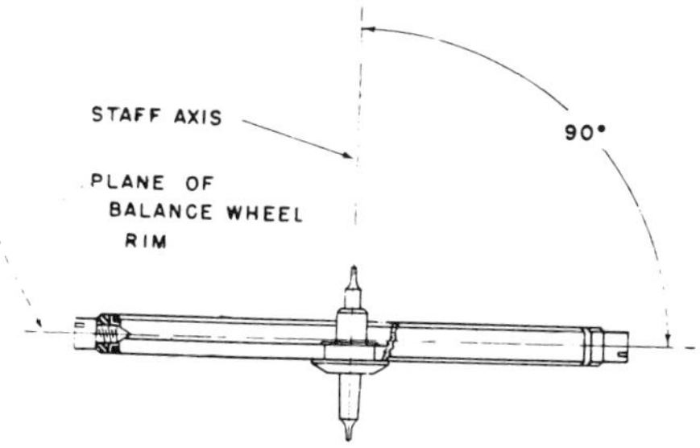

Fig. 10A

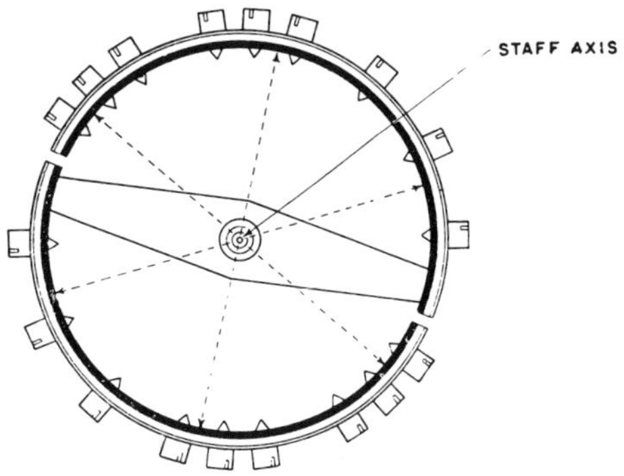

Fig. 10B

II. TRUING BI-METALLIC BALANCE WHEELS

A. TRUING IN THE FLAT

The operation known as truing in the Flat consists of bending the balance wheel rim until it conforms to the first two requirements of a true balance as illustrated in Fig. 10A, namely

1. The entire rim must lay in a single plane.
2. The plane in which the rim lies must be perpendicular to the staff axis.

To perform this operation, place the balance wheel in the calipers in such a manner that the conical portion of the pivots support the balance staff. See Fig. 2.

Adjust the index to a position just above the balance arm and as close to the end of the arm as possible. Fix the space between the index and arm accurately in mind and turn the balance wheel until the other arm is directly under the index. The space between the index and either arm must be exactly the same. If any variation is noted the arms must be adjusted until they are equal in height. This adjustment is accomplished by grasping the rim with the thumb and forefinger and gently forcing the arm up, Fig. 11A, or down, Fig. 11B as required.

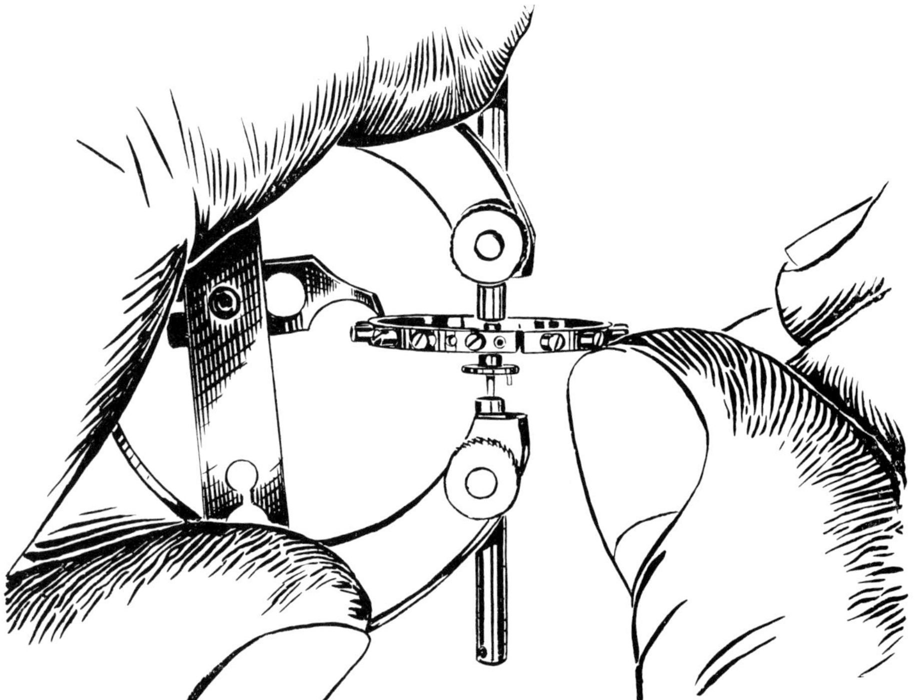

Fig. 11A

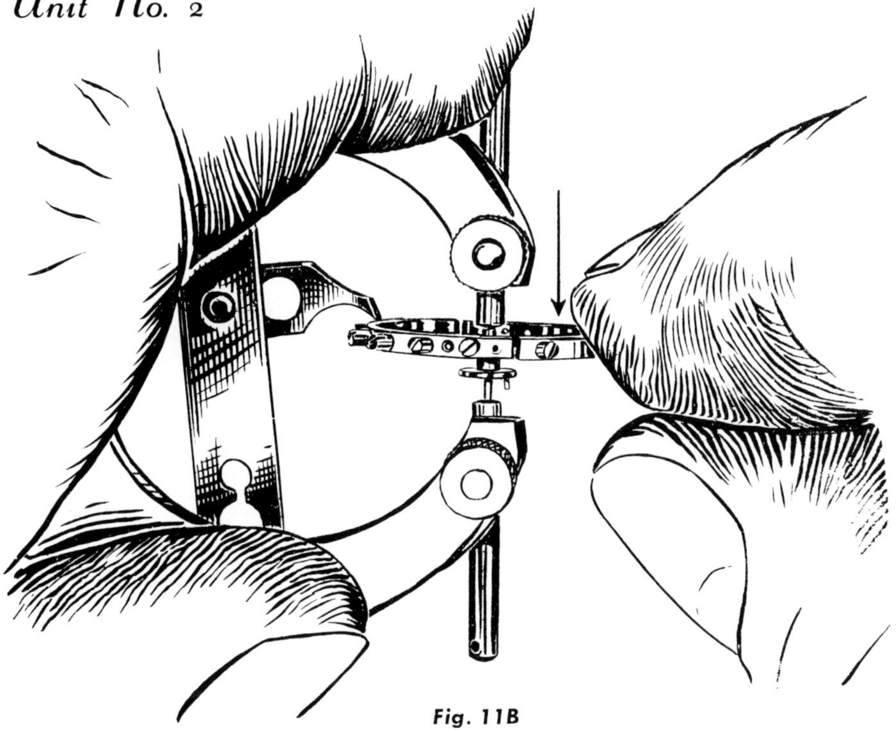

Fig. 11B

Readjust the index to a position directly over the rim at the end of an arm so that a narrow slit of light is visible between the index and rim when the calipers are held in the proper truing position. Fig. 12. Rotate the balance with the side of the right forefinger noting any variation in the width of the slit of light. Since the index serves as a fixed reference point, any increase in the light space indicates a downward bend in the rim. Likewise, any decrease in the light indicates an upward bend.

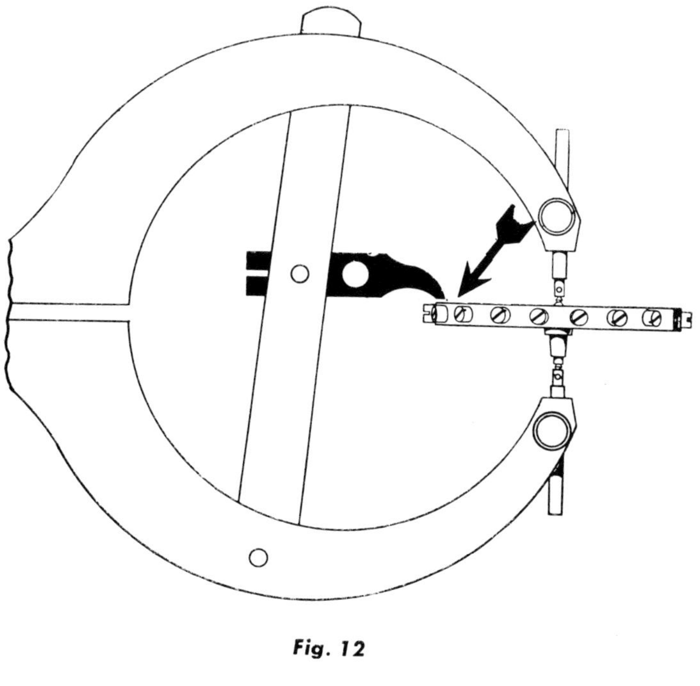

Fig. 12

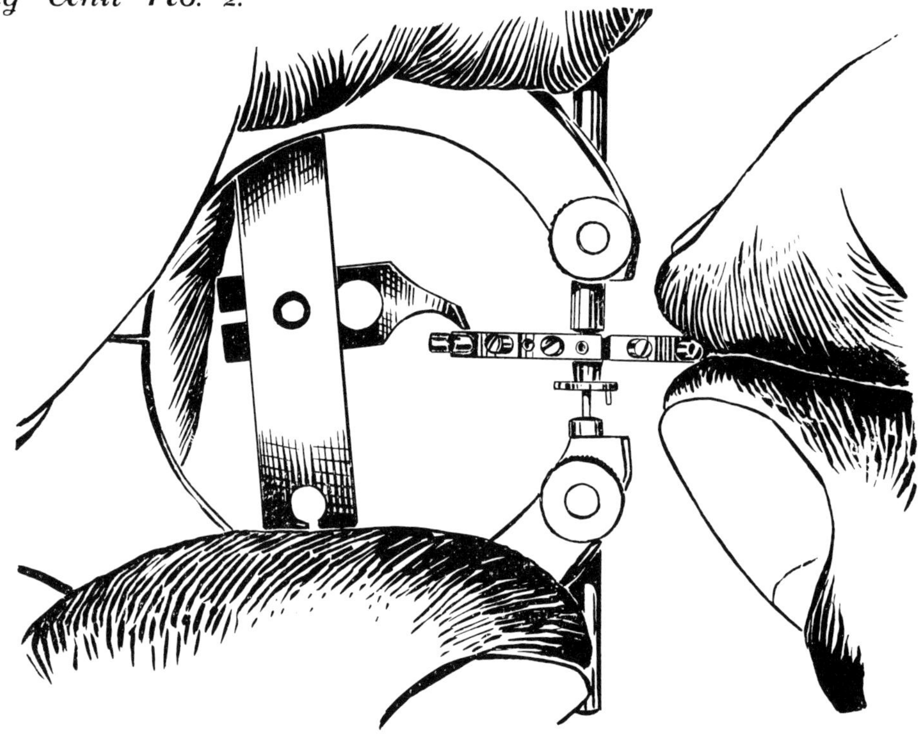

Fig. 13

If a gradual bend is encountered in the rim it may be eliminated by pressing the rim between the soft parts of the thumb and forefinger as shown in Fig. 13.

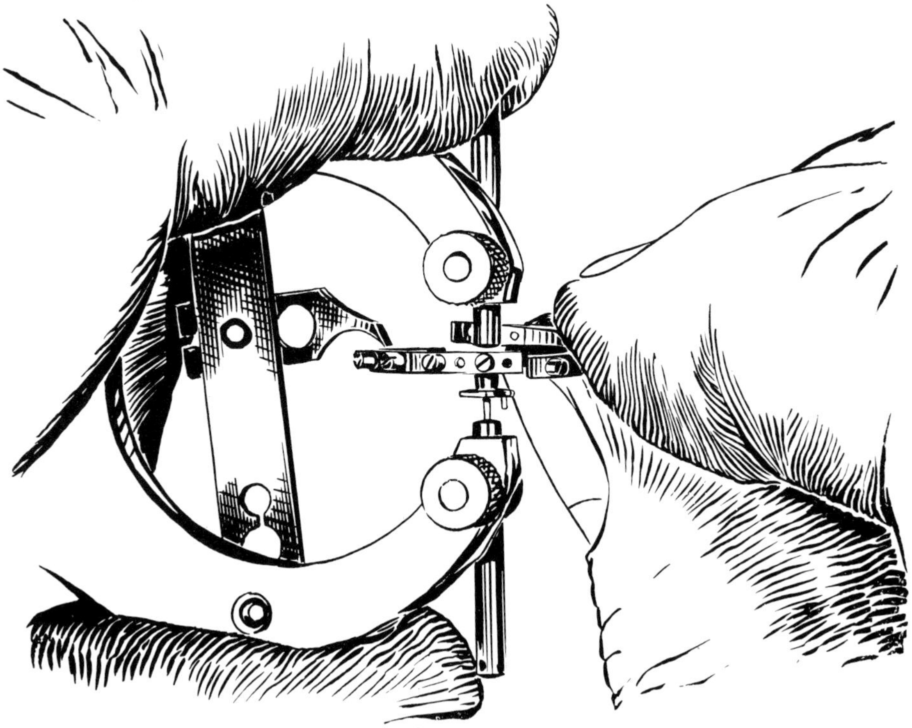

Fig. 14

If the bend is abrupt more pressure must be applied within a limited area of the rim in order to eliminate it. This is best accomplished by using the very ends of the fingers instead of the soft parts to exert pressure on the rim. Fig. 14.

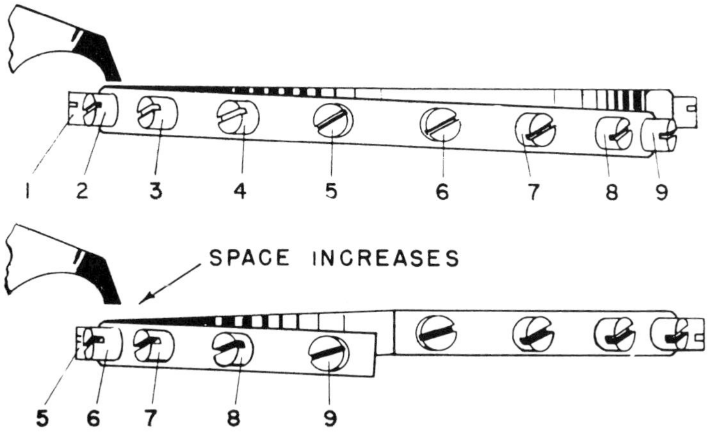

Fig. 15A

In Fig. 15A a downward bend from the arm to the cut is indicated, and in Fig. 15B a similar bend in the opposite direction is shown.

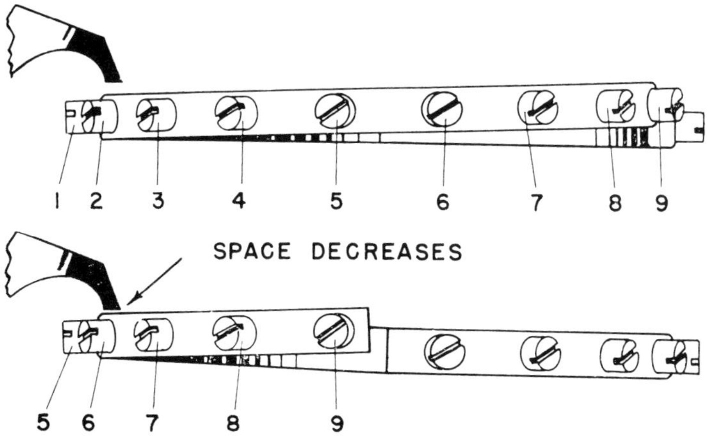

Fig. 15B

If a rim is found to be bent as in Fig. 16A, first straighten the bend nearest the arm, Fig. 16B, using the soft part of the fingers. The remaining bend is straightened by rolling the forefinger over the thumb from the bend to the cut, thus eliminating all bends. Fig. 16C.

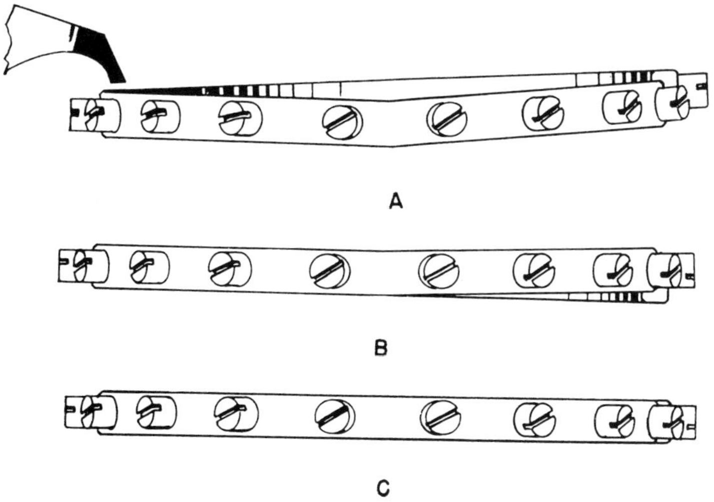

Fig. 16

When a more pronounced bend than that shown in Fig. 16 is encountered it will be necessary to develop the three fingered bend shown in Figs. 17A and 17B. In this bend the first and second fingers are used to support the rim on either side of the bend while the thumb is used to exert the correcting pressure in the opposite direction.

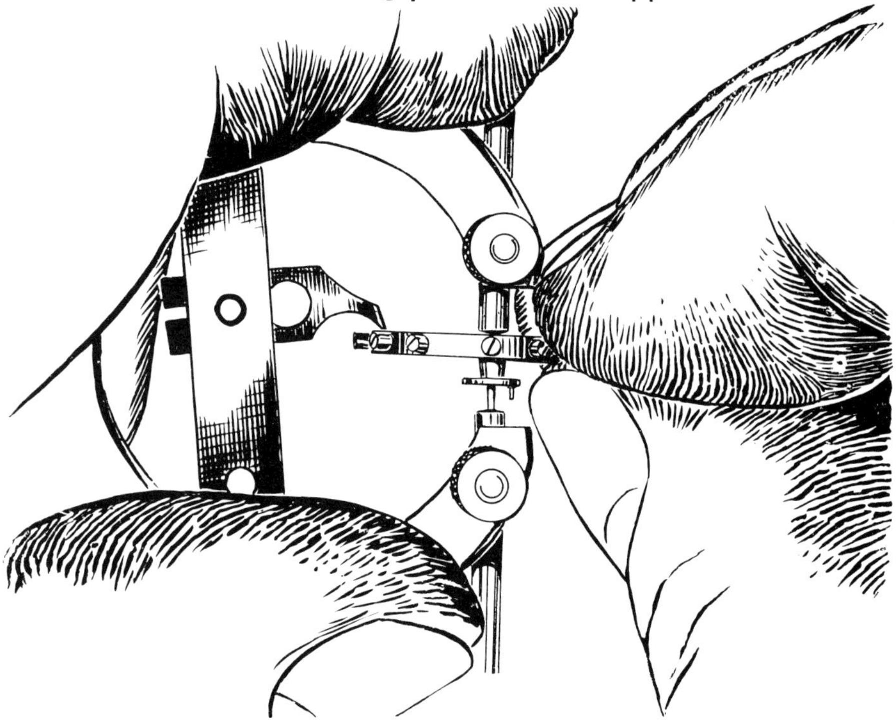

Fig. 17A

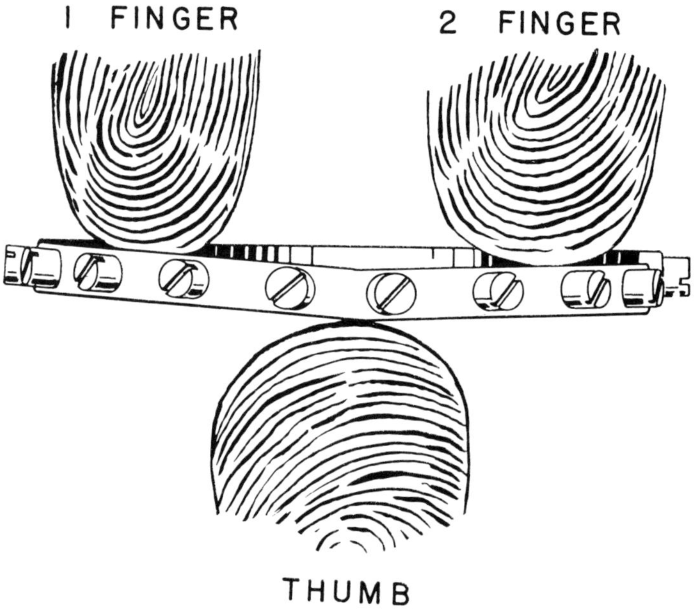

Fig. 17B

There are an infinite variety of bends, and combinations of bends possible in a balance rim, but the principles outlined here will serve as a guide in their correction. Through practice and experience the watch repairer will develop various finger movements to correct any bends that may be encountered.

B. TRUING IN THE ROUND

The operation known as "truing in the round" consists of bending the rim of a balance wheel until it conforms to the third requirement of a true balance as given at the beginning of this training unit. Namely: The entire rim must be concentric with the staff axis.

To perform this operation, readjust the index to lay just above the outside surface of the rim as shown in Fig. 16.

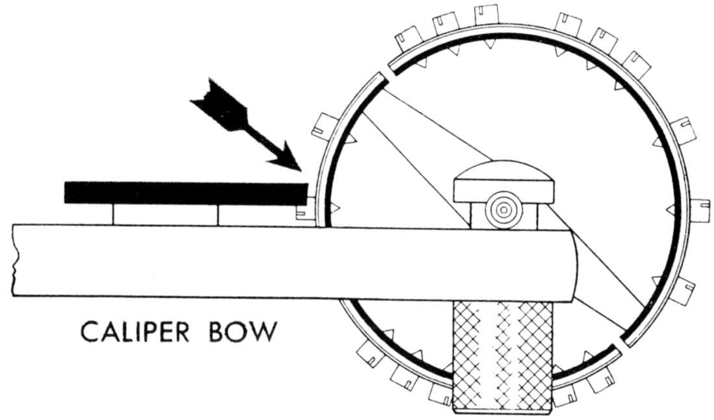

Fig. 18

Slowly rotate the balance, working from an arm to the cut. Any change in the index clearance indicates a bend in the rim. Eliminate all bends as they appear, using a rim wrench to achieve best result.

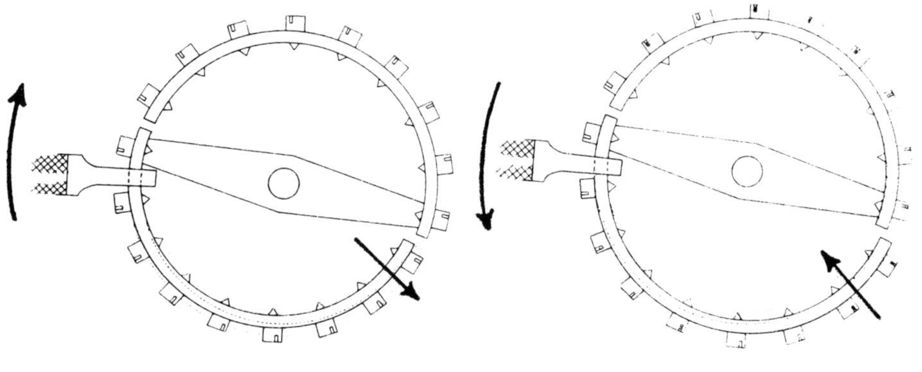

Fig. 19

In Figs. 19A and 19B the rim is bent out of true from the arm to the cut. To correct the bend place the wrench on the rim as close to the arm as possible and bend the rim outward for A and inward for B. The arrow on the wrench indicates the direction of force applied and the arrow at the arm indicates the direction the rim is to be bent.

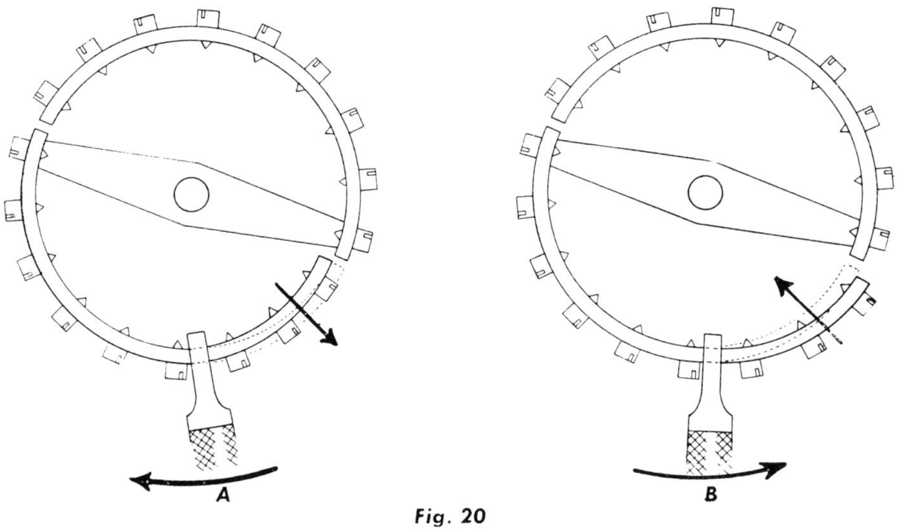

Fig. 20

In Figs. 20A and 20B the rim is bent at a point between the arm and the cut. Note that the wrench is applied to the rim at the point where the bend begins.

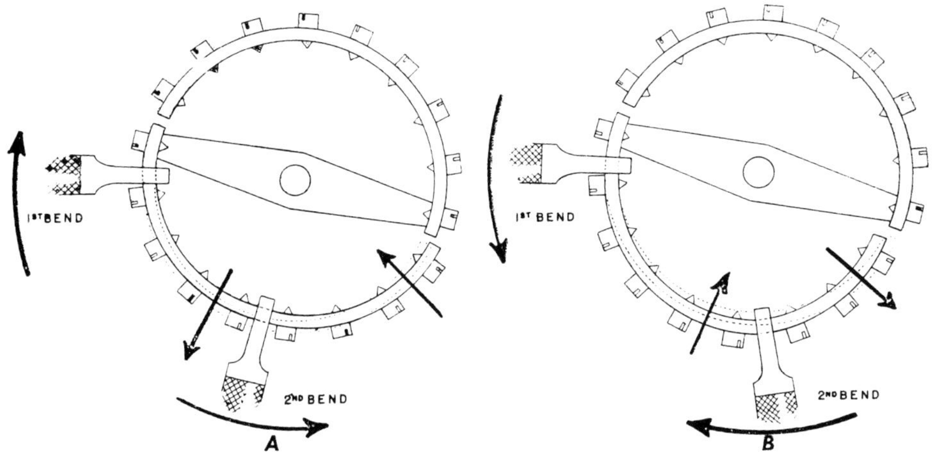

Fig. 21

Examples of multiple bends are shown in Figs. 21A and 21B. Starting from the arm, true the first portion of the rim as shown and then procede to eliminate the second bend as if the first one had not existed.

The rim must not be twisted during this truing operation. Upon completion, the entire rim must be perpendicular to the arms as shown in Fig. 22B. Any tendency of the rim to vary from this condition (see Fig. 22A) must be corrected before the balance wheel can be considered trued.

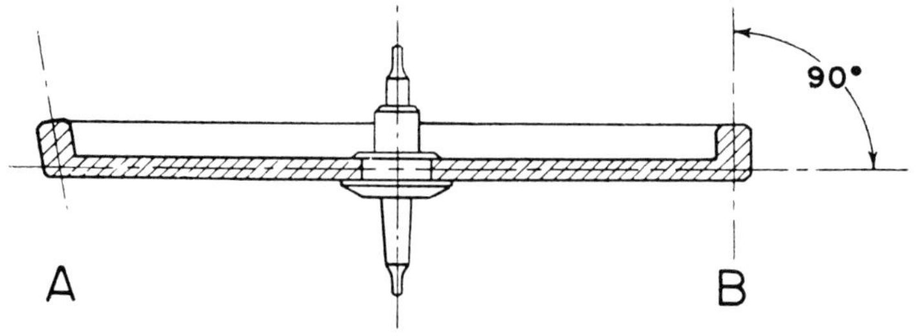

Fig. 22

III. TRUING MONO-METALLIC BALANCE WHEELS

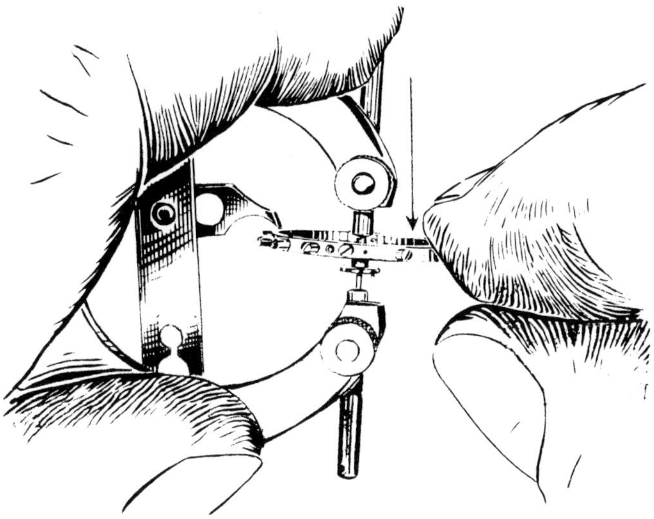

Fig. 23A

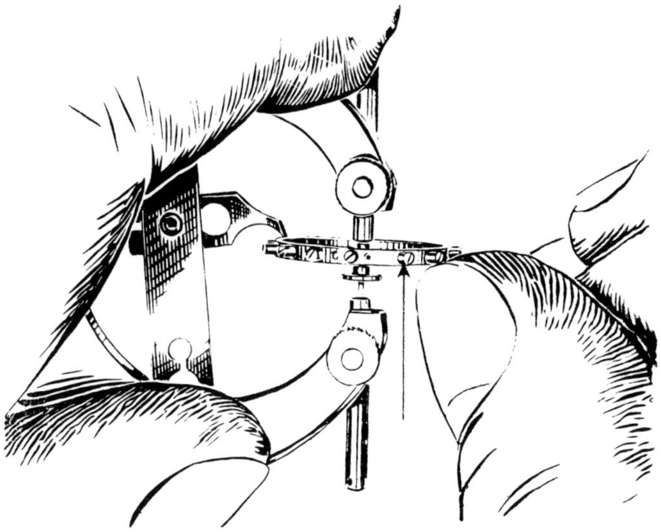

Fig. 23B

TRUING IN THE FLAT

A. Because of the solid rim, the mono-metallic wheel can best be trued in the flat using the three fingered bend discussed under "Truing of Bi-Metallic Balance Wheels" on page 32 See also Figs. 23A and 23B above. If a rim of this type is found to be bent to any appreciable extent it should be replaced.

B. TRUING IN THE ROUND

It is very difficult to correct any errors in the round on a mono-metallic balance wheel. If the wheel is found to be out of true it should be replaced.

TRAINING UNIT NUMBER 3

★

BASIC TURNING

★

REMOVING DAMAGED BALANCE STAFFS

Turning on the watchmaker's lathe is a fundamental part of the watchmaker's required skills.

Whenever a balance staff is to be replaced, good practice demands that the old one be turned off rather than knocked out. New balance staffs may frequently require dimension adjustments which also must be done on the lathe.

Again there will be situations in which the entire staff or stem or some other part may have to be turned by the watch repairman because material is not available.

A thorough knowledge of the principles involved and a high degree of skill in using the watchmaker's lathe, sharpening hand gravers, and measuring with precision instruments are necessary requirements in producing high-grade watch repair work.

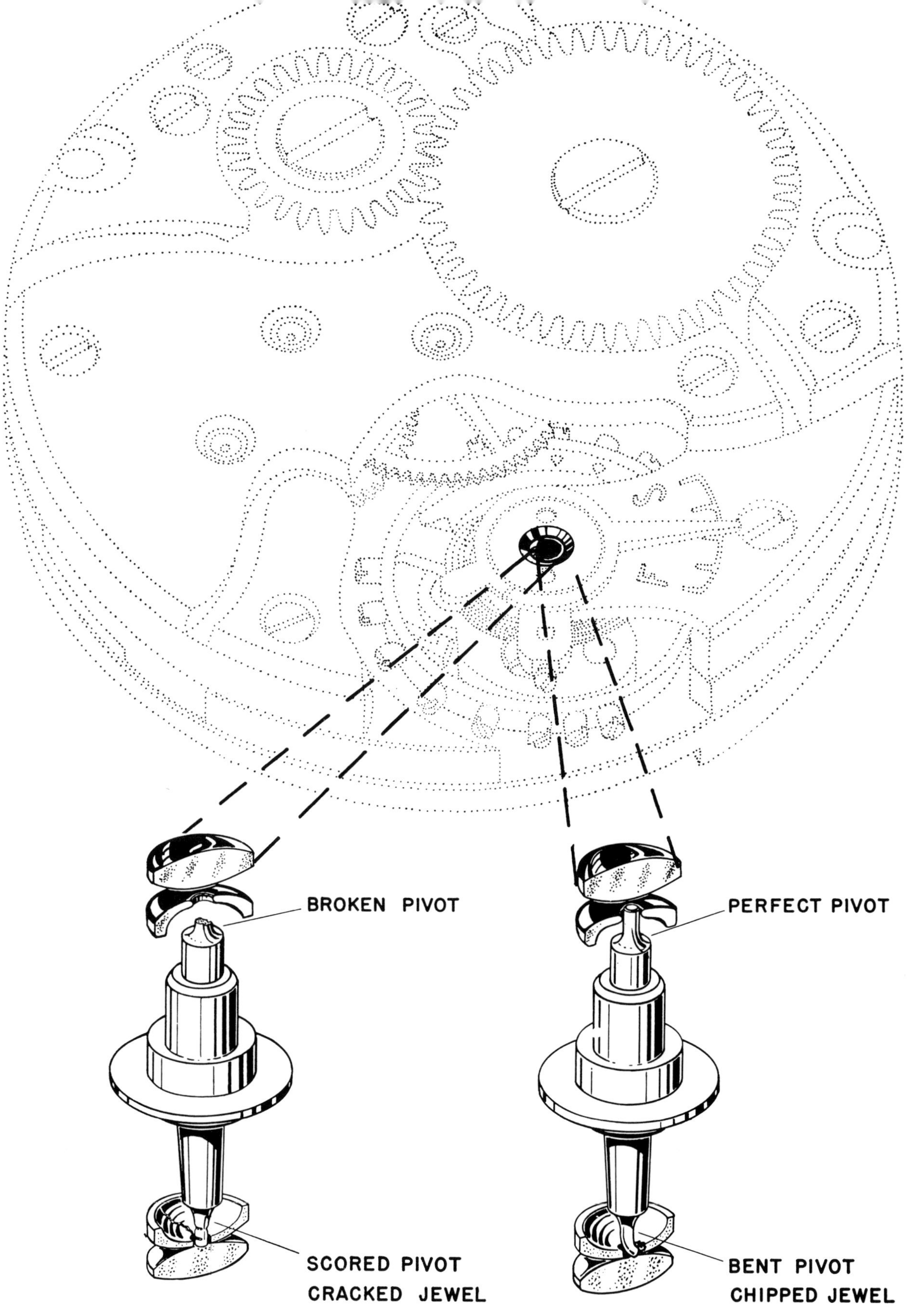
BROKEN PIVOT
PERFECT PIVOT
SCORED PIVOT
CRACKED JEWEL
BENT PIVOT
CHIPPED JEWEL

BULOVA SCHOOL *of* WATCHMAKING

TRAINING UNIT NUMBER 3

SUBJECTS:

Care and Use of Hand Gravers
Reading the Vernier Caliper
Turning Brass and Steel
Removing Damaged Balance Staffs

OBJECTIVES:

To develop skill in the use and maintenance of the tools and equipment required for these operations.

To familiarize the student with the proper methods used to remove various types of balance staffs.

NEW TOOLS REQUIRED:

1. Watchmakers Lathe
2. Powerstat
3. Motor
4. Grinding Wheel
5. Arkansas Oilstone
6. India Oilstone
7. Assorted Hand Gravers
8. Vernier Caliper
9. Set of Chucks

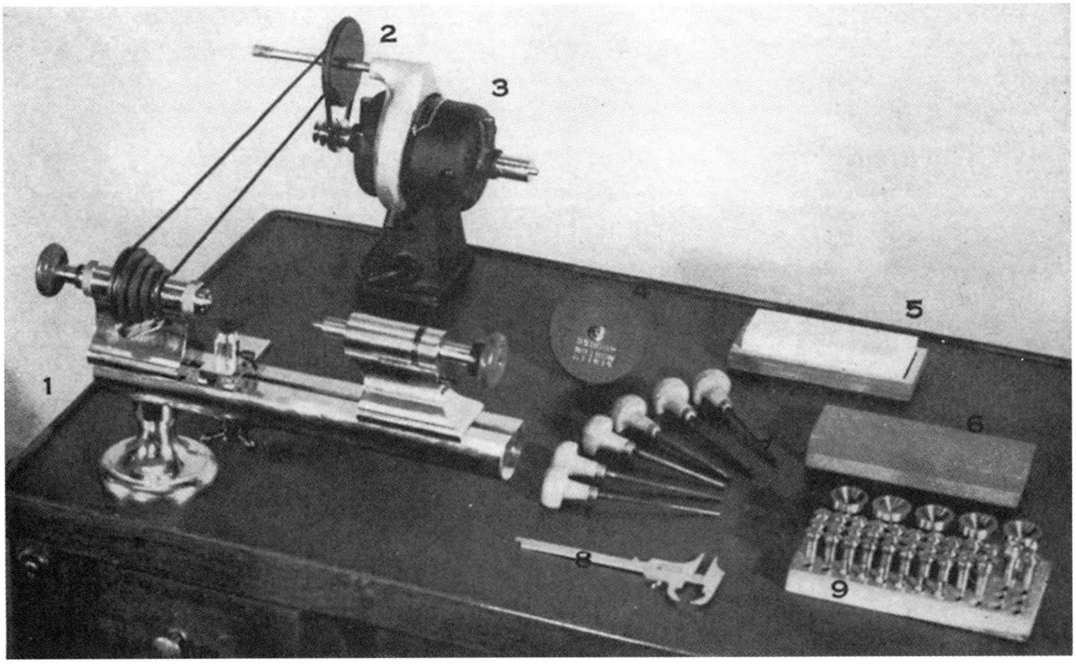

PRINCIPLES AND PROCEDURES
for the
CARE AND USE
of
HAND GRAVERS AND THE VERNIER CALIPER, TURNING BRASS AND STEEL
and
REMOVING DAMAGED BALANCE STAFFS

GRINDING HAND GRAVERS

Hand gravers are hardened steel cutting tools designed to shape metals revolving in a lathe. They are held in their correct cutting position by hand. Gravers are made in many sizes and shapes depending upon the purpose for which they are to be used. Satisfactory results cannot be expected unless a graver is sharpened properly and used correctly.

In case the tool is improperly angled or being shaped from an unfinished blank, grind it to the desired shape with a No. 80 or 100 carborundum wheel.

When using a carborundum wheel, do not over-heat the tool. This will draw the temper of the steel, rendering it soft and unfit for cutting.

Present the graver to the wheel so that a true flat surface of the desired angle is achieved. See Fig. 1.

To prevent the wheel from wearing unevenly, becoming grooved, or glazed, move the graver slightly across the flat surface of the wheel.

Never allow the surface of the wheel to become glazed with imbedded particles of metal.

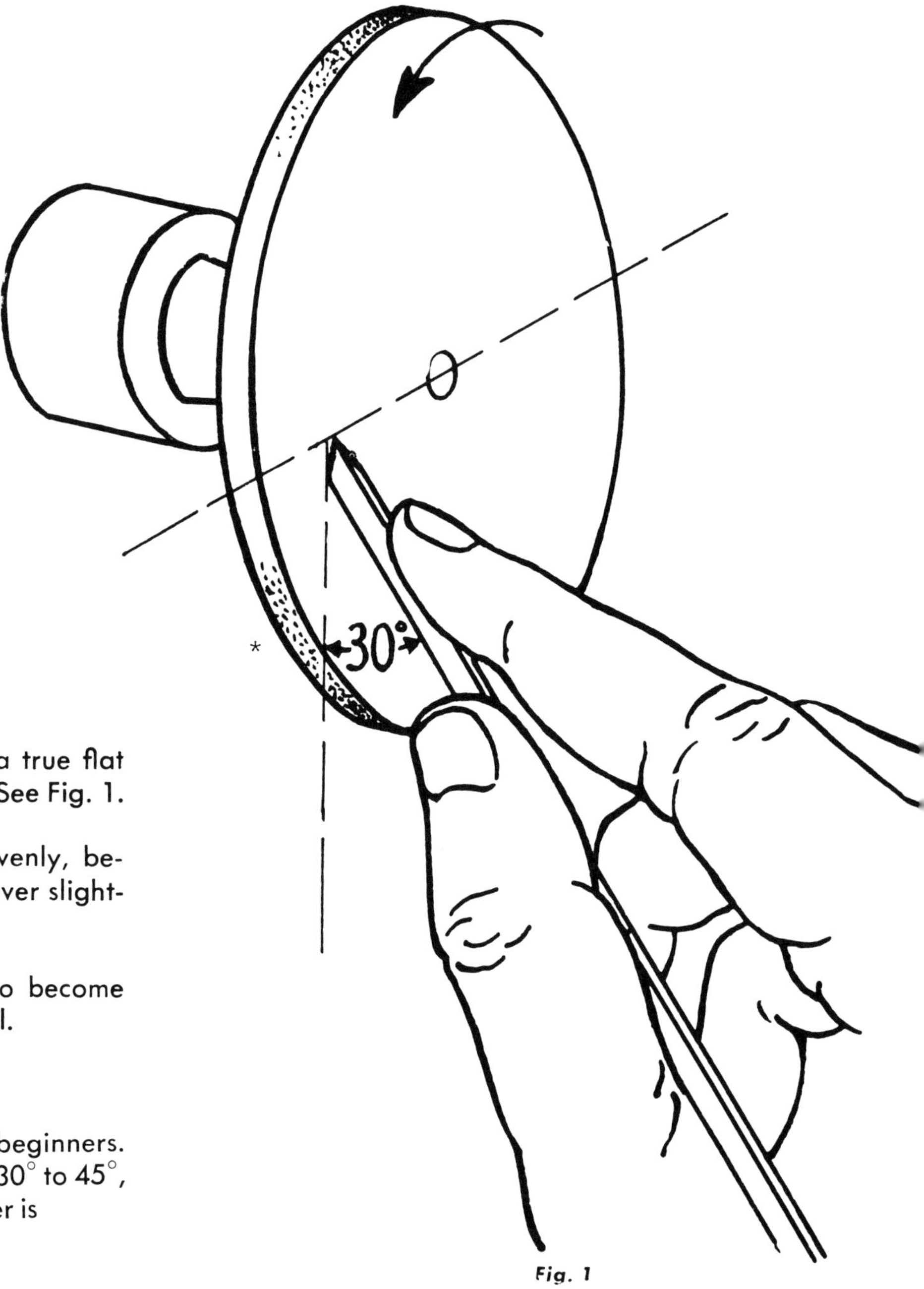

Fig. 1

Note: This method is not recommended for beginners.
*The angle of sharpening may vary from 30° to 45°, depending on the purpose for which graver is to be used.

SHARPENING HAND GRAVERS

When touching up a properly shaped graver, stone the graver lengthwise or sidewise depending on personal choice on a combination India oilstone. Use either the coarse or fine side as the condition of the graver demands. See Fig. 2.

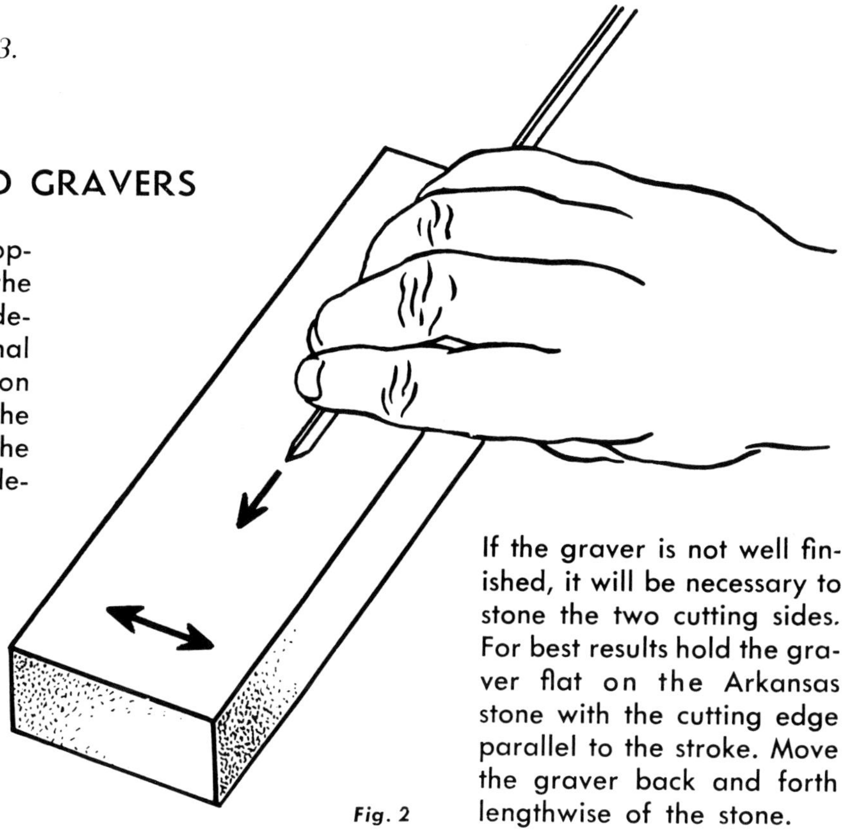

Fig. 2

If the graver is not well finished, it will be necessary to stone the two cutting sides. For best results hold the graver flat on the Arkansas stone with the cutting edge parallel to the stroke. Move the graver back and forth lengthwise of the stone.

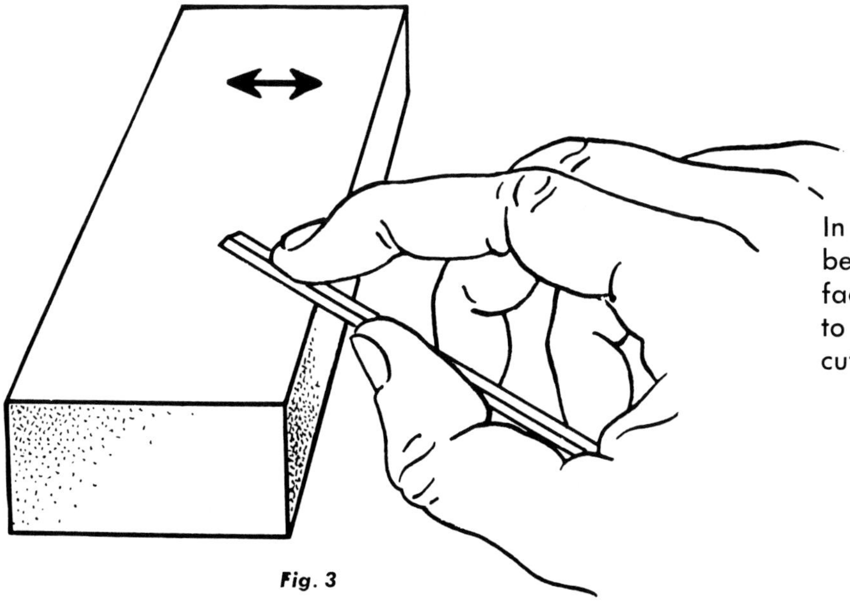

Fig. 3

In Fig. 3 the graver is shown being drawn across the surface of the stone in order to remove burrs from the cutting edge.

The graver must be held at a constant angle to the surface of the stone. A rocking motion will round the cutting edges. Apply a coating of light machine oil to the India oilstone and the Arkansas stone to float the small particles of metal, thus preventing the pores of the stone from becoming clogged. Wipe the stone clean with benzine after using.

The final finishing is done on the Arkansas oilstone. Gravers sharpened in this manner will have a keen cutting edge and will produce smoother work.

READING THE VERNIER CALIPER

The vernier caliper or slide gauge is graduated in millimeters. To be able to read to a tenth of a millimeter, the vernier scale is added to the sliding jaw of the caliper.

It will be seen from Fig. 4 that there are two scales: the upper or millimeter scale (A), on the fixed bar of the gauge and the lower or vernier scale (B), which is on the sliding head. Observation will show that nine divisions on the millimeter scale have been divided into ten equal parts on the vernier scale. Therefore when a division on the vernier scale lines up with a line on the millimeter scale, it indicates that the vernier has moved in tenths of a millimeter to correspond with the matching number.

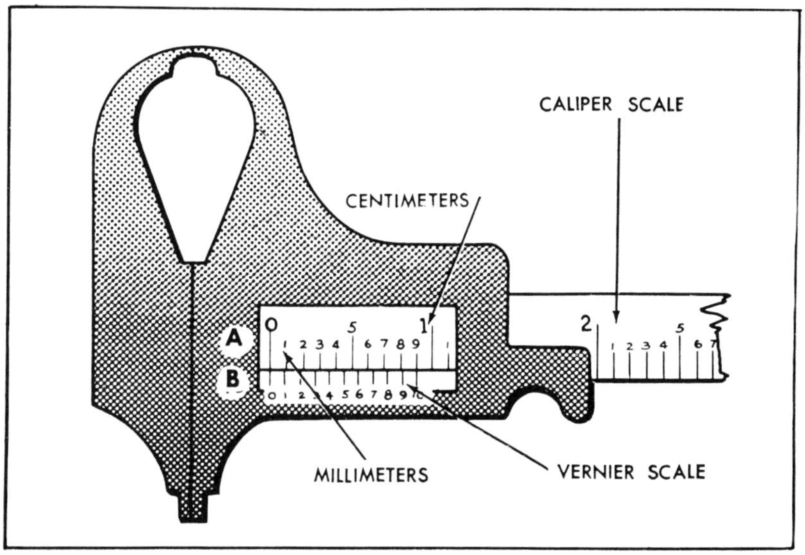

Fig. 4

For example, Fig. 5 shows a setting where the number one division of the vernier scale (B) matches the number one division on the millimeter scale (A).This indicates a reading of one tenth of a millimeter. When the second line on the vernier coincides with the number two division on scale A, the vernier has moved two tenths of a millimeter.

Fig. 6 shows the third division on scale (B) in line with the number three division on Scale (A), indicating a reading of three tenths of a millimeter.

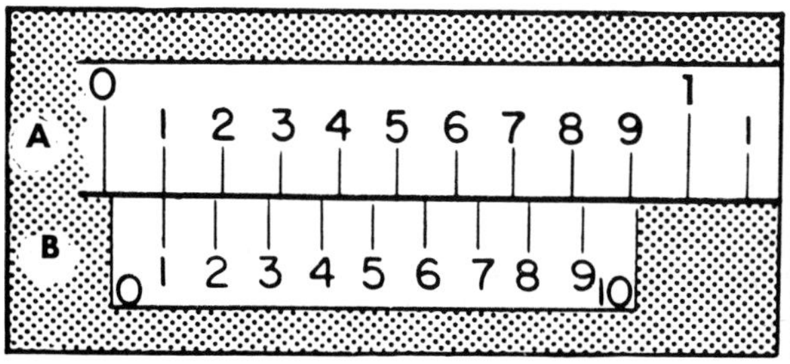

Fig. 5

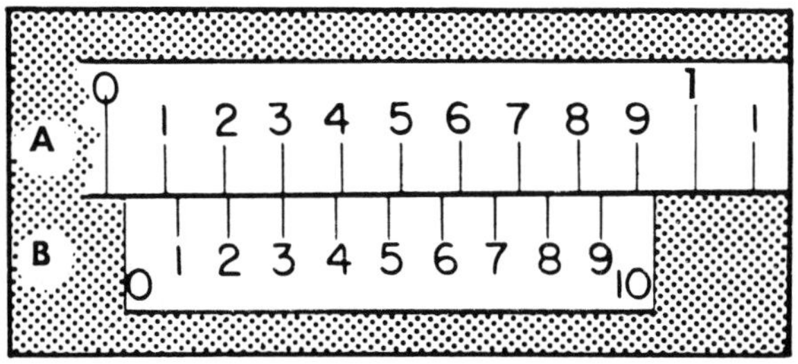

Fig. 6

Therefore, whenever any one division of the vernier (B) becomes lined up with a division of the millimeter scale (A), the number stamped under this vernier division will indicate the distance in tenths of a millimeter that the vernier scale has been moved, corresponding to the same distance that the caliper jaws have been opened.

When the vernier has been moved more than one millimeter on the millimeter scale (A), the position of the zero point will indicate the amount of movement in total millimeters.

In fig. 7, the zero point is located between three and four: therefore the jaws have been opened three millimeters plus a fraction of a millimeter. Since the six of the vernier (B) is lined up with the nine on the millimeter scale (A), then the fraction in this case will be six tenths of a millimeter. Adding the sixth tenths of a millimeter to the three millimeters, we now have a total opening between the caliper jaws of three and six tenths of a millimeter.

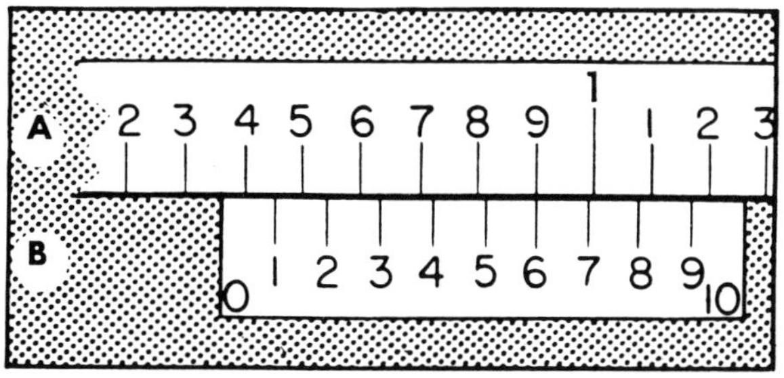

Fig. 7

Depth measurements are obtained by using the opposite end of the caliper. Slide the vernier (B) until it is exactly flush with the extreme end of scale (A) for a 0. mm reading. Since the scale is also calibrated from this end, the sliding vernier is adjusted to the desired depth and the reading made as explained above. In Fig. 8, the reading would be 4.6 mm.

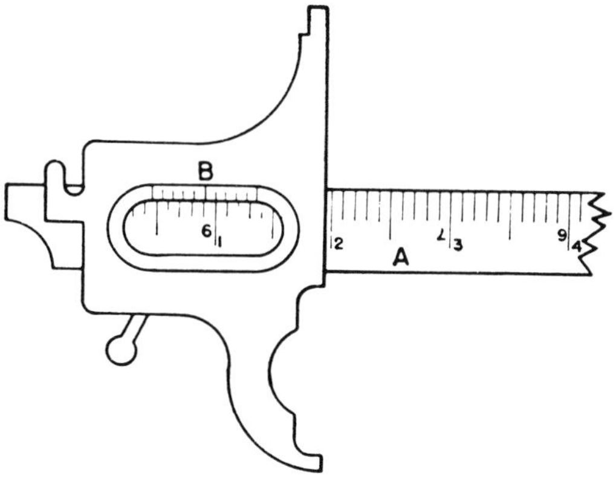

Fig. 8

TURNING BRASS

The first turning operation is taught on brass held in wire chucks. This metal is used because it is softer and easier for the beginning student to work with.

Wire chucks are numbered in tenths of millimeters. The chuck number indicates the size of the opening. For example, No. 32 chuck should be used with 3.2mm stock.

Select the chuck suitable for the wire used. Chuck the wire in the lathe to extend out from the chuck.

Position the tool rest so that it is parallel and close to the wire. The finer the work, the closer the adjustment should be. The height should be adjusted so as to permit the cutting edge of the graver to approach the brass slightly above center. If chatter accompanies cutting, raise the tool rest slightly. See Fig. 9.

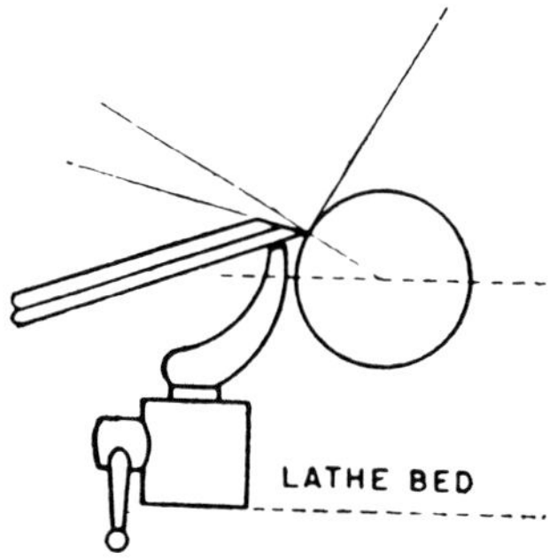

GRAVER AT CORRECT CUTTING POSITION

Fig. 9

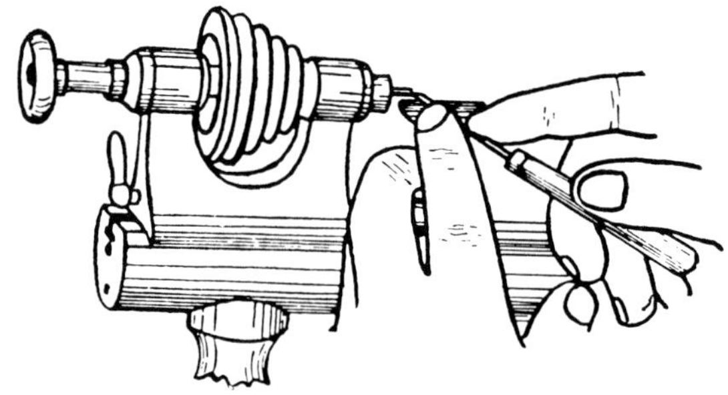

Fig. 10

Hold the graver firmly on the tool rest as shown in Fig. 10.

Cut wire to the length shown in Fig. 11.

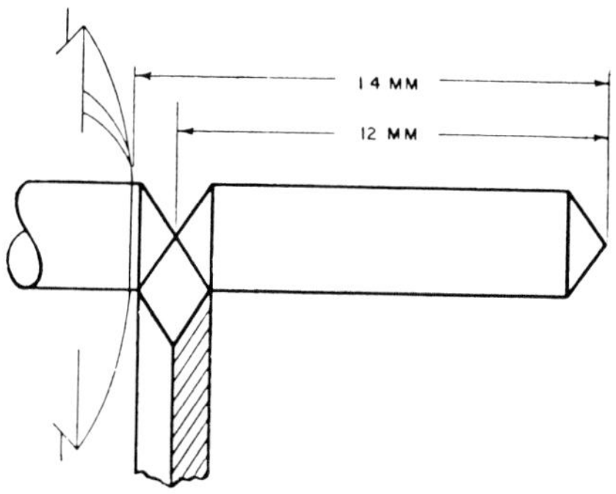

Fig. 11

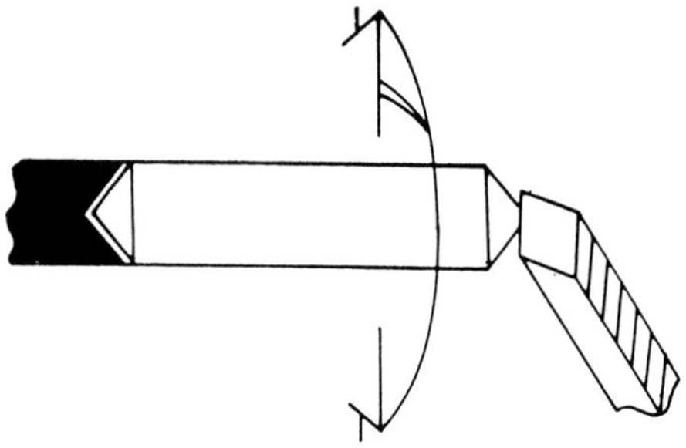

Fig. 12

Rechuck the blank so that about 1mm extends from the chuck.

Lower the tool rest slightly. With the cutting edge of the graver perpendicular to the wire, face off the work with a sweeping action. See Fig. 12.

Remove blank from chuck. Mark off 10 mm from faced-off end with a graver point. Rechuck blank with the mark about 1 mm out from the chuck. With lathe turning slowly, cut a reference ring at the mark. Face off to reference ring as shown in Fig. 13.

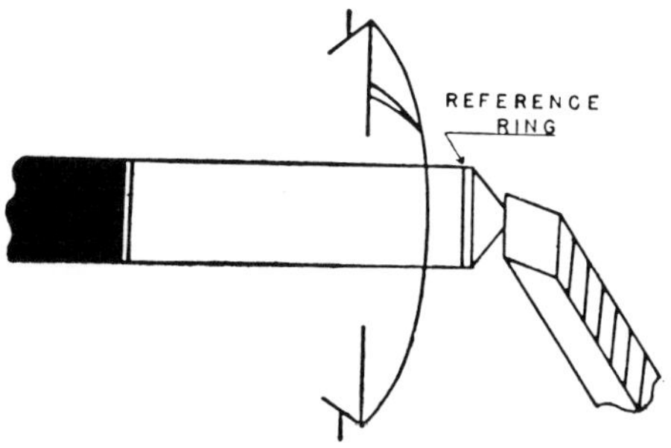

Fig. 13

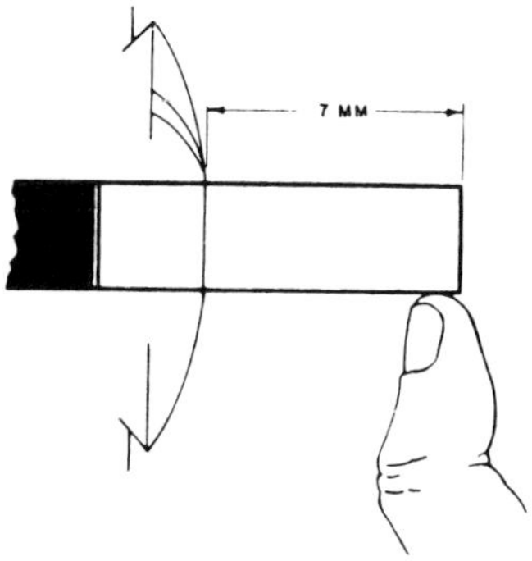

Fig. 14

Rechuck the blank as shown in Fig. 14. Tighten the chuck just enough to prevent the blank from falling out. With the lathe rotating, true the blank using a slight upward pressure of the thumb nail. Tighten chuck and examine trueness with an eye loupe.

Mark off 6mm from the end. Turn to within 0.2 mm of the mark, using the full cutting edge of the graver. Move the graver along the work to prevent ridges until the shoulder is 2.1 mm in diameter. See Fig. 15.

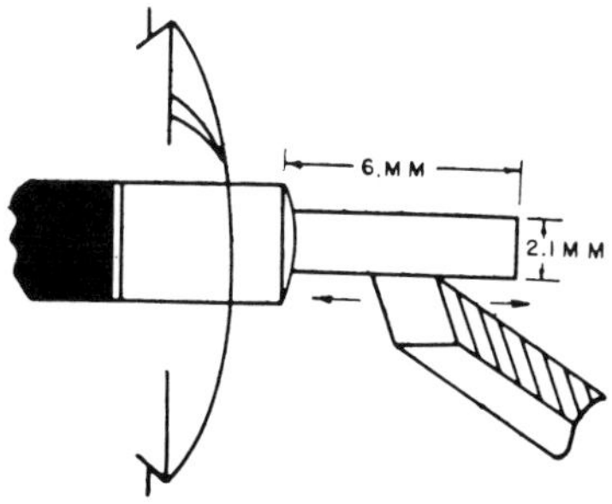

Fig. 15

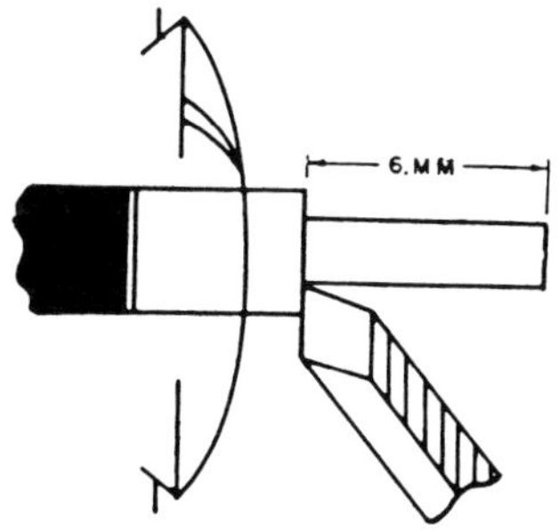

Fig. 16

Square off to 6 mm as shown in Fig. 16.

Using about a No. 3 sharp pointed graver, rivet undercut the square shoulder by presenting the graver to the work as shown in Fig. 17. Feed graver slowly and carefully. Exert only enough pressure to secure free cutting action at the point.

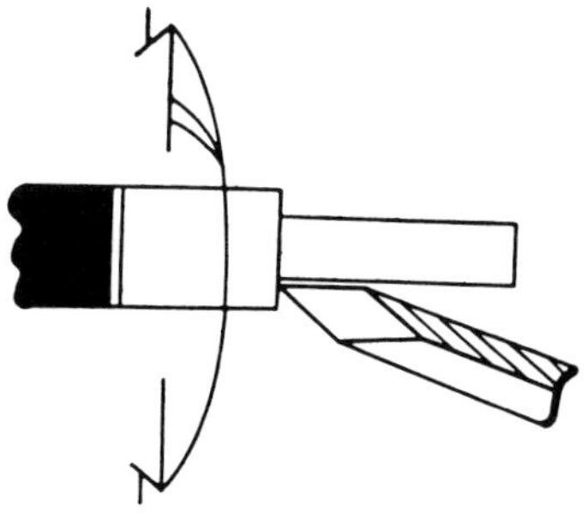

Fig. 17

Gradually cut deeper into the seat until the side cutting edge of the graver is against the shoulder as shown in Fig. 18. This operation requires considerable care and skill to avoid breaking the graver point.

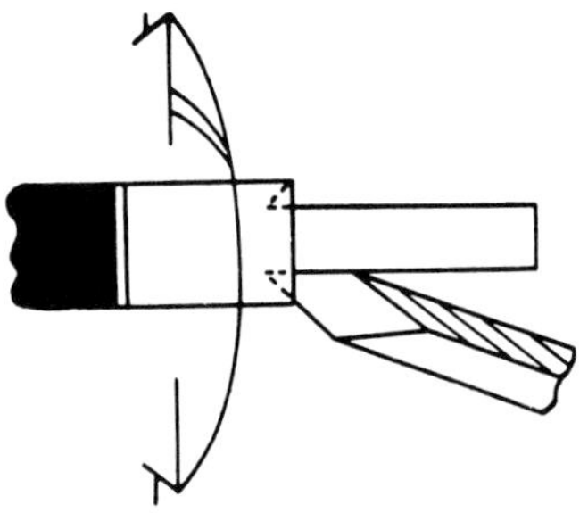

Fig. 18

Mark off 4 mm and turn the next shoulder to 1.4 mm diameter. Square the shoulder to exact length. See Fig. 19.

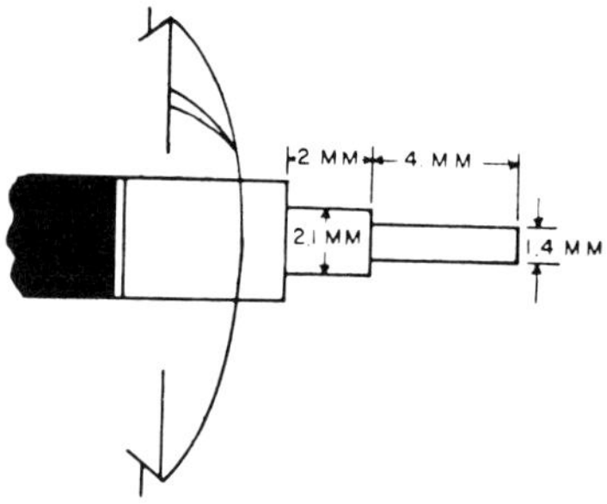

Fig. 19

With the graver previously used in rivet undercutting, undercut the corner slightly. See Fig. 20.

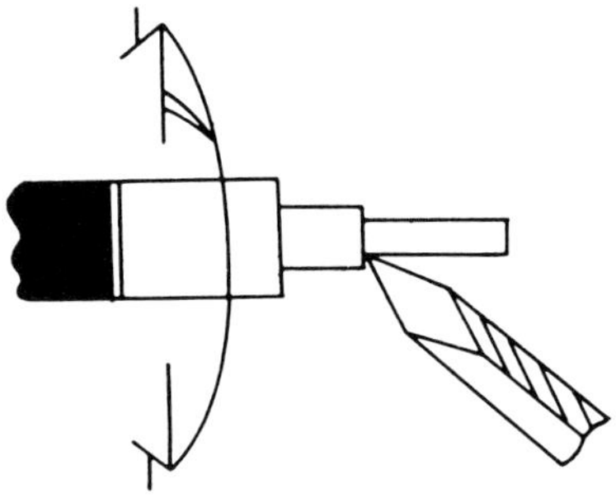

Fig. 20

Mark off the next shoulder 2 mm. Turn this shoulder to 0.7 mm diameter. See Fig. 21. When turning a small diameter, it is best to begin turning at the end and work toward the seat. Slightly undercut the corner. A deep undercut will weaken the pivot.

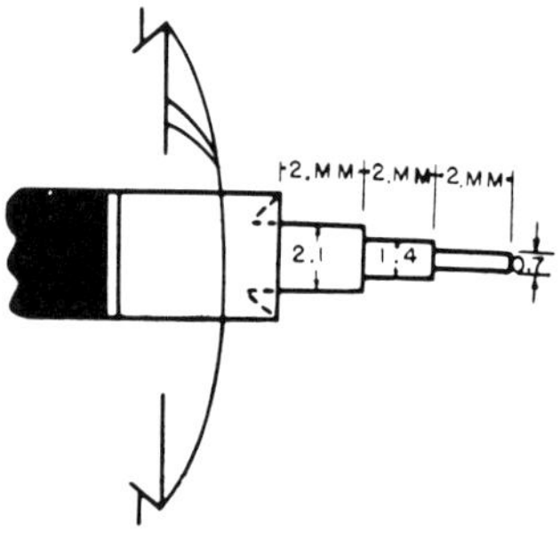

Fig. 21

Rechuck, true, and tighten the work. See Fig. 22.

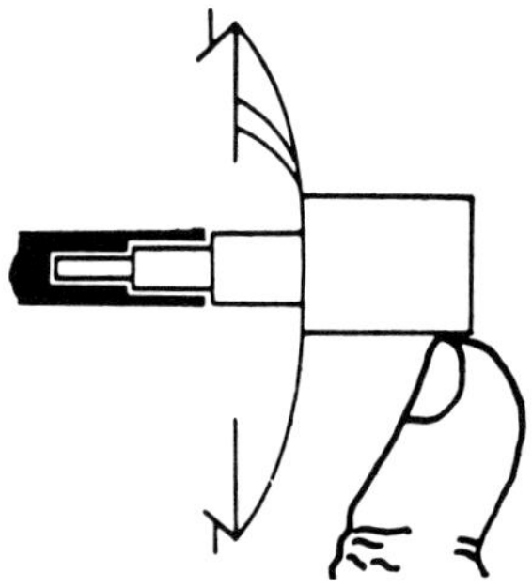

Fig. 22

Mark off 3 mm and turn the taper shown in Fig. 23.

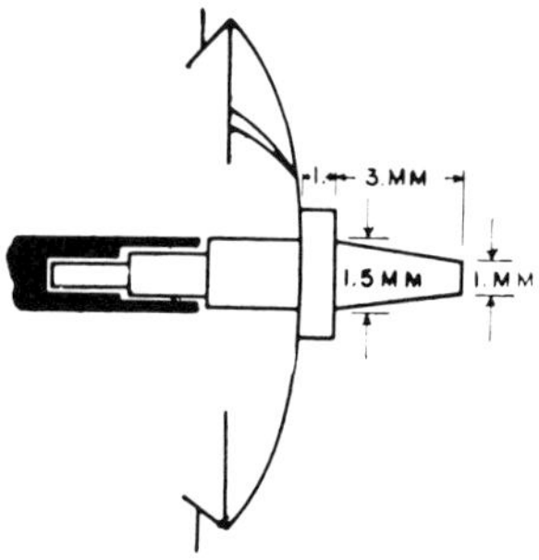

Fig. 23

TURNING STEEL

Steel is more difficult to turn because it is harder and tougher than brass. Turning steel with carbon steel gravers requires slower speeds.

Select a suitable chuck that will fit the steel wire to be used. Turn and face off the steel blank to 10 mm. Refer to Figs. 11, 12 and 13 under Turning Brass.

Chuck and true the blank with approximately 7 mm extending from the chuck.

Mark off 4 mm. Turn a shoulder to the dimensions shown in Fig. 24. Undercut the corner.

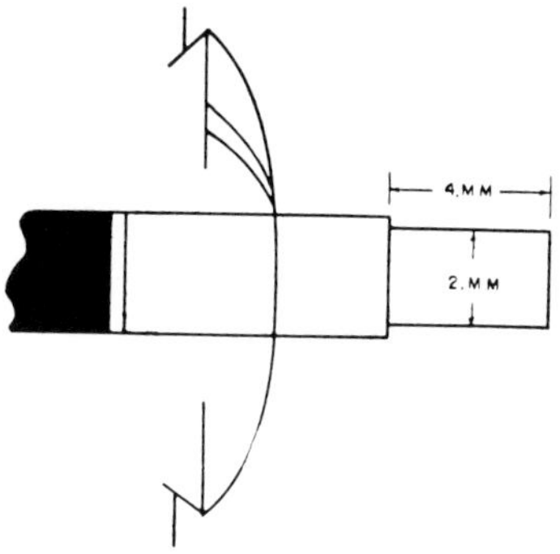

Fig. 24

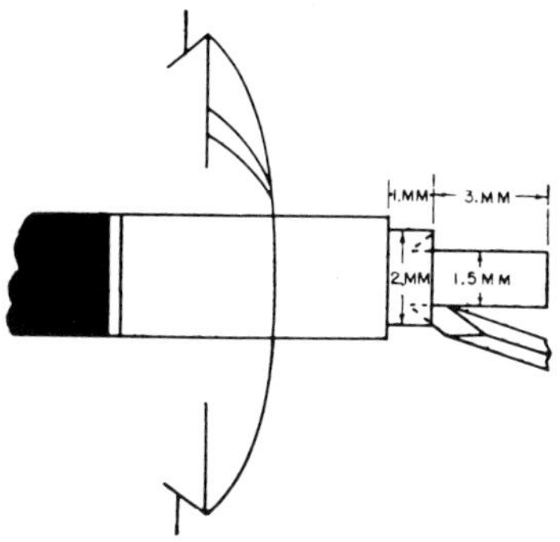

Fig. 25

Mark off and turn the next shoulder 3 mm long and 1.5 mm in diameter. Rivet undercut as shown in Fig. 25.

Mark off 2 mm and turn this shoulder to 0.5 mm diameter. Corner undercut carefully. See Fig. 26.

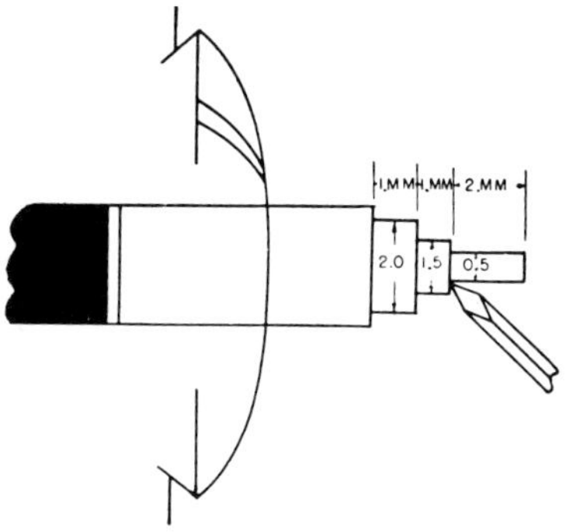

Fig. 26

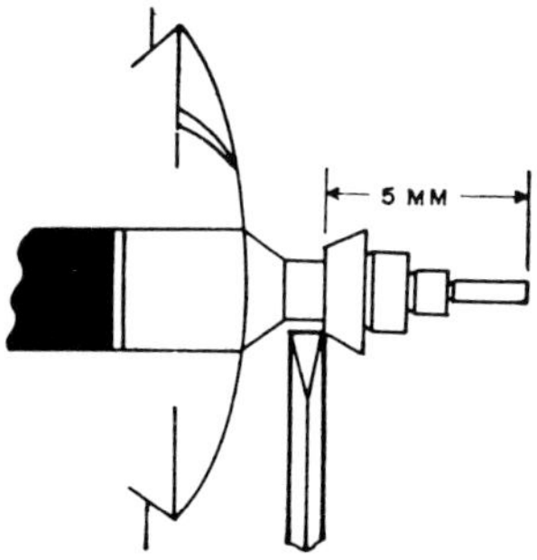

Fig. 27

Mark off 5 mm and rough out the lower part of the practice staff. The taper is turned to form the hub before rechucking. Large diameter of hub should measure 2.8 mm and the small diameter 1.8 mm. See Fig. 27.

Reverse the staff and rechuck. True and tighten with extreme care to secure a perfectly straight staff. See Fig. 28.

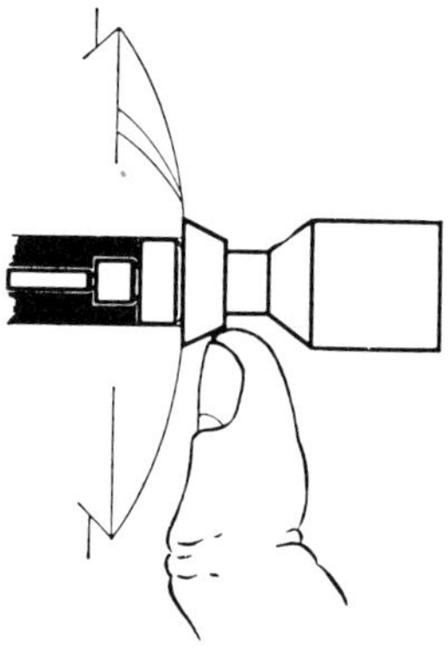

Fig. 28

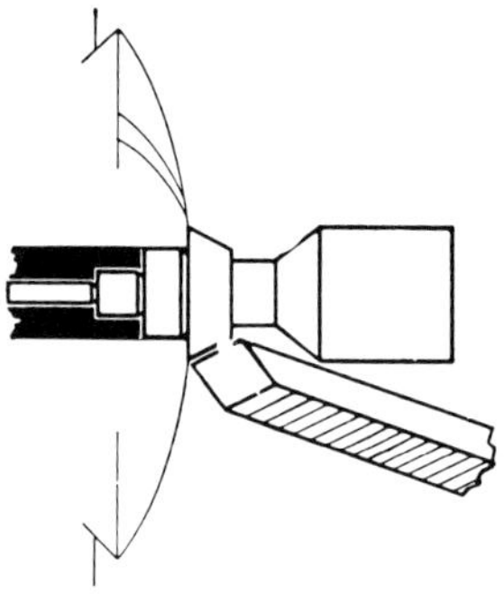

Fig. 29

Check the trueness of the staff by sighting a thin line of light between the full cutting graver edge and the hub. As the work is slowly rotated, this line of light should remain constant, proving that the staff has been chucked true. See Fig. 29.

Turn a taper on the lower part of the staff. The diameter near the hub should measure 1.5 mm and the end of the taper 1 mm. Slightly undercut the corner between the hub and the tapered shoulder. See Fig. 30.

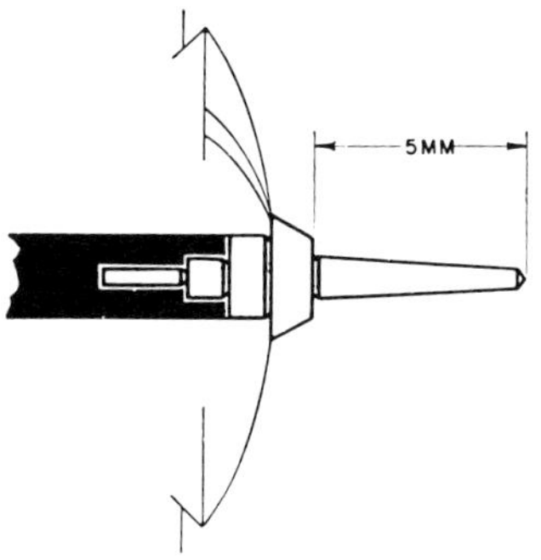

Fig. 30

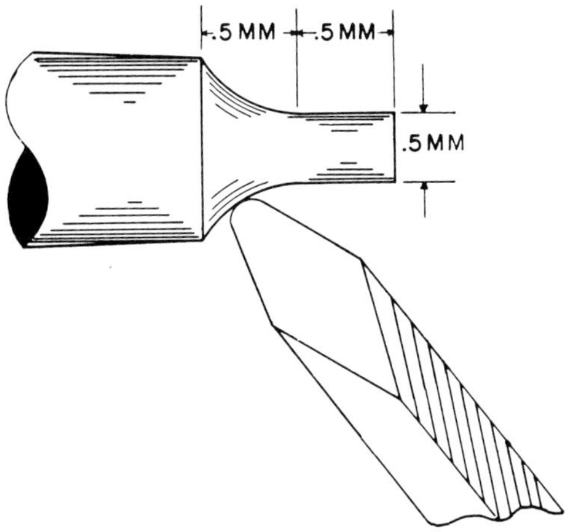

Fig. 31

Turn the cone with a rounded graver. The length of the cone should be 0.5 mm. The length and the diameter of the pivot should be 0.5 mm. See Fig. 31.

If the work has been accurately done, the distance from the roller seat to the end of the pivot should measure 5 mm.

REMOVING DAMAGED BALANCE STAFFS

When the standard rivet type of staff is to be replaced, the old staff may be removed by turning off the hub under the balance arm or turning off the rivet, swaged on top of the balance arm.

In most cases the method of turning off the hub of the staff is preferred and has two advantages. First, the balance shoulder of the staff which may be slightly spread due to staking, will not be forced through the hole of the balance wheel. Second, the hub will be easier to turn than the riveted shoulder because the staking process will have hardened the metal of the rivet.

However, when the damaged staff cannot be replaced with stock material and must be duplicated on the lathe, then the method of turning off the rivet should be used.

If the broken staff should be the top-grooved, side-grooved or friction type (see pages 15 and 16, Training Unit No. 1), the removing process does not require the lathe. These staffs are removed with staking equipment by reversing the staking process.

1. **TURNING OFF THE HUB** of a standard type staff such as Bulova 10 AK.

Approach the hub so that the left cutting edge of the graver tends to shave the roller seat of the hub. The point of the graver is held in the corner of the hub and roller shoulder. See Fig. 32.

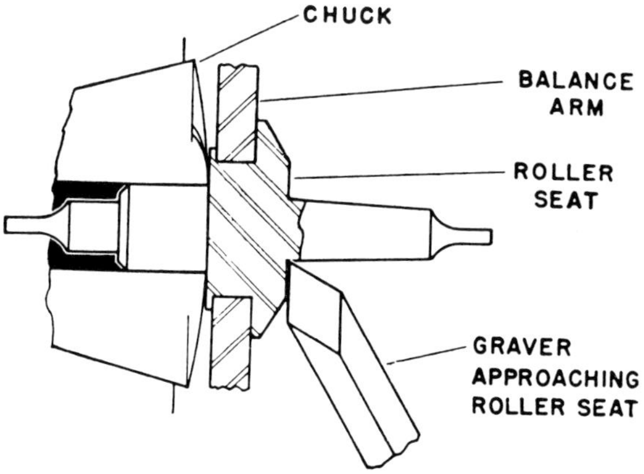

Fig. 32

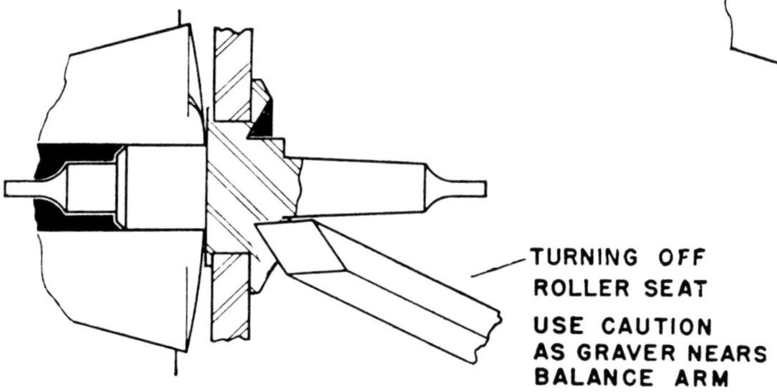

Fig. 33

As the shavings begin to leave the hub, swing the graver slightly to the right. The point will now be cutting deeper than the side cutting edge, forming a shallow V-groove. See Fig. 33.

As the metal of the hub is turned almost to the balance arm, the point of the graver will break through, leaving a loose ring of metal consisting of the thin outer portion of the turned-off hub. Care must be used to avoid damaging the balance arm. See Fig. 34.

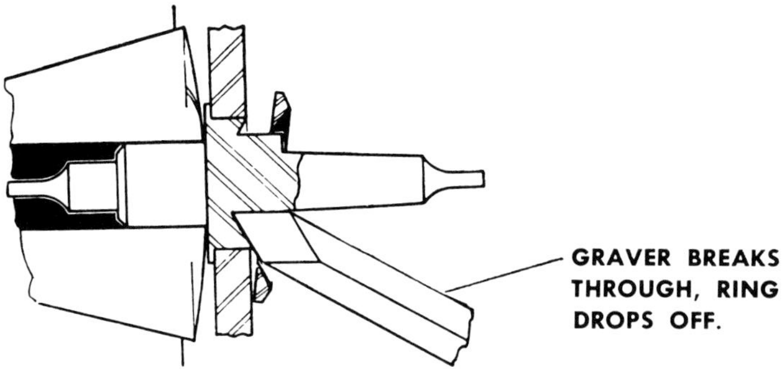

Fig. 34

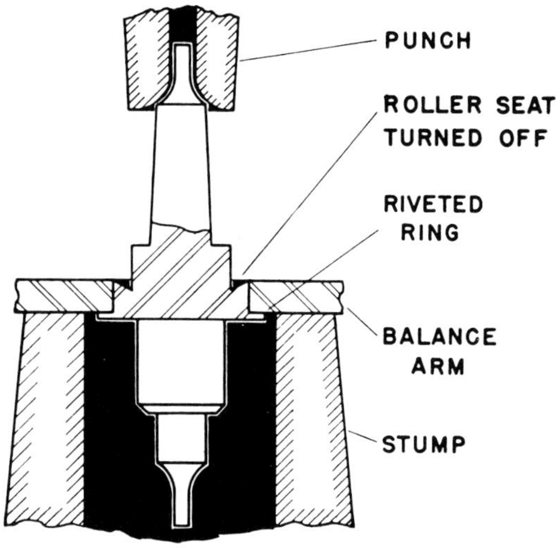

Fig. 35

In most instances the balance wheel can be eased off the staff with the fingers. If the wheel does not come off readily, remove by means of the staking set. See Fig. 35.

2. TURNING OFF THE RIVET of a standard type staff.

Greater care must be used when turning off the rivet. Use the graver in the same manner as in turning off the hub, being careful not to touch the balance arm. See Figs. 36 and 37.

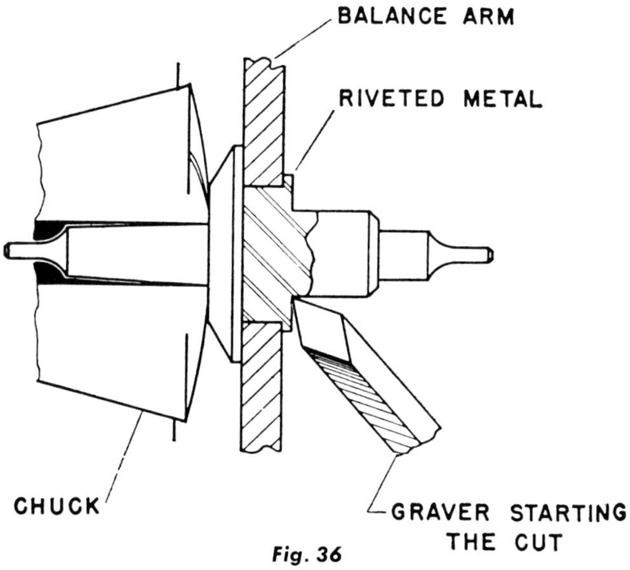

Fig. 36

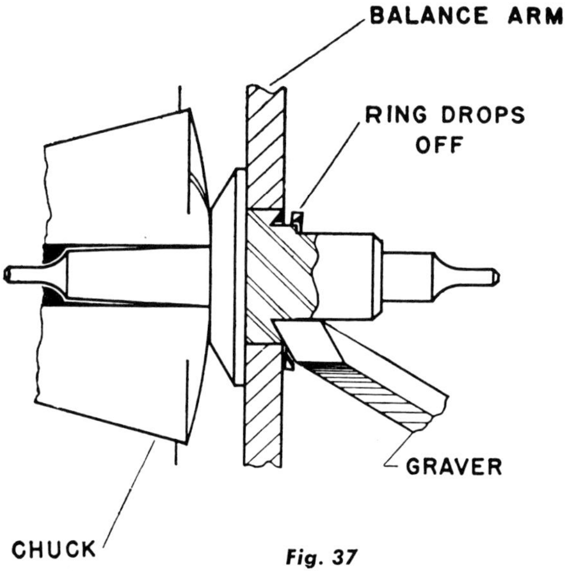

Fig. 37

TRAINING UNIT NUMBER 3A

★

TURNING BALANCE STAFFS

TURNING A BALANCE STAFF

There are many priceless heirlooms handed down from generation to generation and now in the possession of representative people who place high values upon these watches. The largest percentage of these are not in running order and the owners are willing to pay well for the thrill of seeing these watches again tick off the hours, minutes and seconds. Material for these watches can not be purchased.

A watch that had a balance staff replaced by the use of imitation material or poor workmanship, which caused the hole in the balance arms to be enlarged, so that a genuine staff can not be used, must have a balance staff turned to fit the new dimensions.

Herein are the real reasons for the necessity for staff making, and the watchmaker who can turn a balance staff with a sufficient degree of accuracy so that the watch is restored to its original running condition, will establish himself as high grade watchmaker.

TRAINING UNIT NUMBER 3A

Subject:

Balance Staff Turning.

Objective:

To develop skill and proper methods in turning balance staffs.
Practical work — 60 hours.
Related theory — 12 hours.

New Tools Required:

1 Alcohol lamp
2 Millimeter micrometer
3 Polishing Block
4 Bellmetal slip
5 Iron metal slip
6 Pin vise
7 Cement brasses

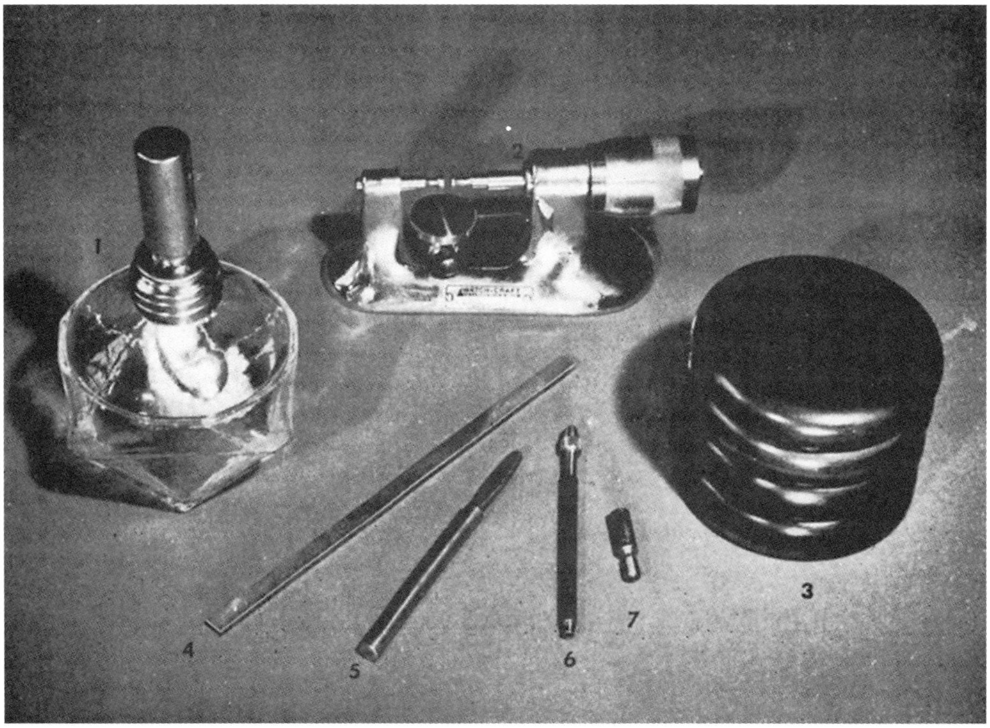

Select a piece of steel wire, with a diameter slightly larger than the largest diameter of the sample staff. In this case, stock of three millimeters thickness will be used. Make a blank one tenth of a millimeter longer than the sample. *(See Fig. 1)* Harden and temper this blank to a full blue color. *(See hardening and tempering)*

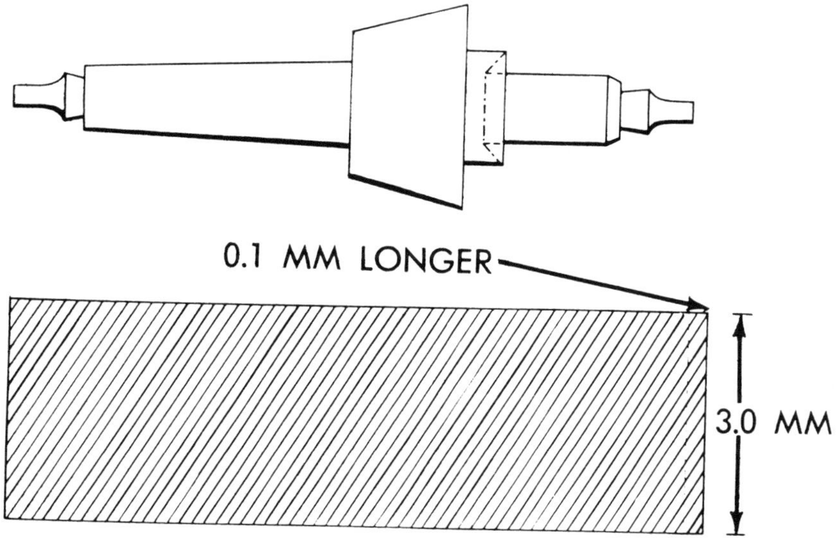

Fig. 1

Select a number 30 chuck which will fit the blank correctly. Chuck the blank in the lathe to extend out from the chuck far enough so that the hub of the staff can be turned. *(See Fig. 2)*

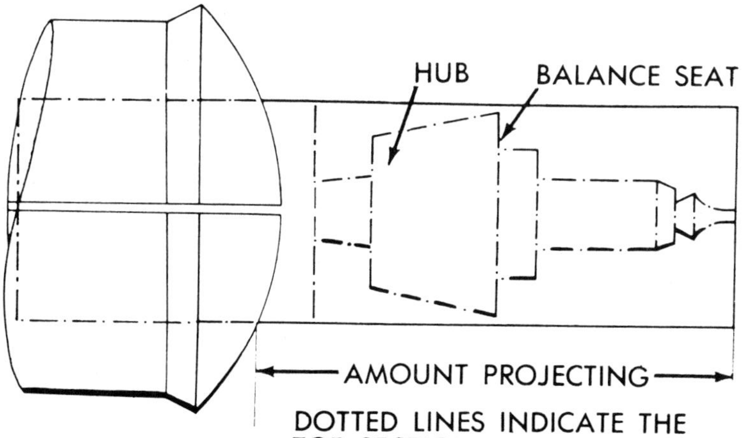

Fig. 2

Tighten the chuck just enough to prevent the staff from falling out while the lathe is rotating. With the thumb nail, true the staff by applying a little upward pressure on the underside of the blank while the lathe is in motion. *(See Fig. 3)* Tighten chuck and with a loupe examine the blank for trueness. This will insure that when the graver is presented to the blank, that very little turning will be required to true it.

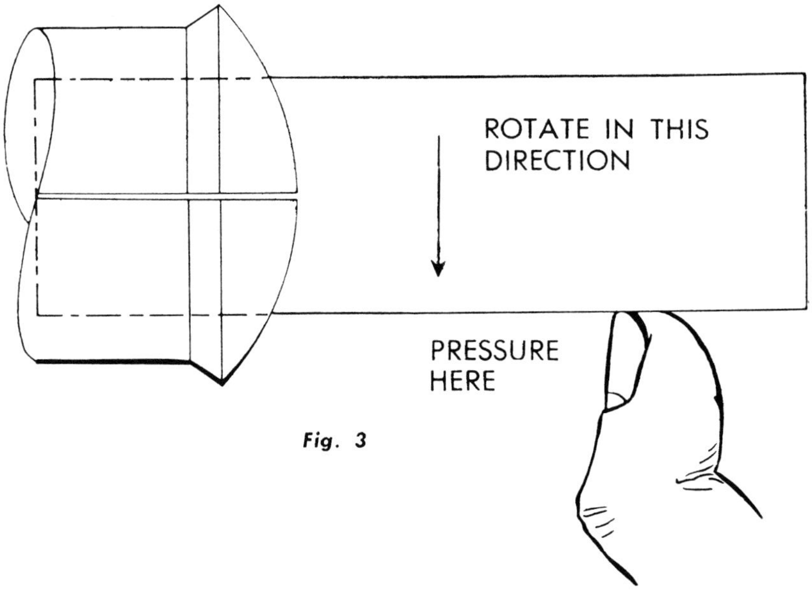

Fig. 3

Clamp the roller shoulder of the sample staff in a pin vise. Place the sample staff parallel with the blank, so that the balance seat can be marked with a graver. *(See Fig. 4)* From this mark proceed turning the balance shoulder.

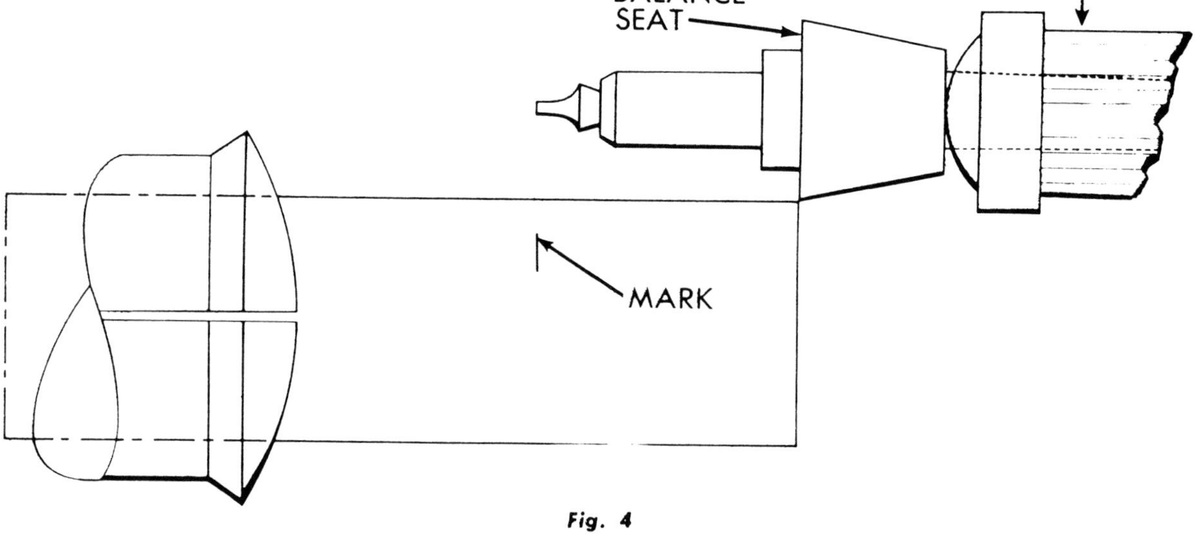

Fig. 4

When you start the turning use the point of the graver to true the steel blank, then use the full cutting edge of the graver moving it along the work to avoid ridges. *(See Fig. 5)*

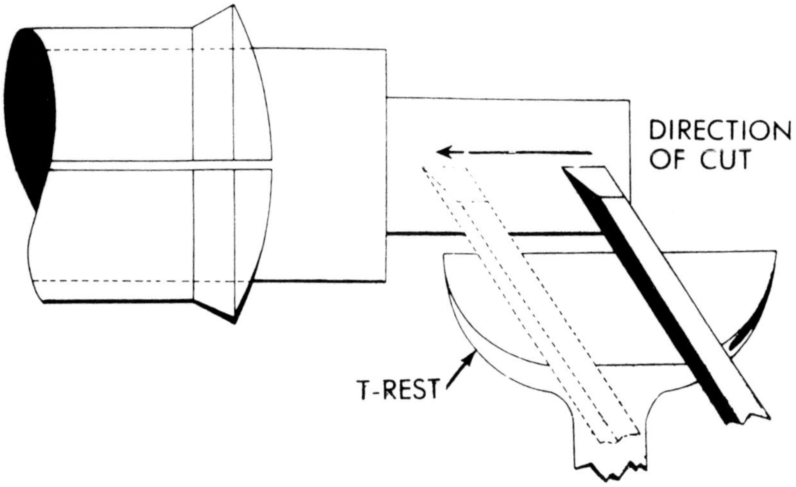

Fig. 5

Care must be used so as not to turn the balance shoulder too small. Turn the balance shoulder on a slight taper until the small end fits into the balance wheel hole. *(See Fig. 6)*

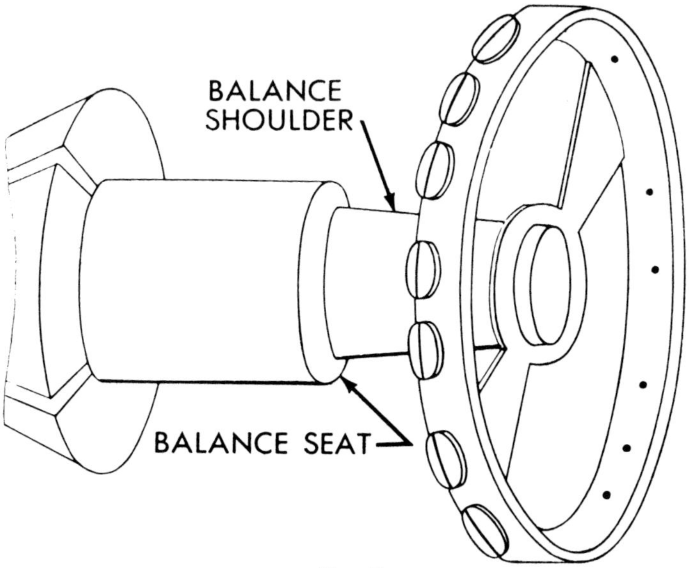

Fig. 6

Gradually reduce the balance shoulder by straightening out the taper until the balance wheel fits. *(See Fig. 7)* The fit should be very close. Too tight a fit will stretch the balance wheel hole and may distort the balance wheel while riveting. Too loose a fit will put the balance wheel out of round when riveted.

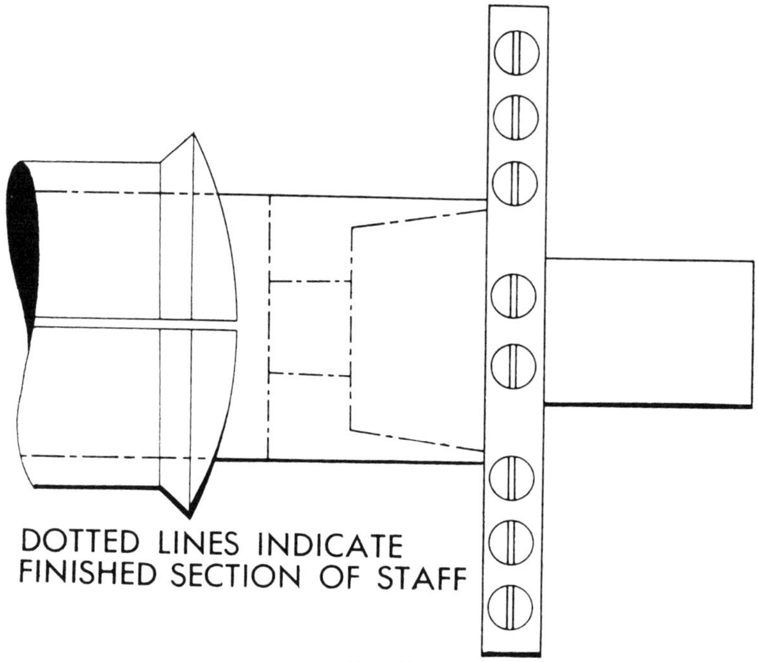

Fig. 7

Turn the hub of the staff, two hundredths of a millimeter larger in diameter than the hub of the sample. *(See Fig. 8)* This operation may also be done after upper part of the staff has been completed.

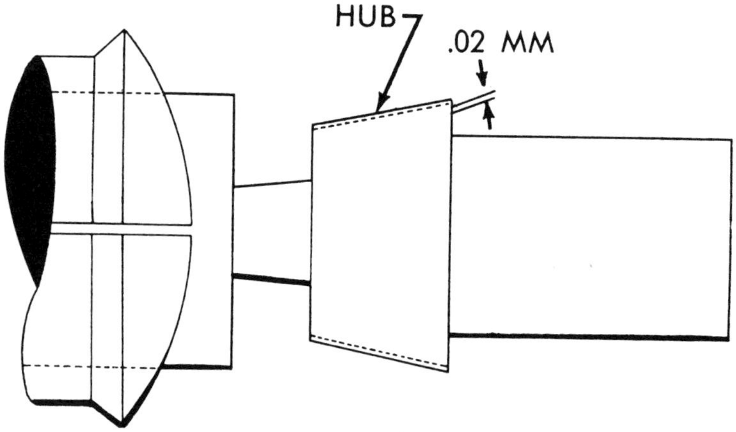

Fig. 8

Put the balance wheel in position on the balance shoulder. Make a mark with the graver approximately one tenth of a millimeter above the balance arms. This determines the length of the balance shoulder. *(See Fig. 9A and 9B)*

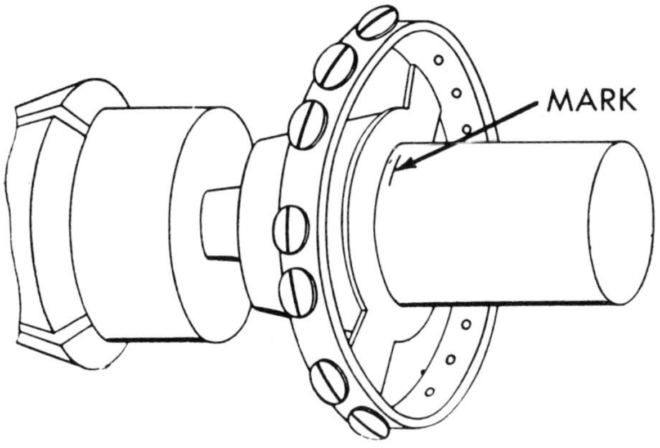

Fig. 9A

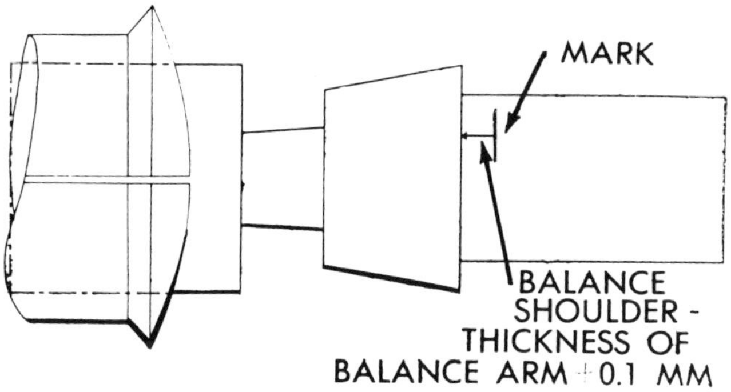

Fig. 9B

From that mark proceed to turn the collet shoulder until the diameter is two hundredths of a millimeter thicker than the sample. *(See Fig. 10)*

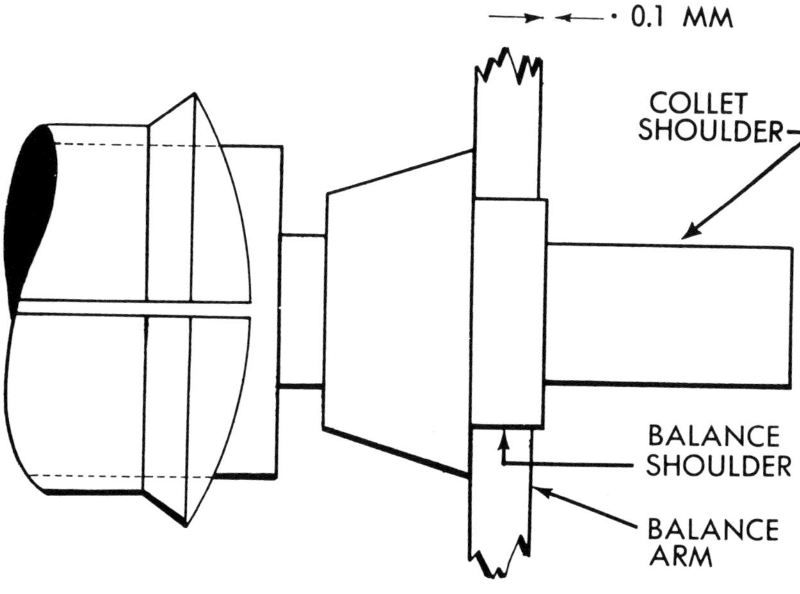

Fig. 10

Use a sharp pointed graver to establish the undercutting for the riveting operation by presenting the graver to the stock as shown in Fig. 11.

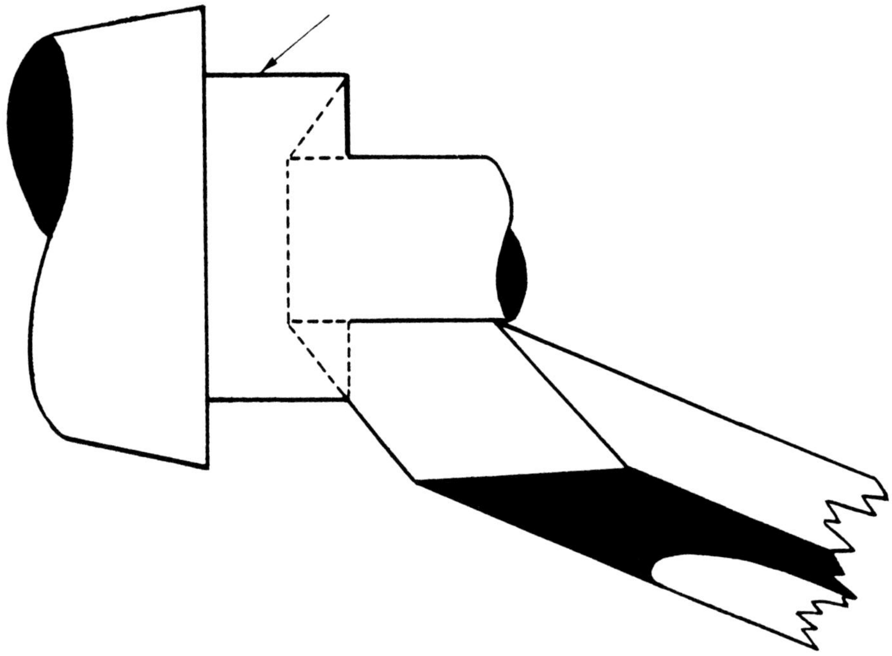

Fig. 11

Place the sample staff parallel with the blank, so that the length of the collet shoulder can be marked on the blank with a graver. *(See Fig. 12)*

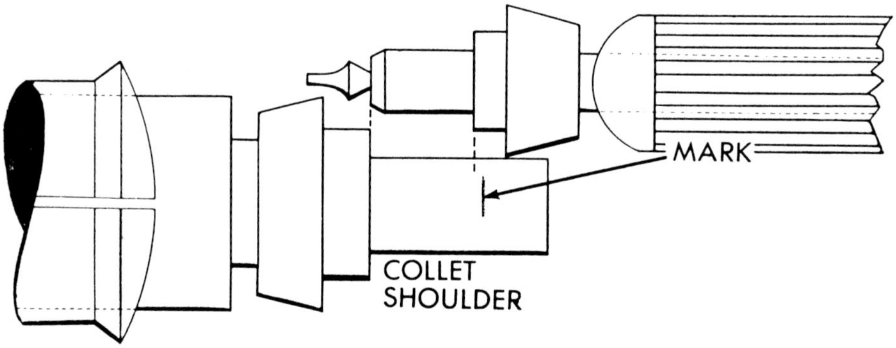

Fig. 12

From that mark, turn the upper cone and pivot. *(See Fig. 13)* Try the jewel on the pivot frequently, using great care so as not to break the pivot or jewel, until the pivot starts to fit into the jewel. This will leave enough metal for grinding and polishing.

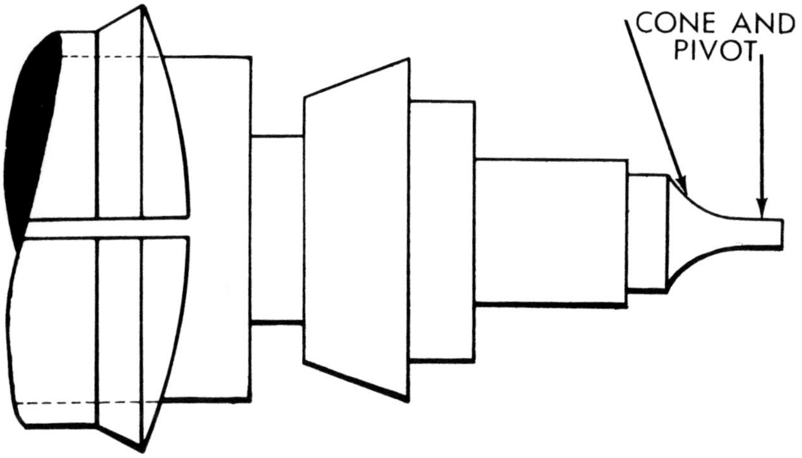

Fig. 13

Turn the cone with a rounded graver as shown in Fig. 14.

A correctly formed conical pivot consists of approximately two thirds cone and one third pivot. *(See Fig. 15)*

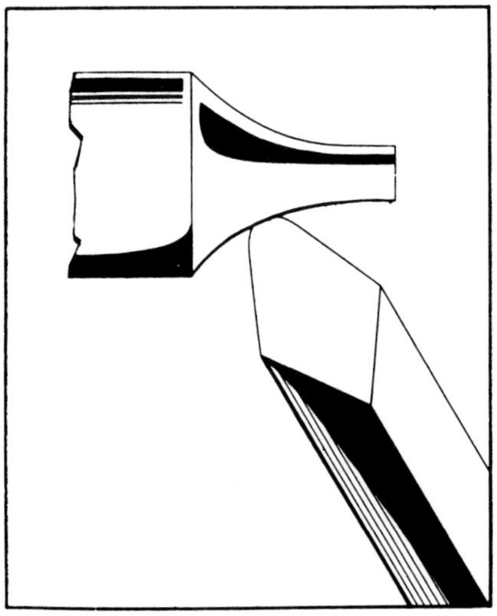

Fig. 14

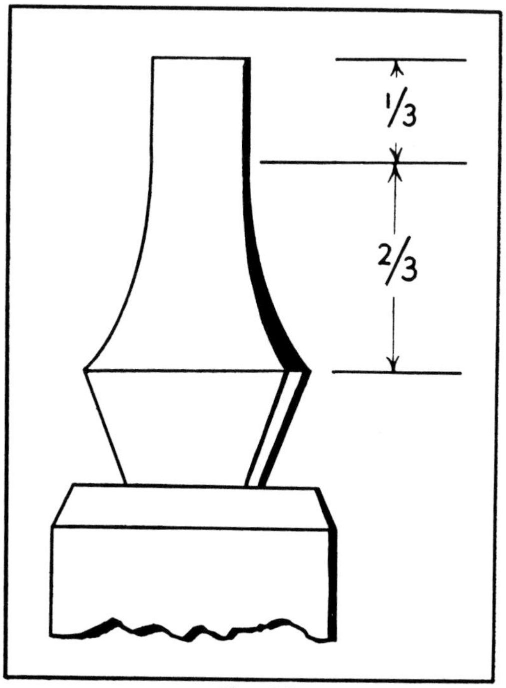

Fig. 15

With a sharp pointed graver turn a back taper on the cone as shown in Fig. 16. This operation requires considerable care and practice, so as not to break off the fine point of the graver.

Turn a bevel on the upper end of the collet shoulder as shown in Fig. 17.

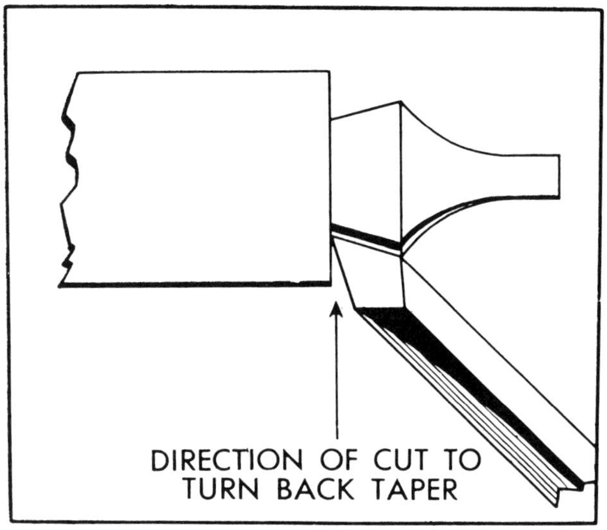

Fig. 16

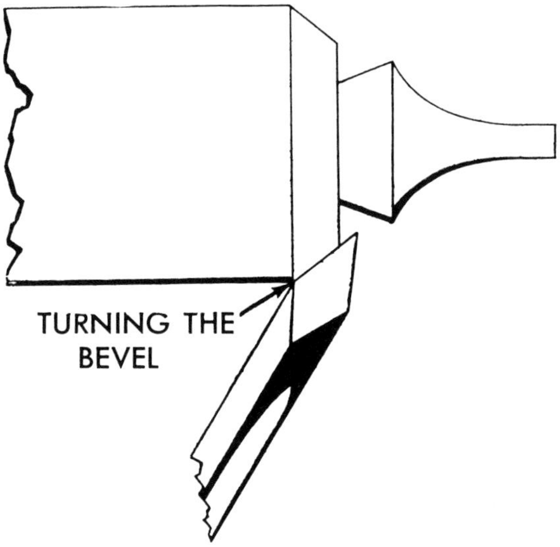

Fig. 17

Use an iron metal slip *(See Fig. 18)* charged with oilstone powder and oil, grind the hub and collet shoulder *(See Fig. 19)* in order to remove all traces of turning marks and to get the correct diameter. Clean off with pith wood to remove all traces of oilstone. Absolute cleanliness is essential before starting the polishing.

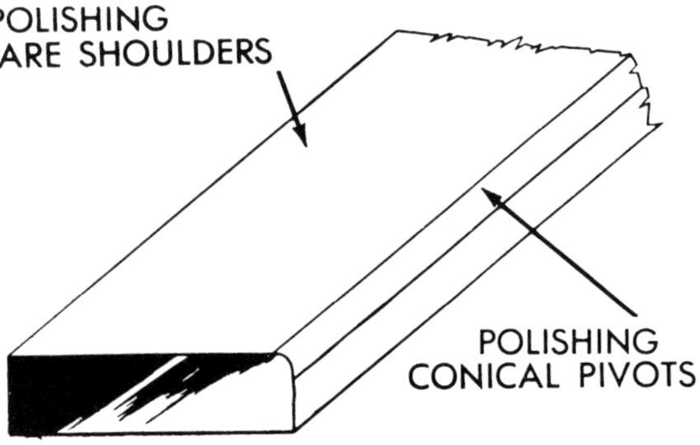

Fig. 18

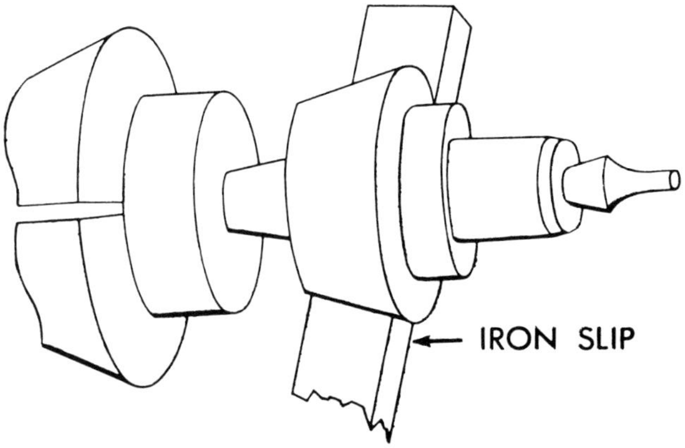

Fig. 19

Use a bell-metal slip shaped the same as the iron slip and charged with diamantine and oil, polish the collet shoulder and hub.

Grind and polish the upper cone and pivot to the correct size using the rounded side of the oilstone and bell-metal slips. Try the pivot in the jewel hole frequently to insure that the pivot is not made too thin.

Reverse the position of the unfinished staff in the lathe and chuck it on the balance shoulder. *(See Fig. 20)*

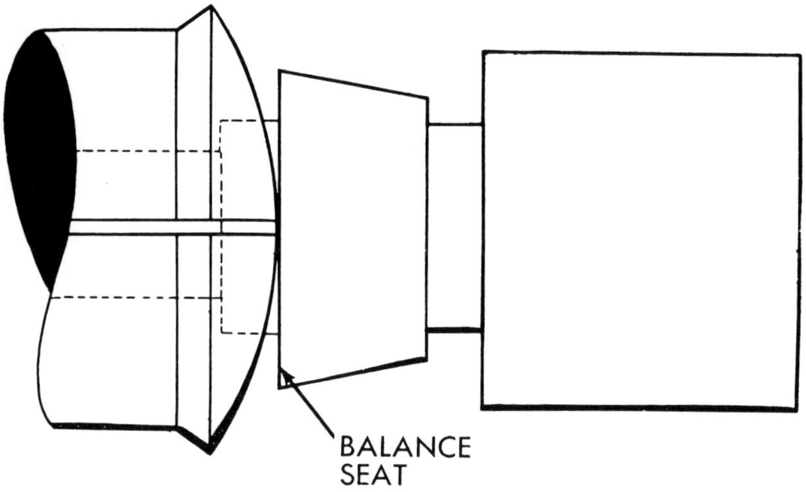

Fig. 20

Tighten the chuck just enough to prevent the staff from falling out while the lathe is rotating. With the thumb nail, true the staff by applying a little upward pressure on the underside of the hub while the lathe is in motion. *(See Fig. 21)*

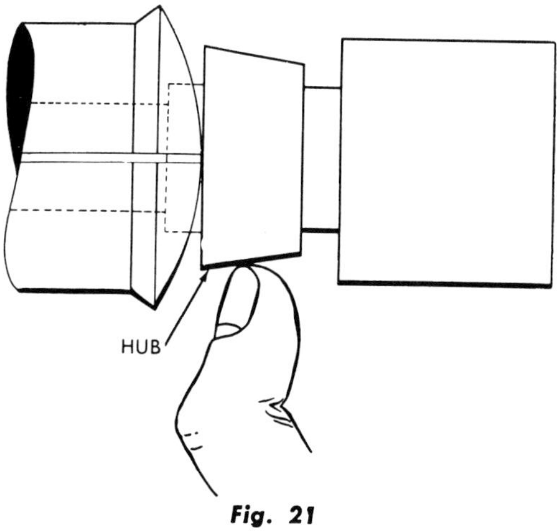

Fig. 21

Tighten chuck and with a double loupe examine the hub for trueness. Place a graver on the underside of the hub by resting it on the T-rest, allowing a thin line of light between the full cutting edge of the graver and the hub. Rotate the lathe slowly. If the space between the graver and the hub of the balance staff does not vary, the staff has been chucked true. *(See Fig. 22)*

Use the sample staff as a gauge. Check the length of the hub and correct, if necessary. *(See Fig. 23)*

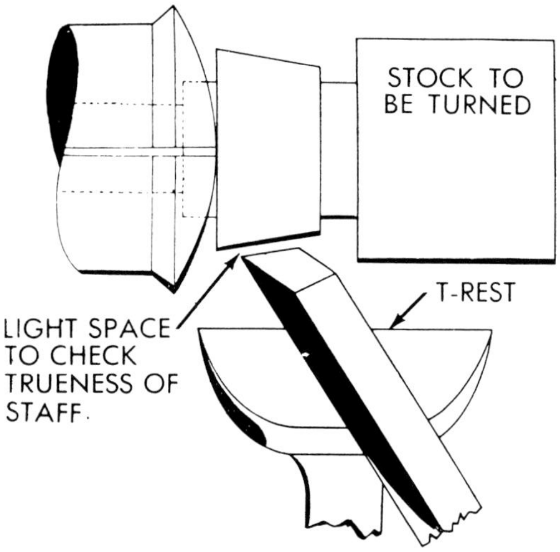

Fig. 22

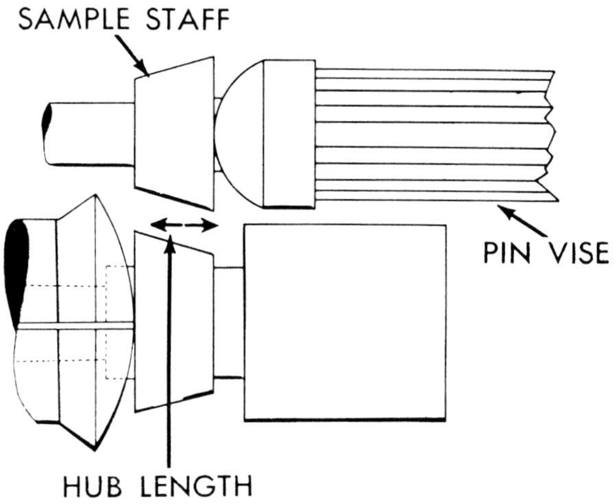

Fig. 23

Turn the roller shoulder with a slight taper until its diameter is two hundredths of a millimeter thicker than the sample. *(See Fig. 24)*

Undercut the roller seat very slightly so that when the roller shoulder is ground and polished it will still remain a square shoulder. *(See Fig. 25)*

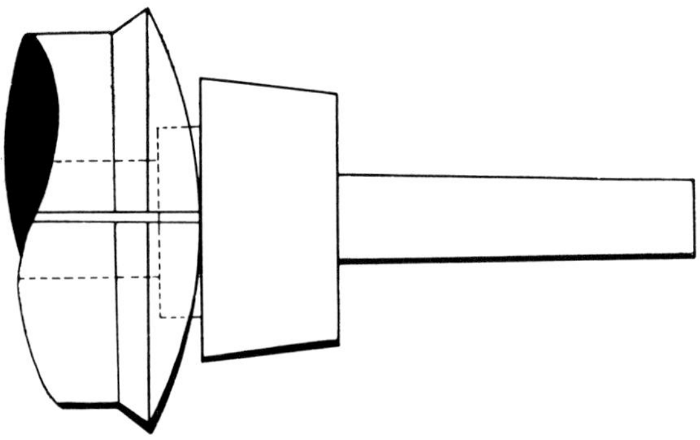

Fig. 24

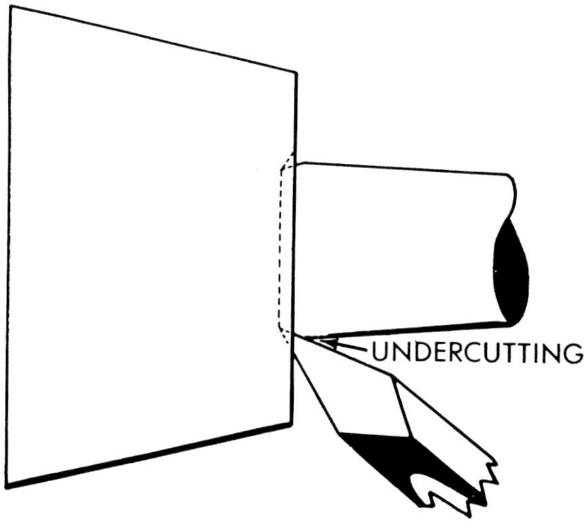

Fig. 25

Turn the lower cone and pivot using the old sample staff as a gauge. *(See Fig. 26)* Use the same precautions as that used in turning the upper pivot. Also see figure 35.

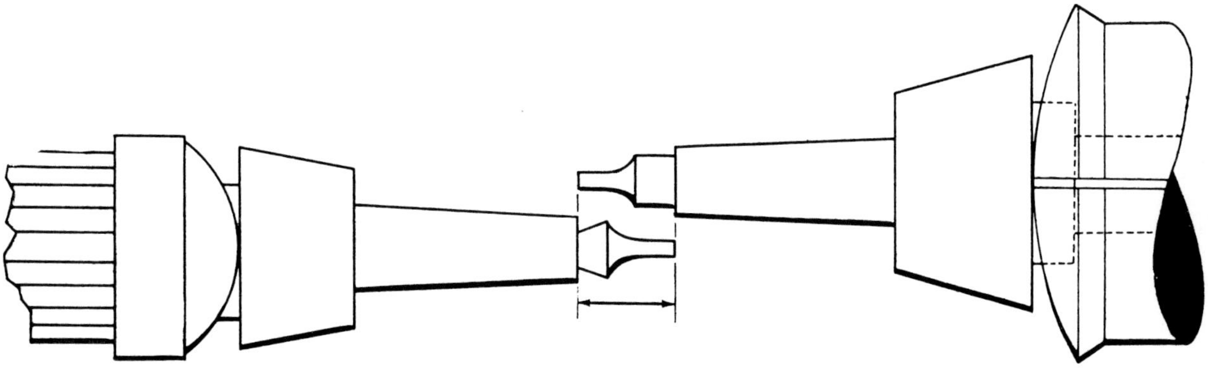

Fig. 26

With a sharp pointed graver turn a back taper on the cone as shown in Fig. 27. Many staffs are now made without this back taper. *(See Fig. 27A)*

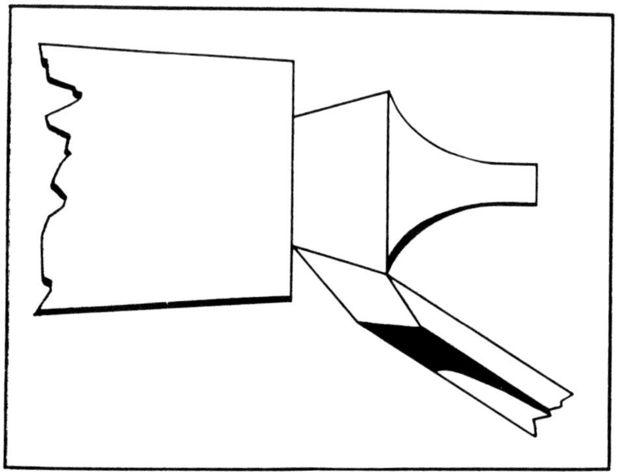

Fig. 27

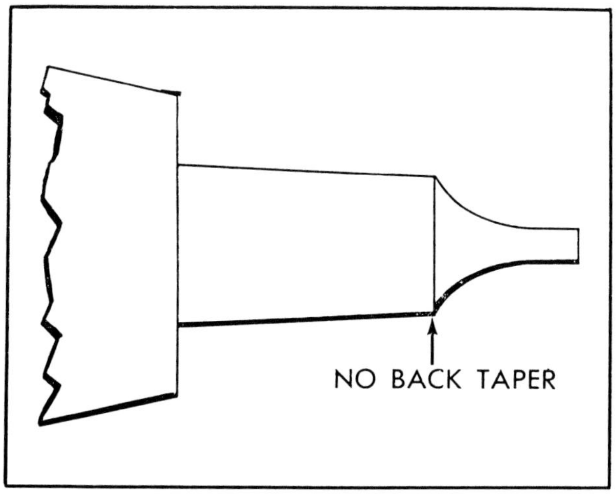

Fig. 27A

Use oilstone powder for grinding and diamantine to polish the lower balance staff to the correct diameters.

Stake balance wheel to staff and then burnish the ends and sides of the pivots.

Another method of turning a balance staff is as follows:

Chuck the steel wire in the lathe allowing it to extend out from the chuck to include the full length of the staff and about two millimeters additional.

Using the previous method proceed to turn the upper staff. *(See Fig. 28)*

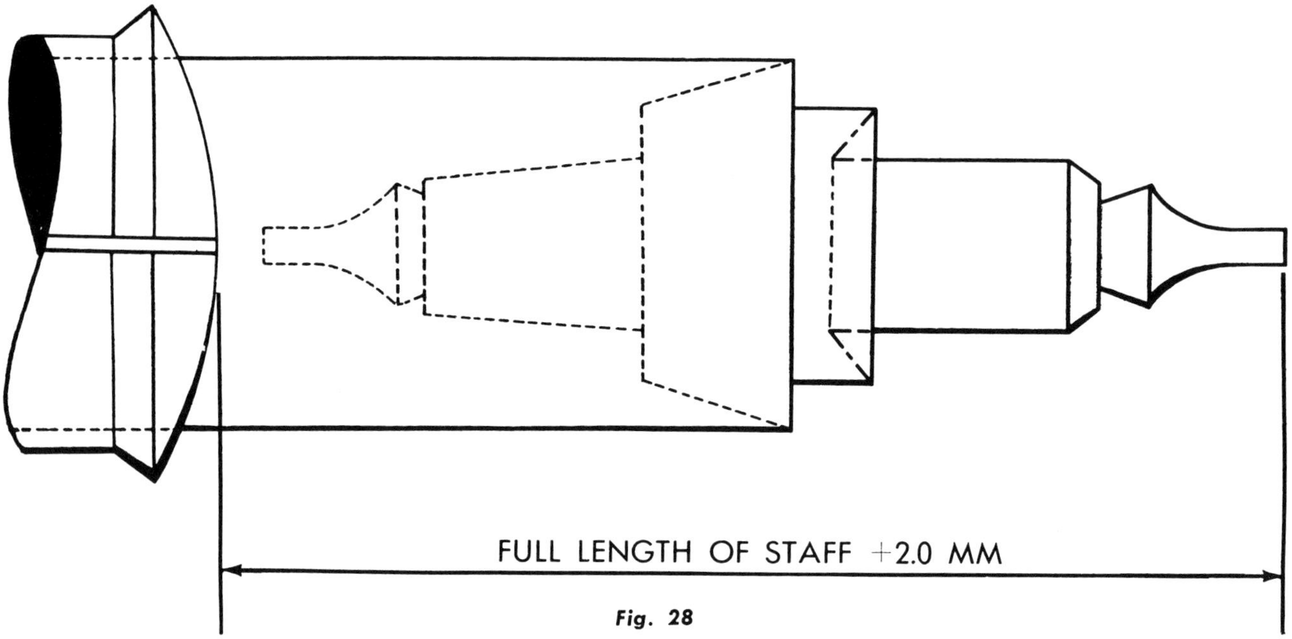

Fig. 28

Rough out the lower part of the staff leaving it long enough to include the lower pivot. *(See Fig. 29)*

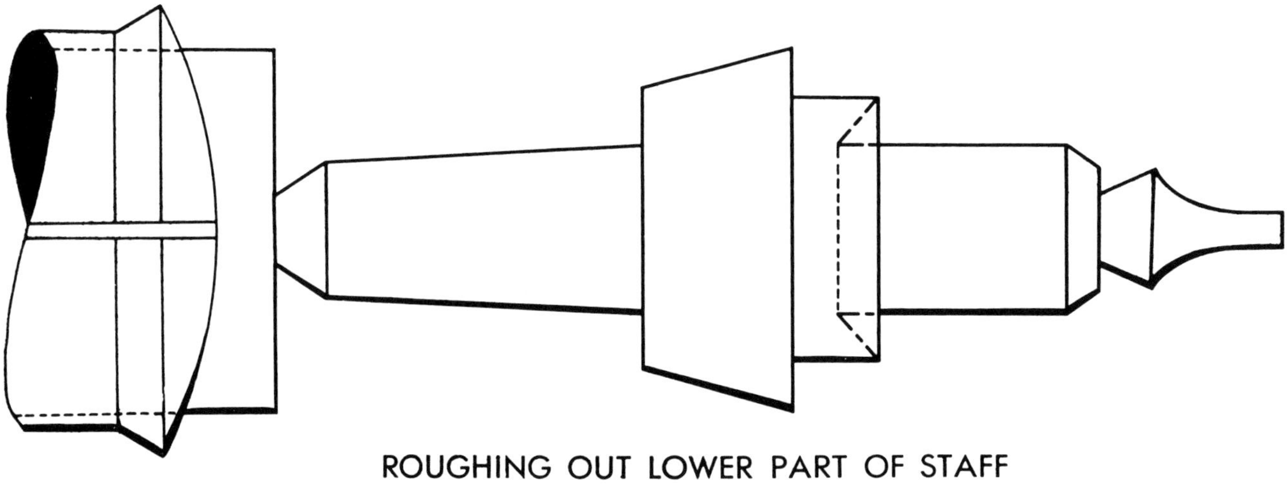

ROUGHING OUT LOWER PART OF STAFF

Fig. 29

Insert a brass chuck into the lathe. Turn the end flat and true. Cut a shellac reservoir into the end of the brass by resting the graver on the T-rest facing the work. *(See Fig. 30)*

It is important that this shellac reservoir should be turned very accurately with a sharp pointed graver, so as to remove any excess material which would hold or locate the pivot off center and cause the staff to run out of true. *(See Fig. 31)*

Fig. 30

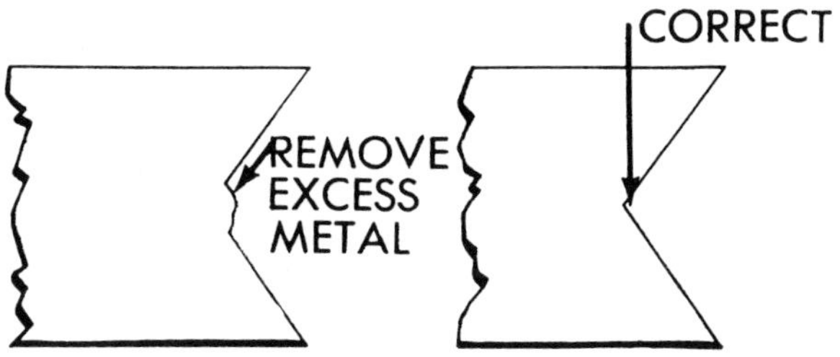

Fig. 31

Start the cutting in the center or as near as you can judge it. If a circle is formed, move the graver to the center of the circle to insure a dead center point. Cut the shellac reservoir deep enough to take in the upper part of the staff. *(See Fig. 32)*

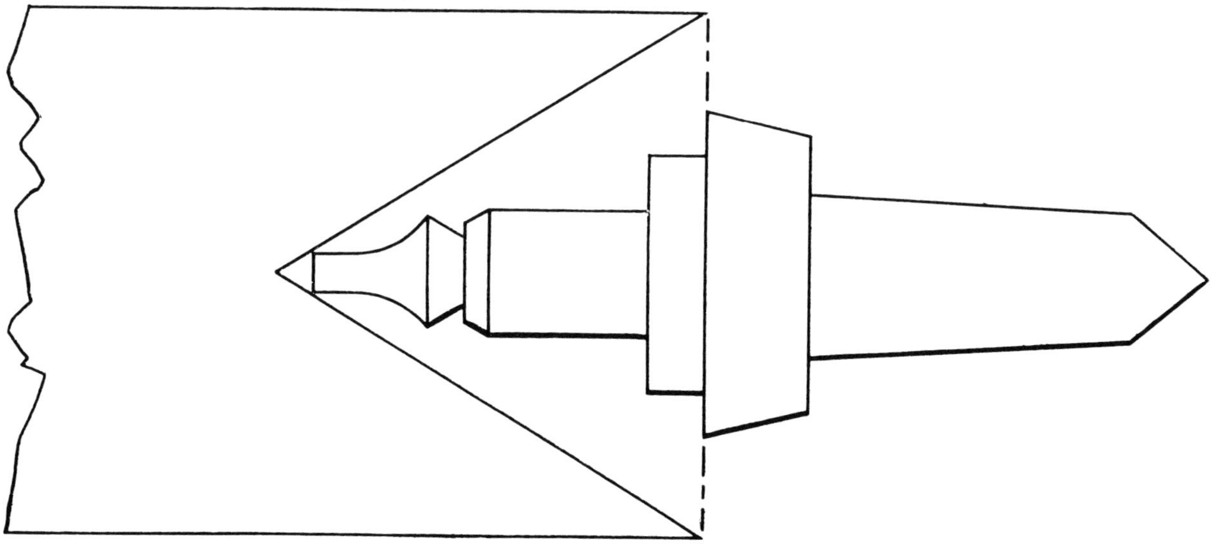

Fig. 32

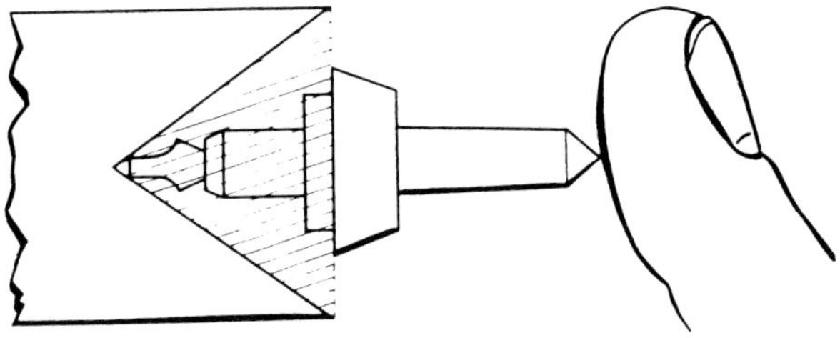

Fig. 33

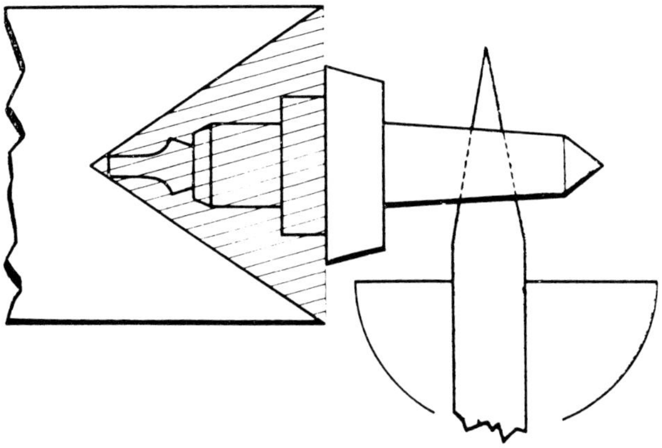

Fig. 34

While the brass chuck is in the lathe, warm the brass and pack the hollow with shellac and then apply more heat so that the shellac is quite soft. While the shellac is still soft, place the half finished staff in position. With the pad of the finger tip and turning the lathe, apply a little pressure to the staff, so that the pivot will be pushed into the point of the hollow. *(See Fig. 33)* Keep rotating the lathe until the shellac is hard.

To true the staff, apply a little heat, just sufficient to allow the staff to move. While the shellac is slightly softened, rest a piece of peg wood on the underside of the staff. Rotate the lathe until the staff turns true. Allow the peg wood to barely touch the roller seat during the truing operation so as to insure that the pivot is held firmly in position in the point of the shellac reservoir. Use the previous method to check if the staff is true. *(See Fig. 34)*

Using the old staff as a gauge reduce the lower part of the staff to the correct length as shown in *Fig. 35*.

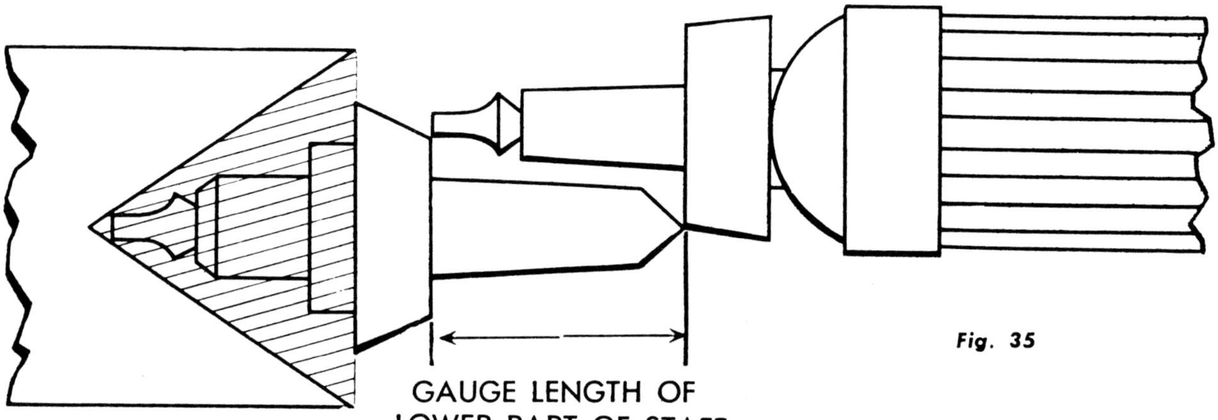

Fig. 35

Using the previous method turn, grind and polish the lower staff.

To remove staff from brass chuck, heat brass until shellac is soft. Remove shellac from staff by boiling it in alcohol.

If for some reason the old staff can not be used as a sample or is lost, the following method is used for taking measurements of a balance staff.

Remove the upper and lower cap jewels of the balance cock and screw the balance cock in position on the pillar plate. See that the balance cock and pillar plate are parallel. If the balance cock is bent, it should be straightened before any measurements are taken. *(See Fig. 36)*

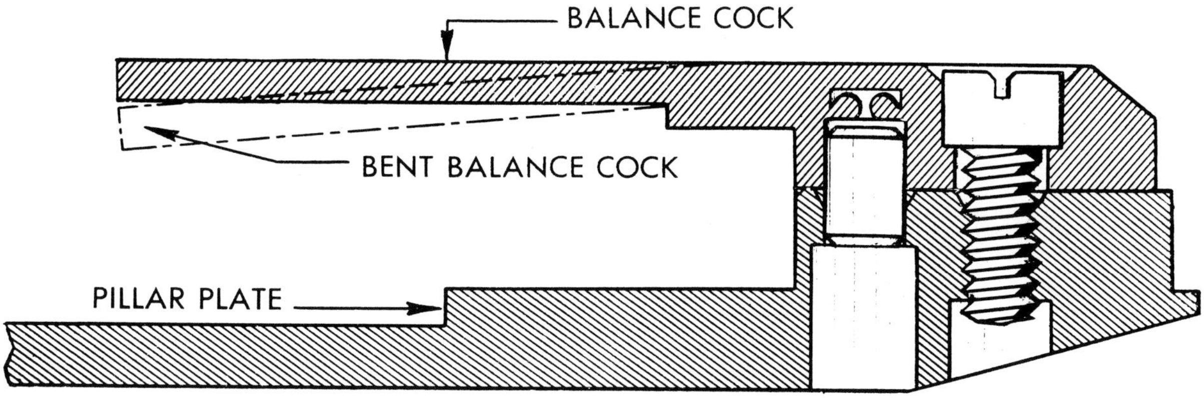

Fig. 36

Measure the distance from the top of the upper hole jewel to the bottom of the lower hole jewel and allow one tenth of a millimeter additional for the protrusion of the balance pivots. *(See Fig. 37)*

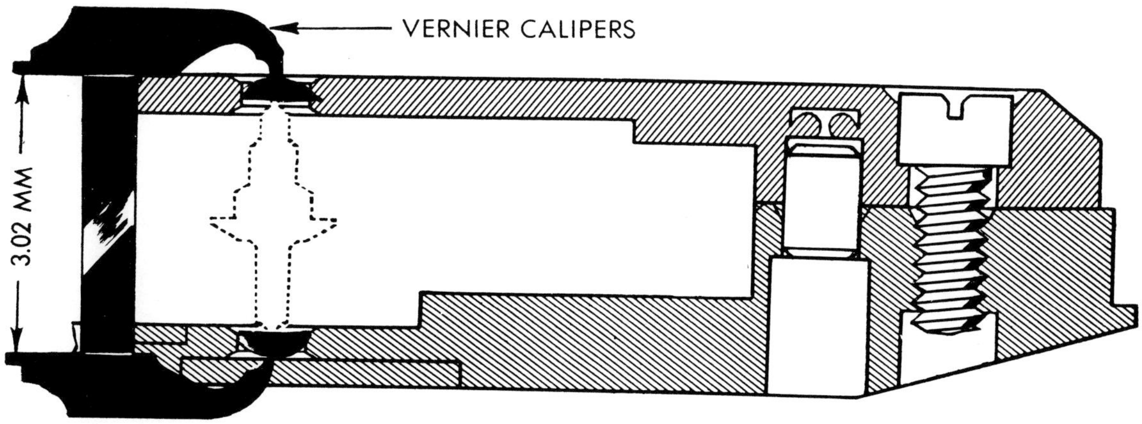

Fig. 37

The position of the balance wheel is measured from the lower balance hole jewel to the top of the pallet bridge. Add to this measurement the freedom necessary between the balance wheel and the pallet bridge. This will usually locate the balance wheel, which will permit the required space between the top of the balance arms and the balance cock for the hairspring collet and hairspring. Special consideration should be given to the correct location of the balance wheel so as to allow for the hairspring and collet. *(See Fig. 38)*

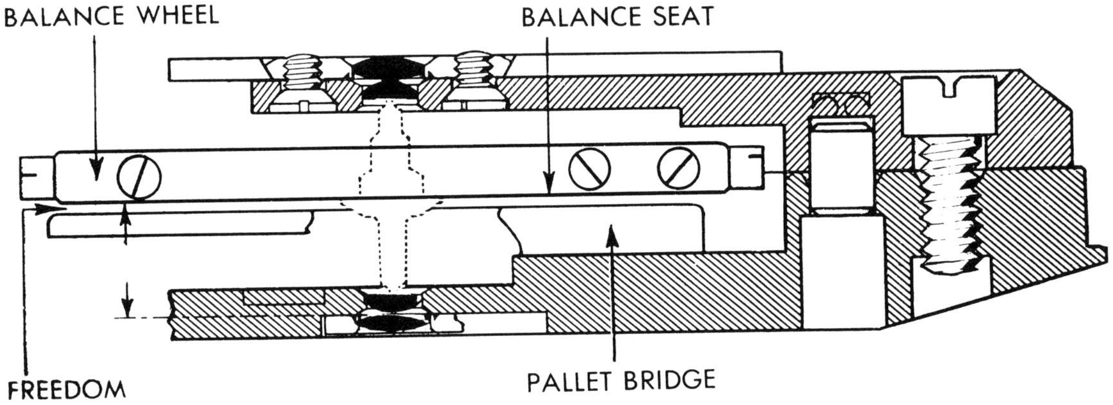

Fig. 38

Next measure from the lower balance hole jewel to the top of the fork and add the thickness of the impulse roller. To this measurement add the freedom of the impulse roller. This will locate the position of the roller seat. The key point of consideration here is that the fork and guard pin is parallel with the pillar plate and that the end of the guard pin coincides with the center of the safety roller. *(See Fig. 39)*

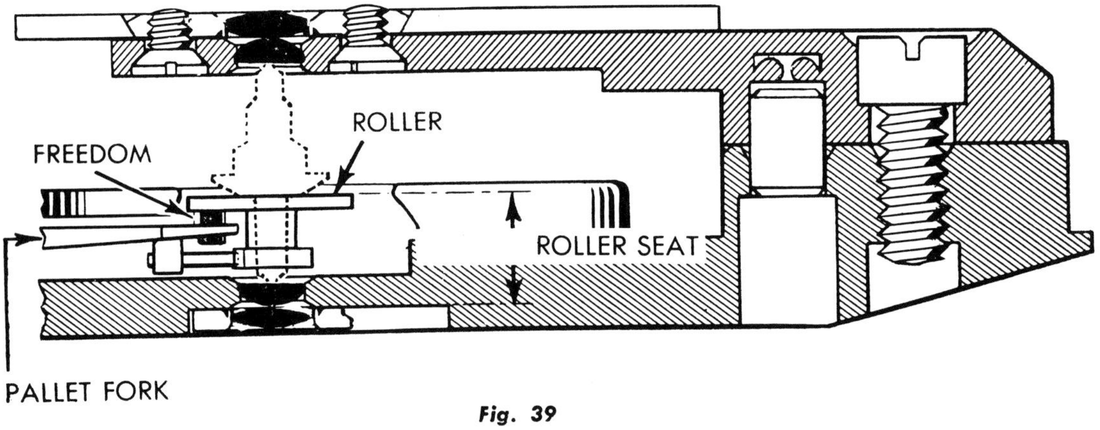

Fig. 39

The length of the collet shoulder is determined by the height of the hairspring collet. The rest of the staff is left for the upper and lower balance cones and pivots.

To determine the diameter of the collet shoulder, use the following method.

Select a brass wire, with a diameter slightly larger than the hole in the collet. Chuck the wire in the lathe. Turn the wire with a very slight taper, until the collet fits friction tight on the brass wire. *(See Fig. 40).* The position of the collet on the large side of the tapered wire determines the diameter of the collet shoulder.

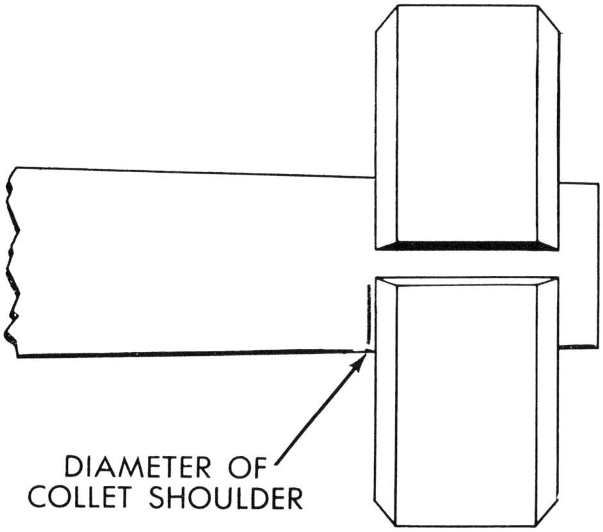

Fig. 40

To determine the diameter of the roller shoulder, turn the roller shoulder on a slight taper until the roller table fits on the small end of the taper. *(See Fig. 41A)*

Use the iron metal slip charged with oilstone powder and oil to grind the roller shoulder until the roller table almost fits up to the roller seat. The space between the roller seat and the roller table should not exceed double the thickness of the impulse roller. *(See Fig. 41B)* Clean off with pith wood all traces of oilstone. With the bell-metal slip charged with diamantine, polish the roller shoulder.

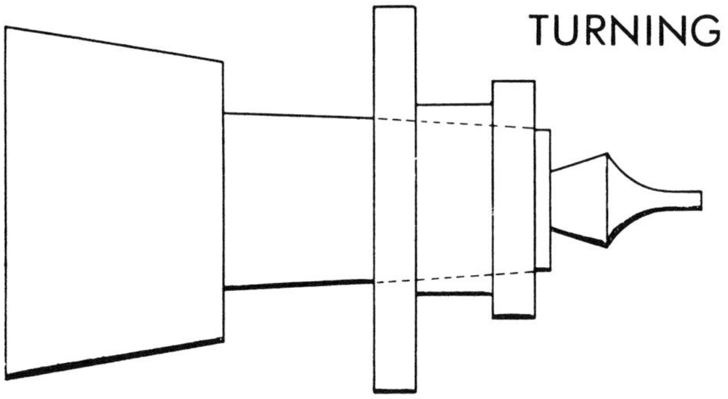

Fig. 41A

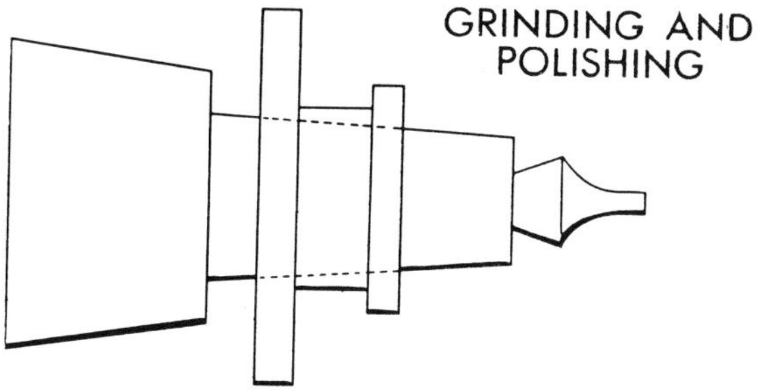

Fig. 41B

THE METRIC MICROMETER

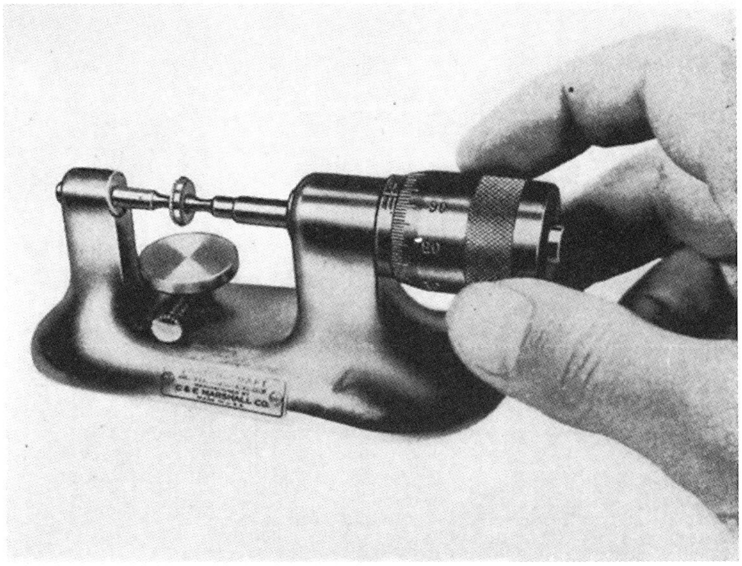

Fig. 42

Fig. 42 shows a metric micrometer for measuring watch parts. The table under the spindle is adjustable for height. The thimble is divided into one hundred parts so that the micrometer can be read to a hundredth of a millimeter.

The marks on the sleeve or barrel show the number of turns the thimble has made. As each division on the thimble is moved across the index of the sleeve, the space between the spindle and the anvil has opened one hundredth of a millimeter. Therefore, a complete turn of the thimble represents one millimeter and each division on the thimble as it passes the index on the sleeve represents a movement of 1/100 of a millimeter.

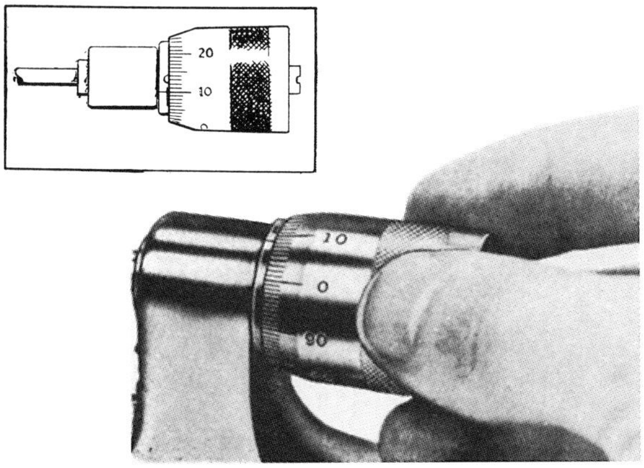

Fig. 43

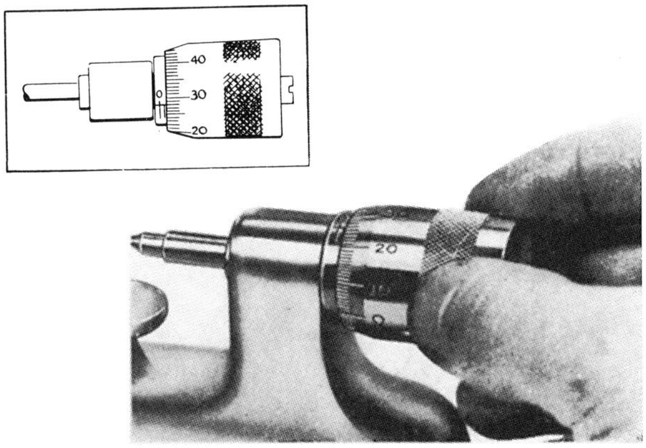

Fig. 44

Fig. 43 shows a setting where the tenth division coincides with the index on the sleeve, indicating a reading of ten hundredths of a millimeter.

Fig. 44 shows a setting where one line on the sleeve is visible and the twenty-eighth division on the thimble coincides with the index on the sleeve, indicating a reading of one and twenty-eight hundredths of a millimeter (1.28 M.M.)

TRAINING UNIT
NUMBER 3B

★

STEM MAKING

STEM MAKING

It is necessary for a watchmaker to know how to make a stem when the stem bearing is so worn that the old stem cannot be used and a new one must be made to fit the new dimensions, or when a stem cannot be purchased from the material house, or when a stem is needed for an old model watch and parts are no longer obtainable.

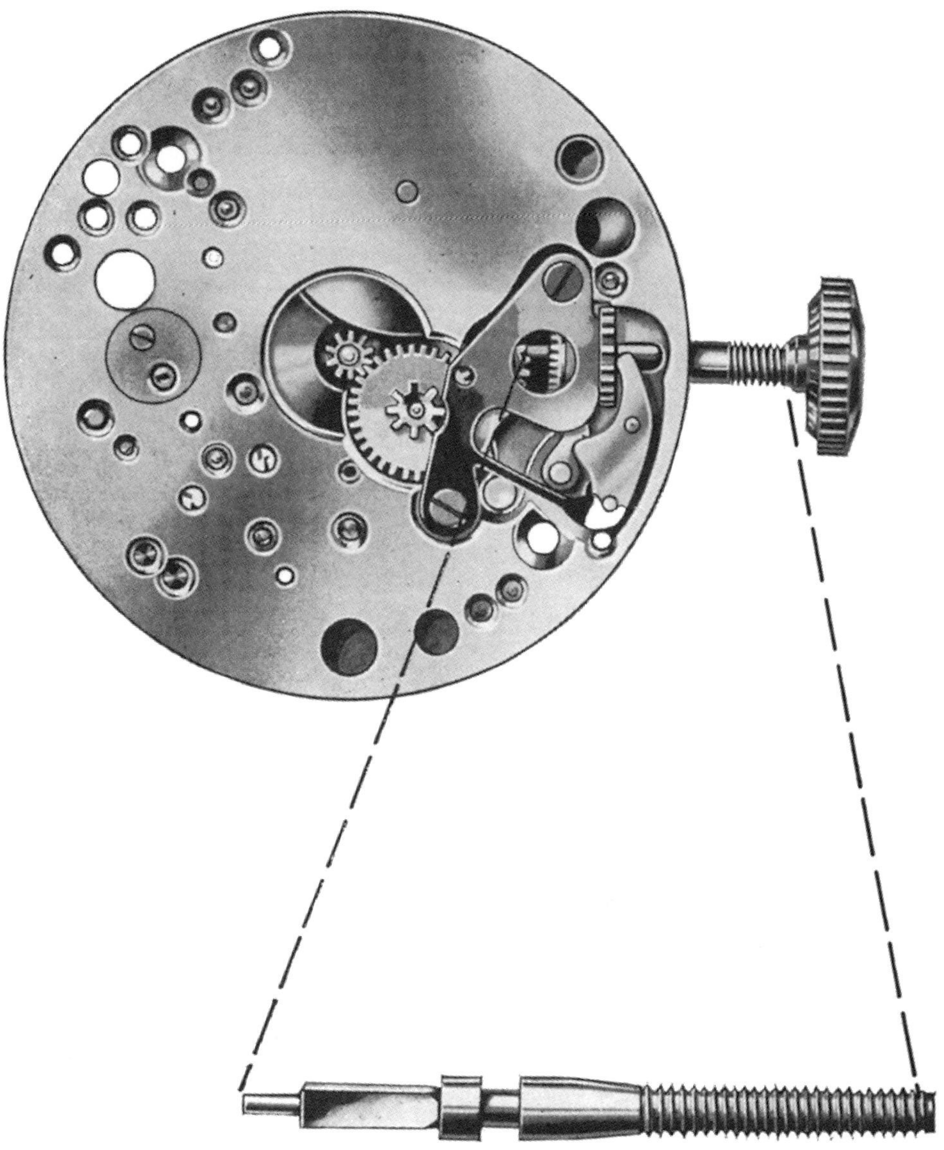

BULOVA SCHOOL *of* WATCHMAKING

TRAINING UNIT NUMBER 3B

Subject:

Making a stem

Objective:

To develop skill and proper methods in making stems.

New Tools:

1. Screwplate
2. Taps
3. Twist drills
4. Filing rest

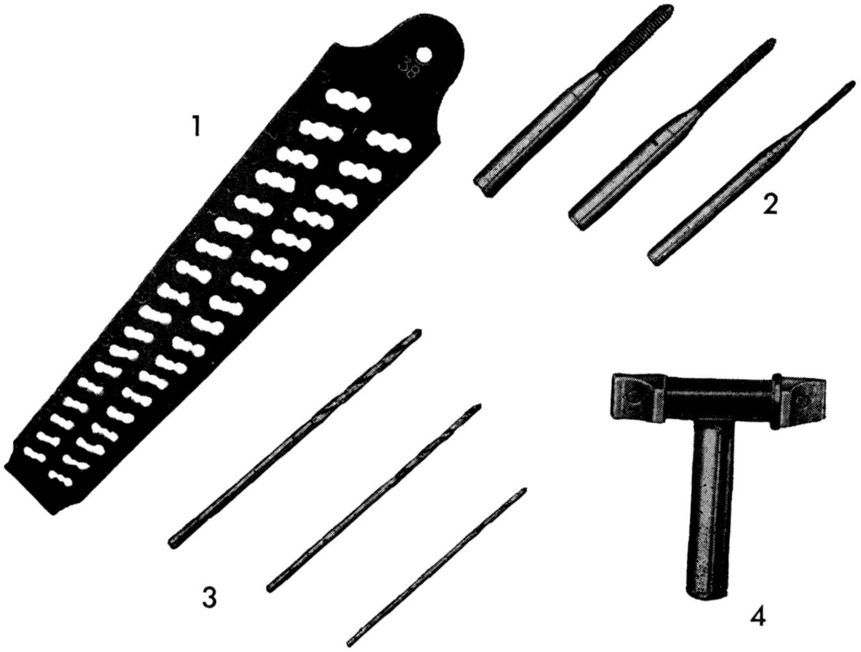

MAKING A STEM FROM SAMPLE

Select a piece of soft steel wire with a diameter slightly larger than the largest diameter of the sample stem. Chuck the steel in the lathe to extend out from the chuck so that the pilot, winding pinion and clutch wheel shoulders can be turned (see Fig. 1)

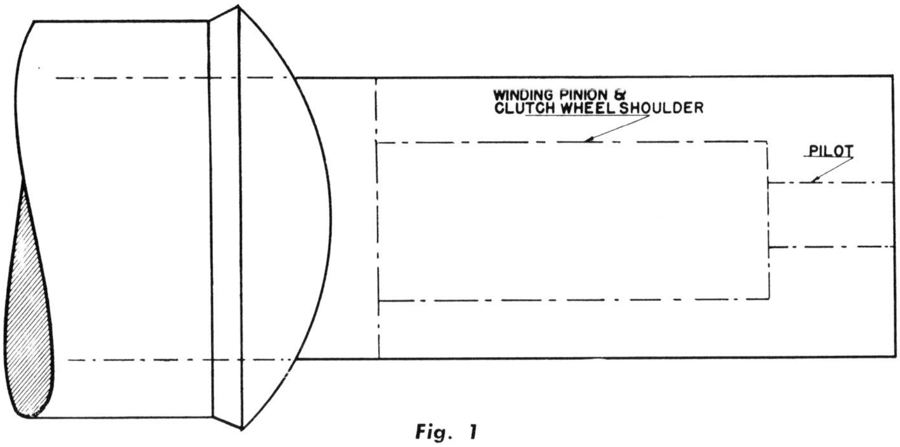

Fig. 1

Place the sample stem parallel with the blank so that the pilot seat can be marked with the graver. (see Fig. 2) From this mark proceed to turn the pilot two hundredths of a millimeter. (.02MM) larger in diameter than the sample, then corner undercut the square shoulder.

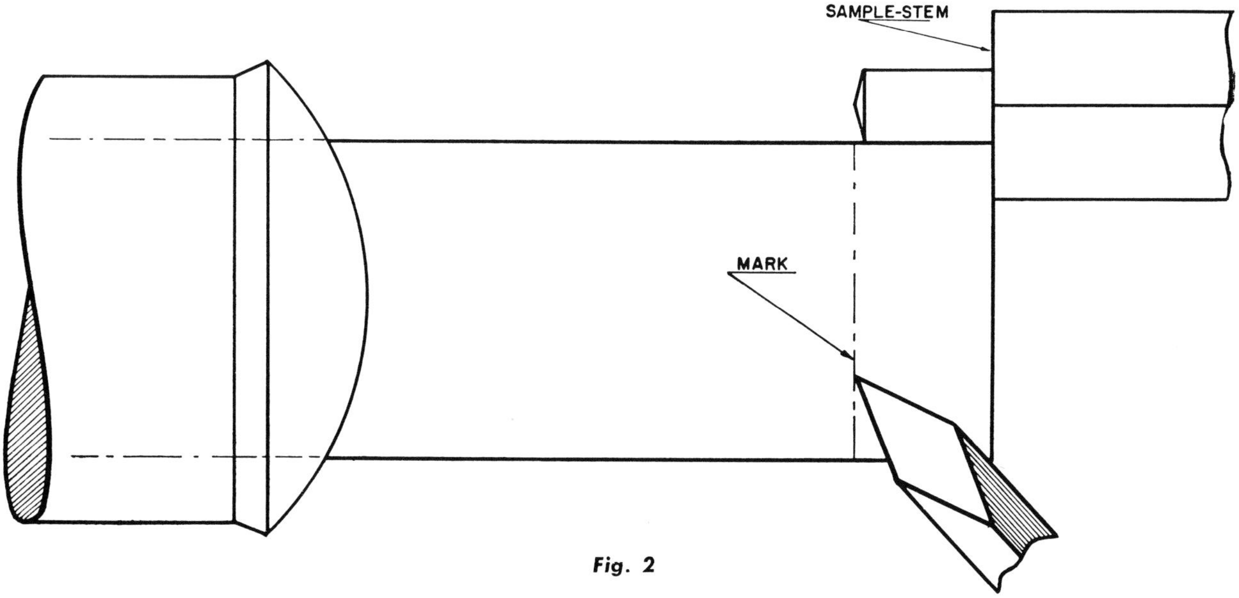

Fig. 2

Turn the rest of the blank which extends out from the chuck two hundredths of a millimeter (.02MM) larger in diameter than the bearing of the sample. (see Fig. 3)

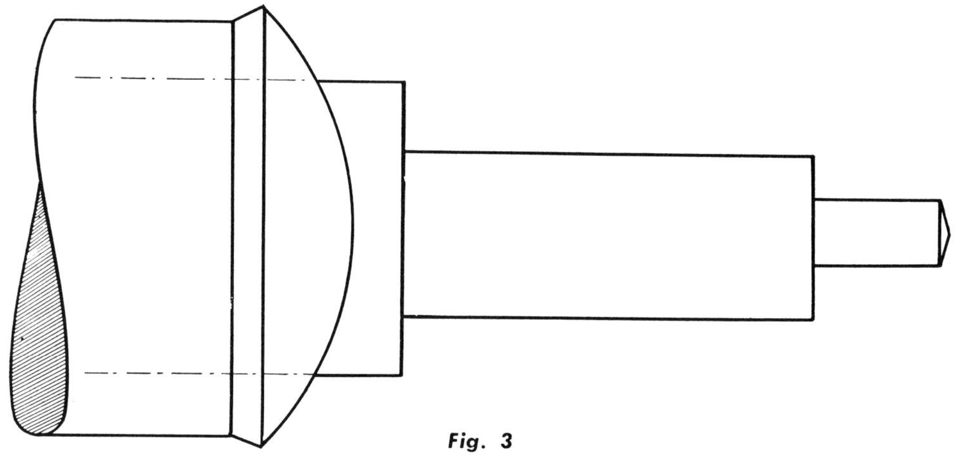

Fig. 3

Place the sample stem parallel with the blank so that the seat for the clutch wheel can be marked with the graver (see Fig. 4)

If the sample is broken at the threads or setting lever slot, (this is the usual breaking point) the section with the clutch wheel shoulder should be used to locate the seat for the clutch wheel.

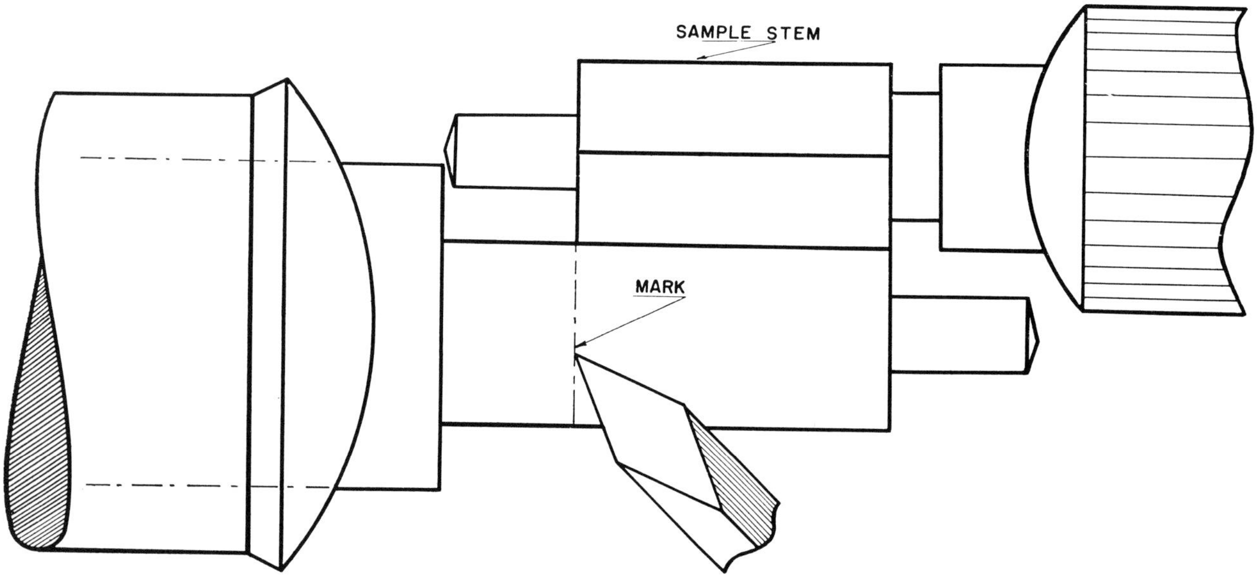

Fig. 4

Into this mark turn a slight groove by presenting the graver to the work so as to turn a square shoulder for the seat of the clutch wheel. (see Fig. 5)

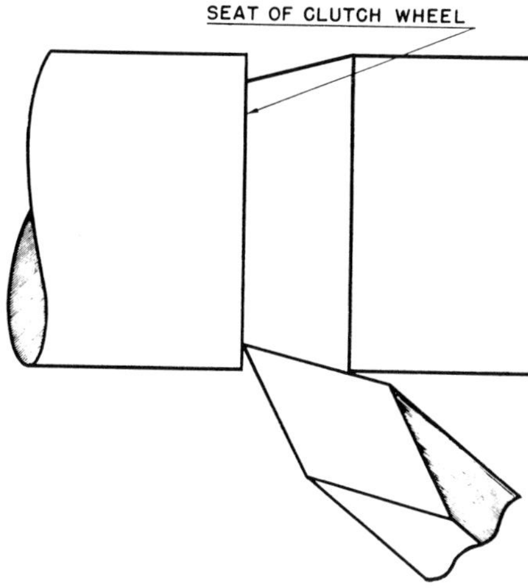

Fig. 5

From this square shoulder, the square must be filed in the following manner. Remove the "T" rest and replace with the filing rest. Adjust the filing rest so that the top of the roller is level with the blank and approximately one-half inch away from the work.

Lock the head on the lathe with the index pin (which fits into the holes of the index plate). Place a number three (No. 3) file, having a safe or non-cutting edge firmly on the roller, with the safety edge of the file against the shoulder of the roller and on the back taper of the groove as shown in Fig. 6.

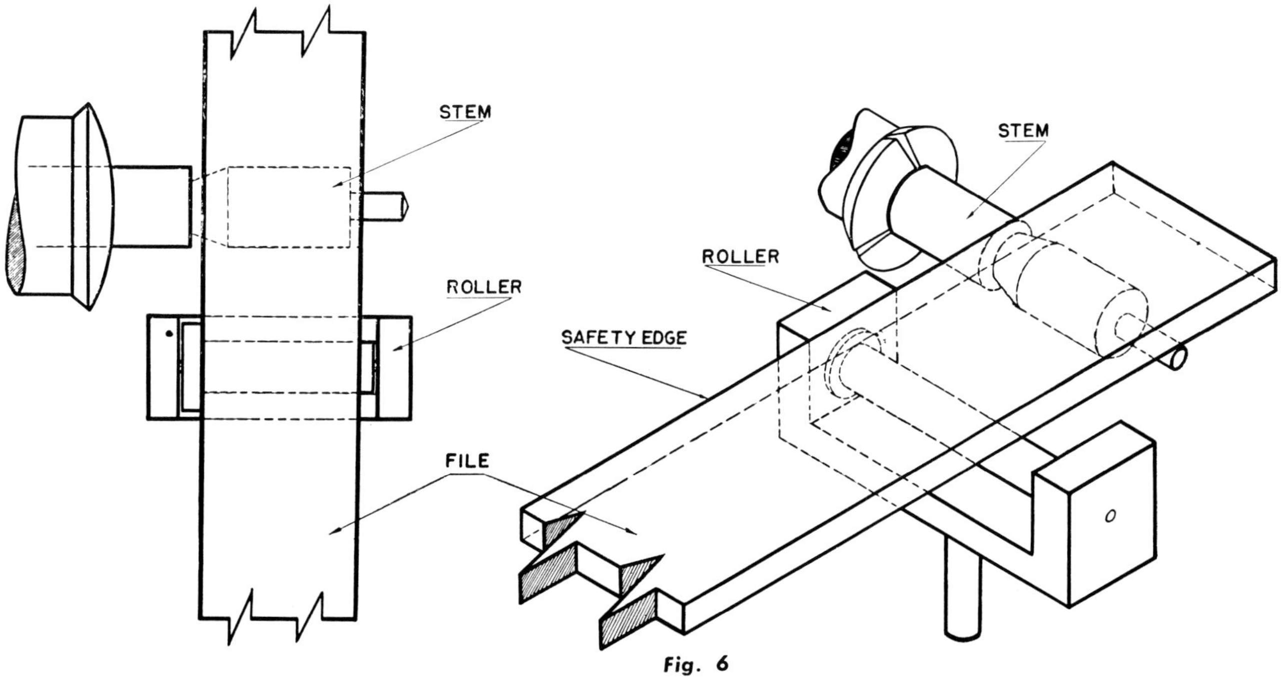

Fig. 6

Apply pressure on the roller as you move the file back and forth, gradually tipping the file until it starts cutting the blank. Care must be used to see that the file is held flat on the roller and against the shoulder of the roller. Take three or four strokes with the file. Make sure that the safety edge of the file does not touch the square shoulder of the groove during these first strokes. After this the safety edge of the file may be permitted to rest against the square shoulder of the groove so that the filing can be continued without slipping past the square shoulder.

Take two more strokes with the file, then turn the headstock ½ turn (180 degrees). Care must be taken to lock the headstock with the index pin before continuing to file. Repeat the filing on this side of the blank using the same procedure as before. Check with a micrometer the diameter of the flat sides with the square of the sample. Continue filing the same amount on each side until the diameter of the flat sides is equal to the square of the sample. (see Fig. 7)

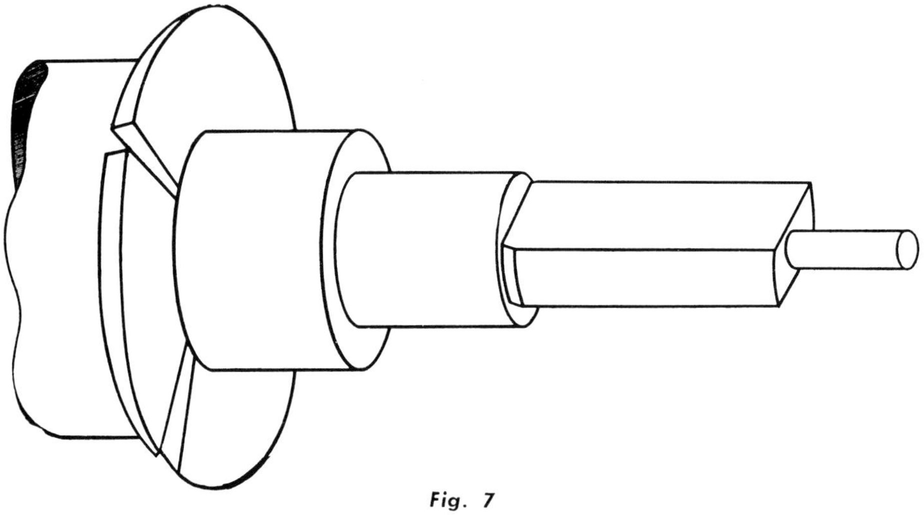

Fig. 7

Turn the headstock ¼ turn (90 degrees) and proceed to file the third side of the square using the same procedure as that used in filing the first side. (see Fig. 8)

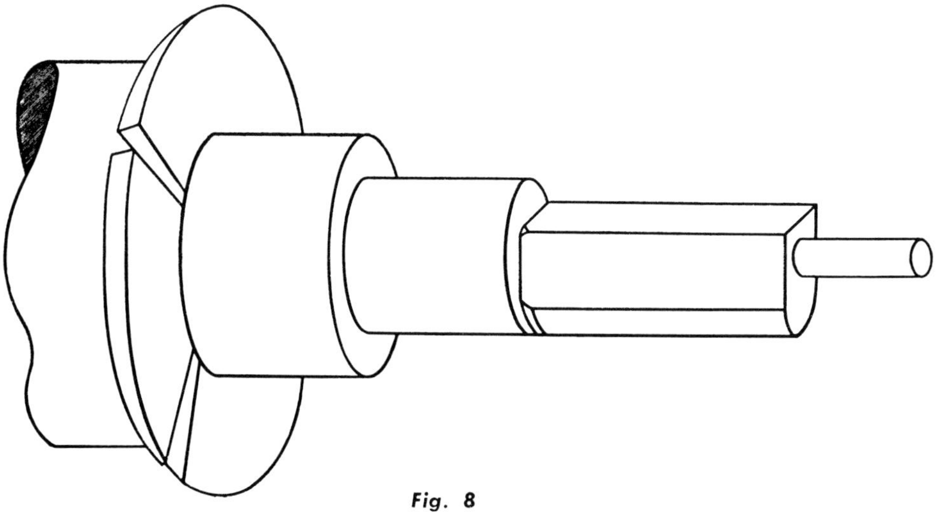

Fig. 8

Turn the headstock ½ turn (180 degrees) and proceed filing the final side of the square using the same precaution in checking the diameter as used when filing the second side. Continue as before; filing the same amount on both sides (3rd and 4th side) until the square is complete. (see Fig. 9)

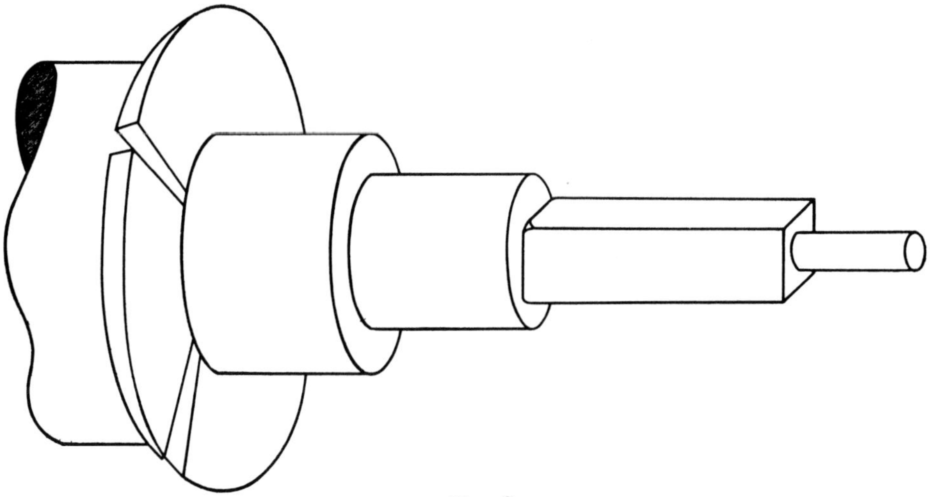

Fig. 9

Place the unfinished stem on the bench and lay the sample stem parallel with the blank. Mark the unfinished stem to indicate the length of the threaded portion (see Fig. 10) and cut off the excess metal.

If the sample is broken at the setting lever slot, the section with the threads should be used to mark the length of the threaded portion of the stem. If the thread of the sample is broken the threaded portion must be made long enough to extend out through the watch case to permit the crown to be screwed on.

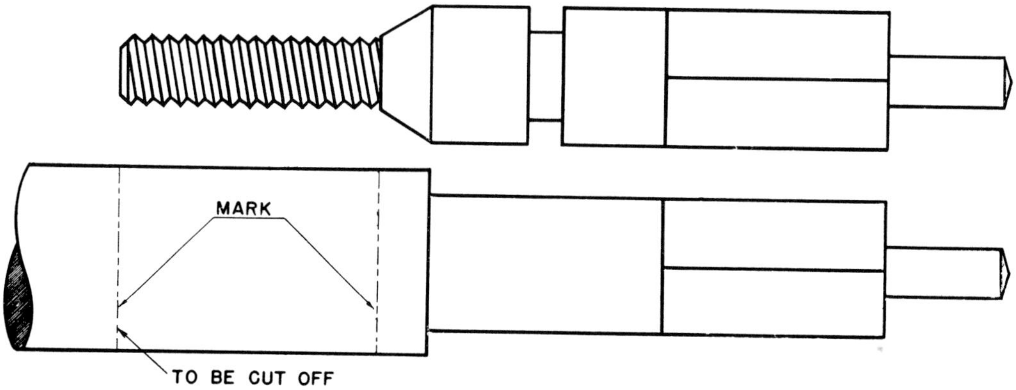

Fig. 10

Reverse the blank in the lathe (use another chuck to hold the bearing of the stem, see Fig. 11) and turn the shoulder for the threads. The diameter of this shoulder is determined by fitting it into the screwplate two holes larger than the one to be used in cutting the threads. (in some screwplates one hole larger than the one to be threaded must be used to determine the diameter of the shoulder for the threads. Experience with your own screw plate will enable you to determine the correct hole to use.)

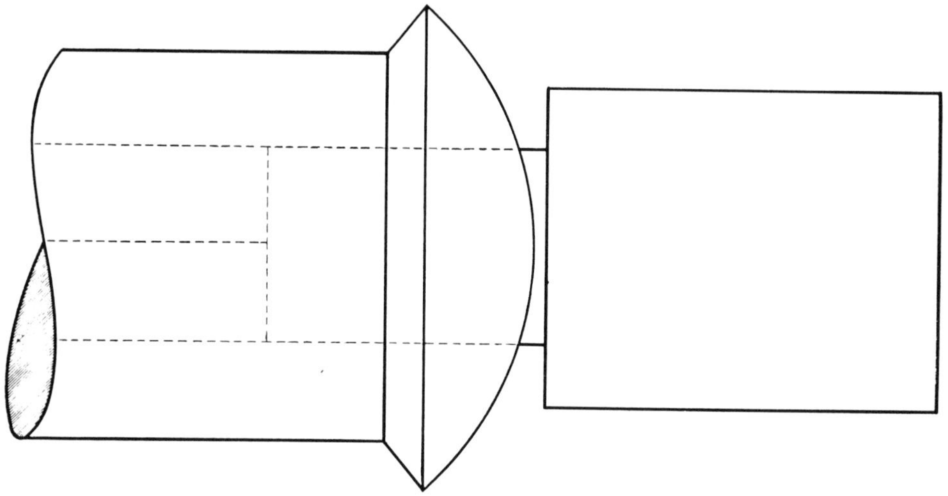

Fig. 11

Turn a small taper at the end of the shoulder to permit the screwplate to start cutting the threads as shown in Fig. 12.

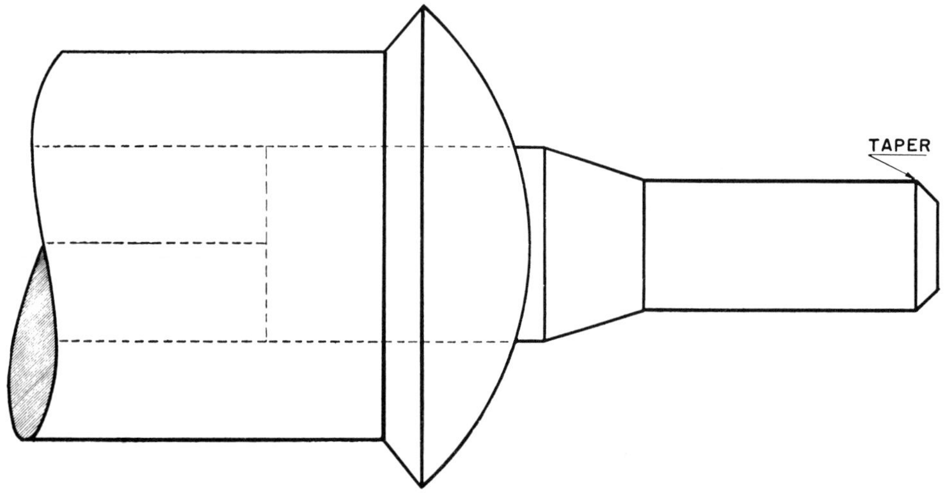

Fig. 12

Cut the threads on the stem by turning the screwplate with the right hand and the headstock with the other. Using plenty of oil, work the headstock backward and forward, at the same time work the screwplate in the opposite direction, gradually cutting the threads to the required length. Care must be taken not to cut the threads too rapidly as the shoulder might be twisted off (see Fig. 13)

Fig. 13

Before turning the slot for the setting lever, the unfinished stem must be hardened and tempered to a blue color.

Chuck the stem in the lathe and place the sample stem parallel with the unfinished stem, (as shown in Fig. 14 method A) so that the slot for the setting lever can be marked with a graver. If the sample stem is broken at the setting lever slot, the sample stem should be held parallel with the unfinished stem as shown in Fig. 14 method B.

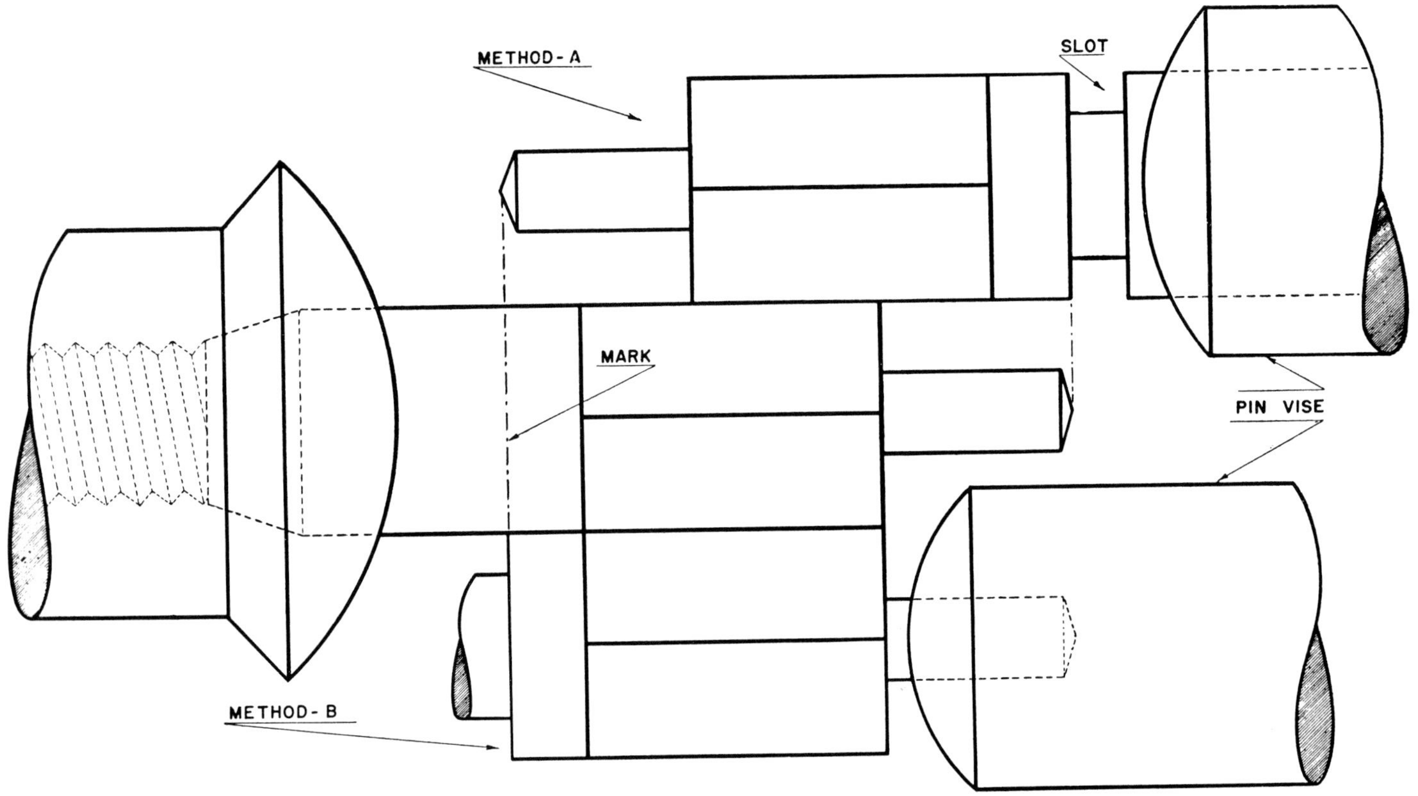

Fig. 14

Use a special graver, (as shown in Fig. 15) turn the slot for the setting lever to the same width and depth as the sample stem. (see Fig. 16)

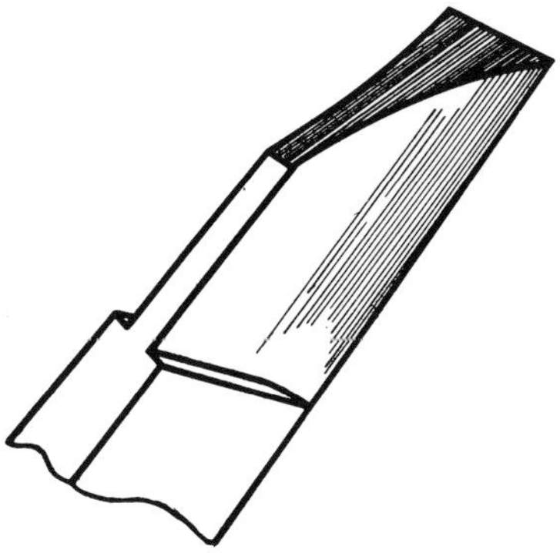

Fig. 15

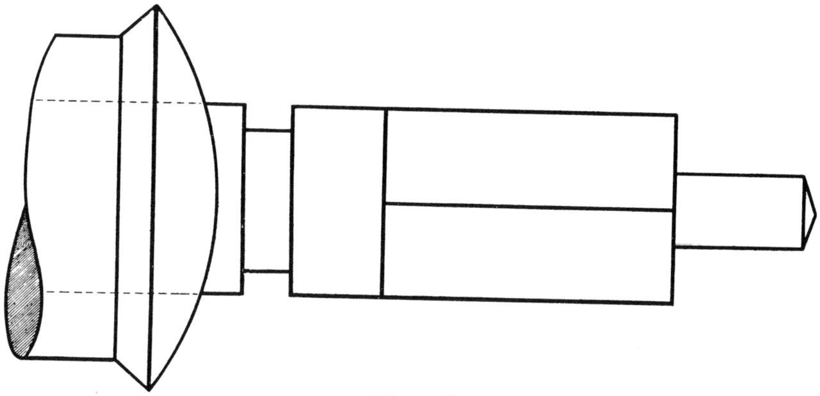

Fig. 16

Use a triangle iron slip charged with oilstone powder and oil, grind the bearing and pilot, in order to remove all traces of turning marks. Clean the stem to remove all traces of oilstone. Absolute cleanliness is essential before starting the polishing.

Use a bell-metal slip charged with diamantine and oil, polish the bearing and pilot.

Chuck a female center in the headstock and a female center into the tailstock and fit the stem between these centers. Care must be taken that the stem moves freely so that when the grinding or polishing slips are used with a little pressure, the stem will find its own level. (see Fig. 17) Grind and polish the square using the same procedure as that used in grinding and polishing the bearing and pilot. This completes the stem made to match a sample.

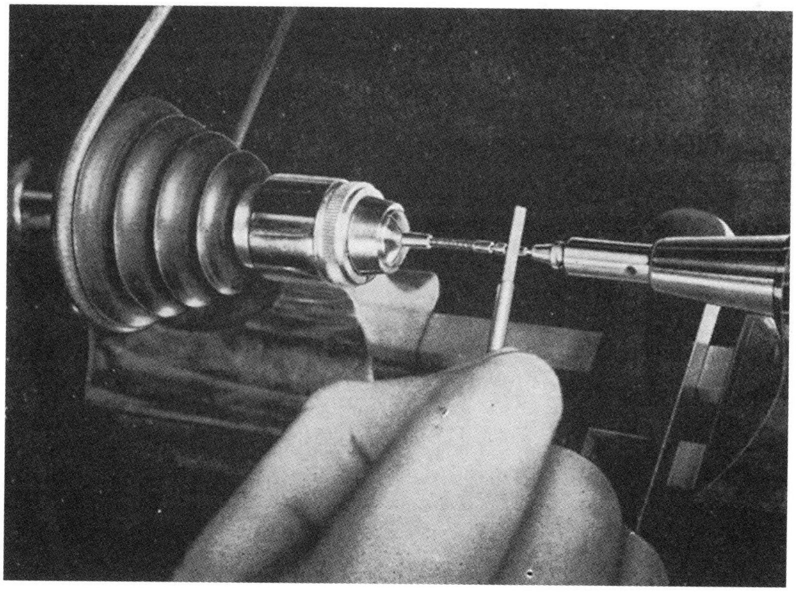

Fig. 17

MAKING A STEM WITHOUT A SAMPLE

Remove all the winding and setting parts of the movement, so that the diameters of the pilot and bearing can be determined.

To determine the diameter of the pilot, select a twist drill so that the shank will fit snugly into the pilot hole of the pillar plate (see Fig. 18)

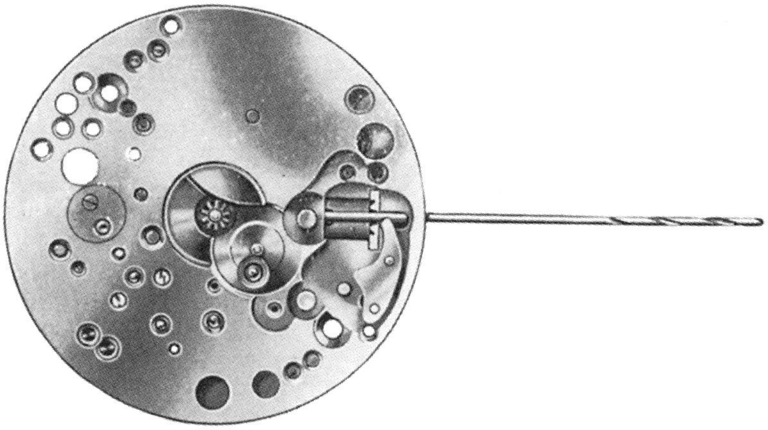

Fig. 18

By selecting a larger twist drill so that the shank will fit into the opening in the pillar plate, the diameter of the stem bearing can be determined, as shown in Fig. 19.

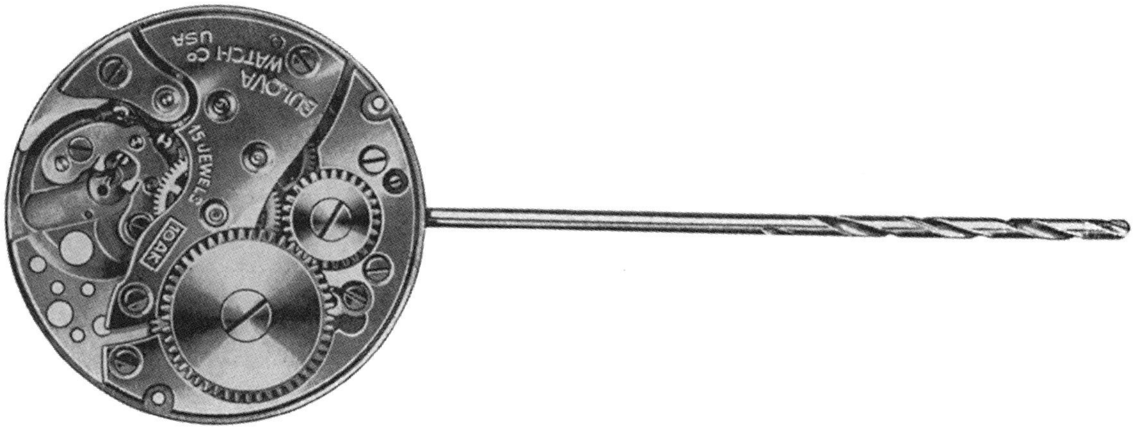

Fig. 19

Select a piece of soft steel wire with a diameter slightly larger than the shank of the larger drill. Chuck the steel in the lathe to extend out from the chuck so that the end of the metal reaches the inside edge of the setting wheel teeth. (see Fig. 20)

Holding the pillar plate in this position, mark the seat of the pilot with a graver as shown in Fig. 20. From this mark proceed to turn the pilot two hundredths of a millimeter (.02 MM) larger in diameter than the shank of the smaller drill.

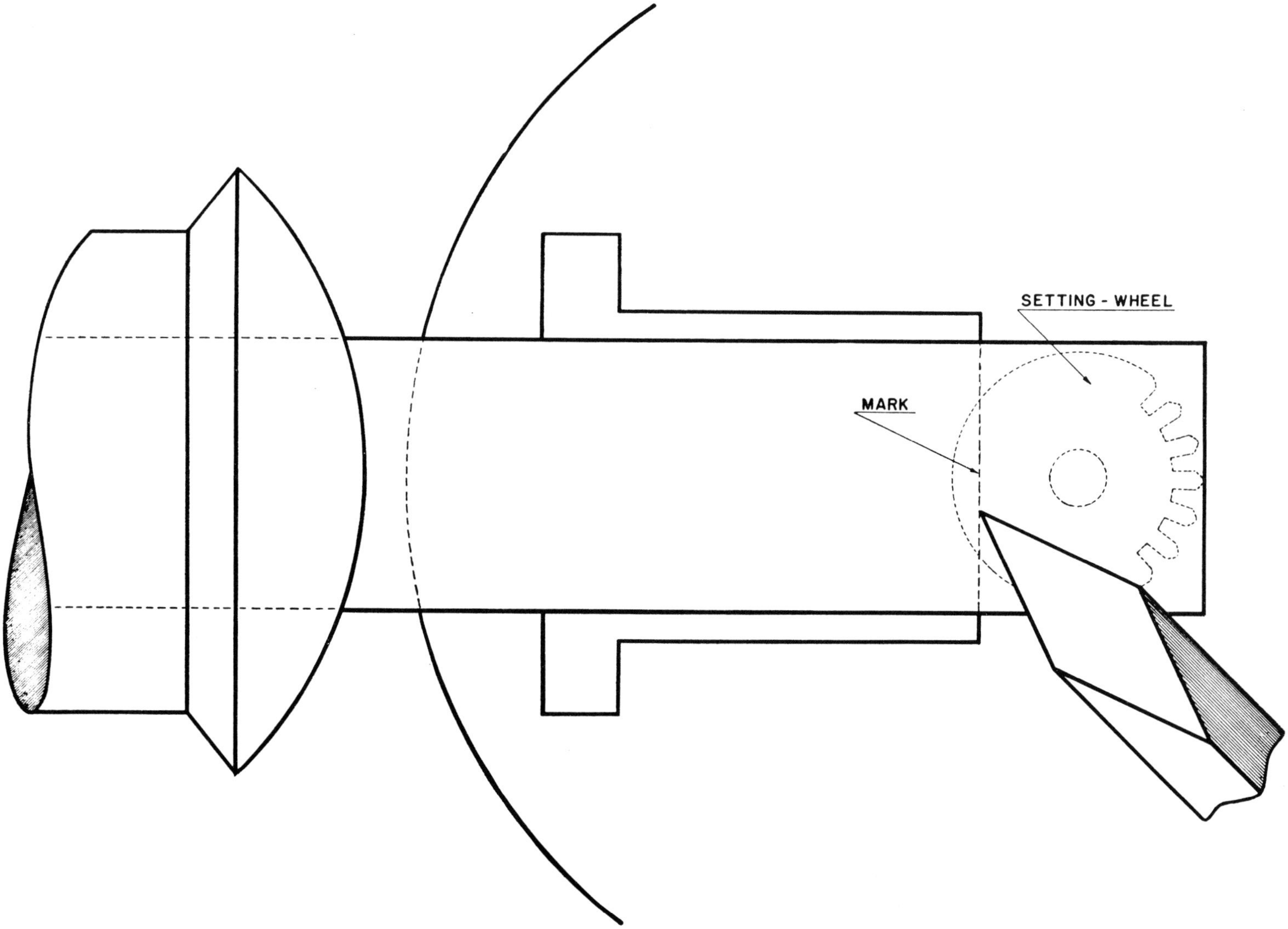

Fig. 20

Turn the rest of the blank which extends out from the chuck two-hundredths of a millimeter (.02MM) larger in diameter than the shank of the larger drill. (see Fig. 21)

Using a triangle iron slip grind the bearing and pilot with oilstone powder and oil to remove all turning marks, then clean the stem.

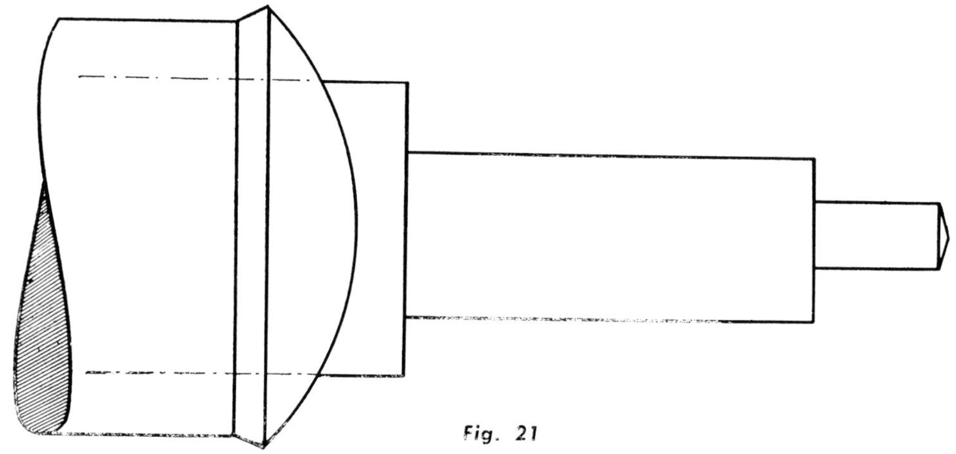

Fig. 21

The stem blank may now be tried in the pillar plate to determine the proper length for the pilot. This should be done without removing the blank from the lathe chuck.

If the hole for the pilot has been drilled through, care must be taken not to make the pilot too long, or the pilot might interfere with a moving part and stop the watch. If the pilot hole has not been drilled through, shorten the pilot until the pilot seat rests on the pillar plate as shown in Fig. 22.

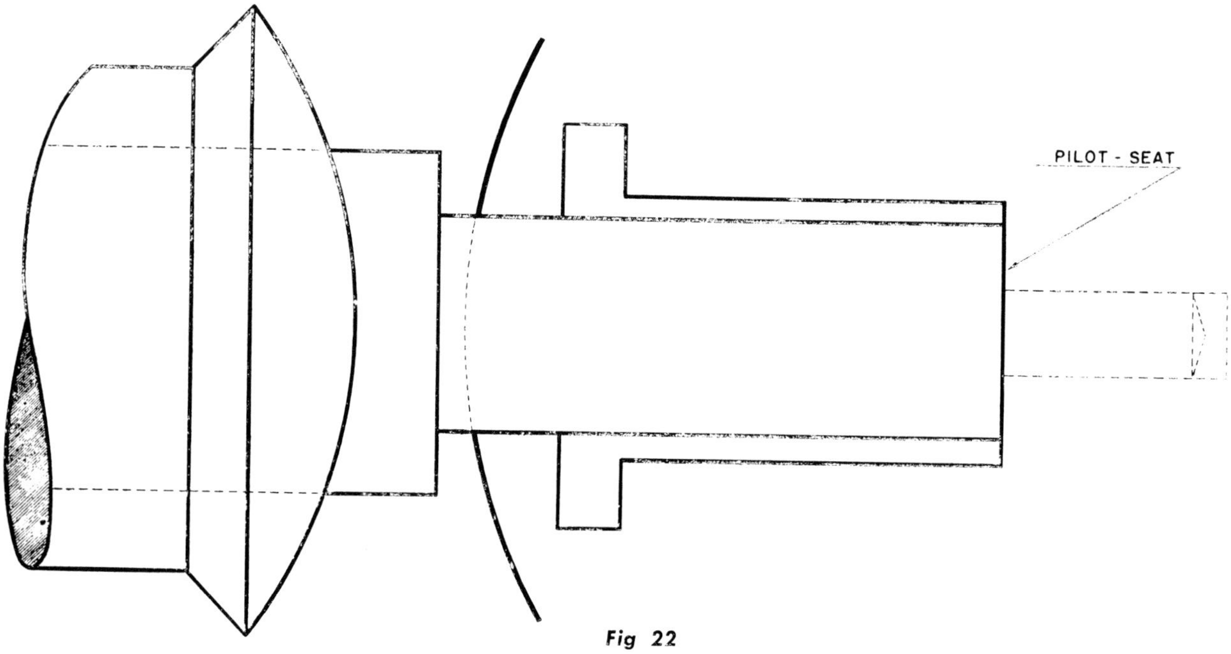

Fig 22

Mark off the seat for the winding pinion. (see Fig. 23-A) Should the winding pinion have a counterbore, the mark must be made at point B in Fig. 23, to allow the bearing to enter the counterbore as shown in Fig. 24. If the bearing is the same diameter as the winding pinion shoulder, there is no need for marking off the seat for the winding pinion.

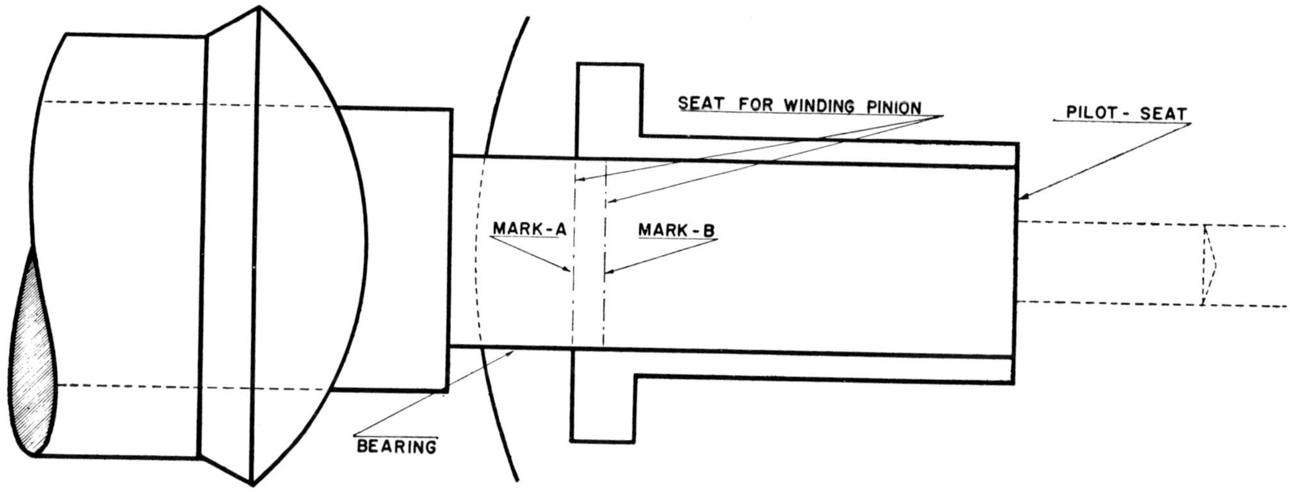

Fig. 23

From this mark turn the shoulder of the winding pinion, until the winding pinion fits on its shoulder. (If the bearing is the same diameter as the winding pinion shoulder this operation is not necessary).

Mark off the seat for the clutch wheel (see Fig. 24) a little above the ratchet teeth of the winding pinion. This will insure the meshing of the ratchet teeth of the clutch wheel with the winding pinion after the square is filed.

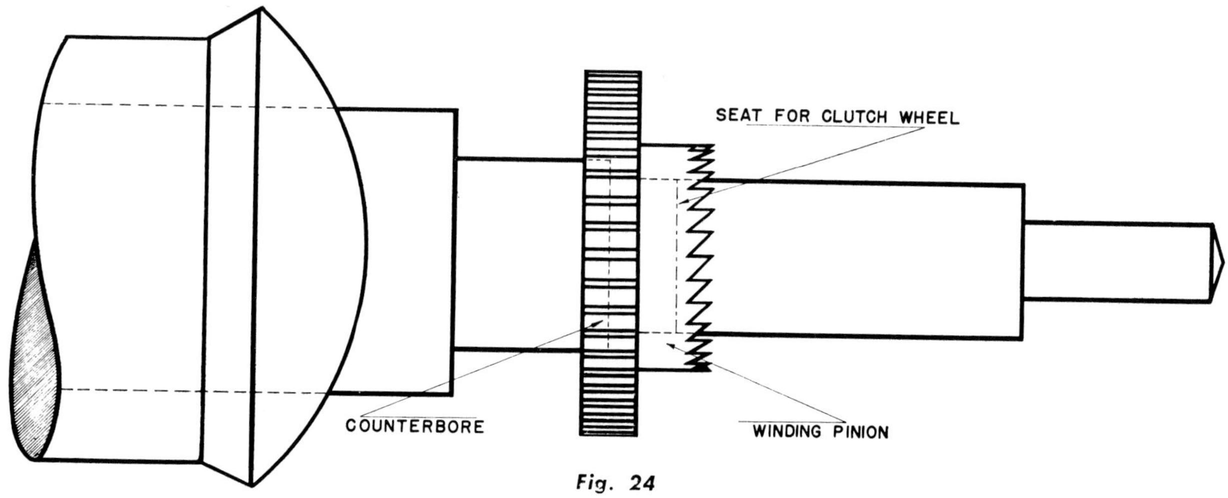

Fig. 24

Using the previous method proceed to turn a slight groove into this mark, then file the square.

Select a tap that will fit snugly into the crown to be used. This will determine the number of the hole on the screwplate for cutting the threads of the stem. HARDEN AND TEMPER the unfinished stem before turning the slot for the setting lever.

To locate the slot for the setting lever, insert the stem into the pillar plate and place a mark where the pin of the setting lever touches the stem in the winding position. (See Fig. 25) (have winding and setting mechanism assembled for this operation). Turn the slot deep enough so that the setting lever pin will be free from the bottom of the slot and wide enough to give it a little freedom. Finish the stem by polishing it, using the previous method.

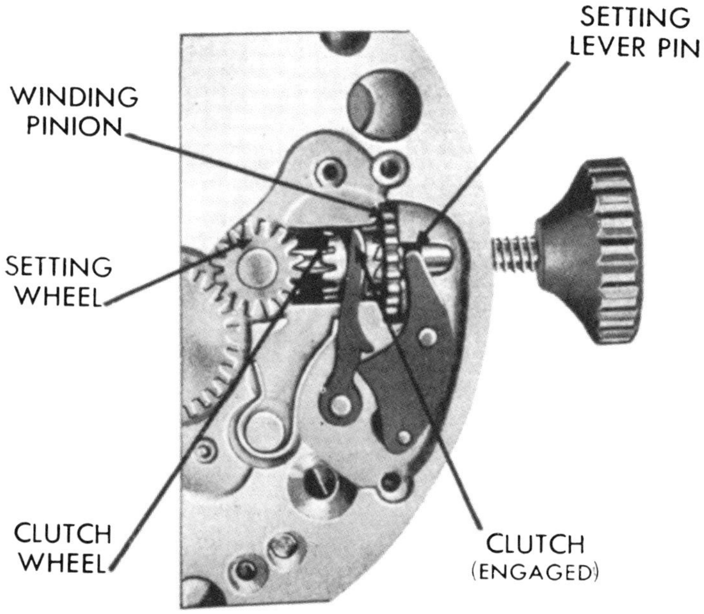

Fig. 25

NOMENCLATURE FOR WINDING & SETTING PARTS

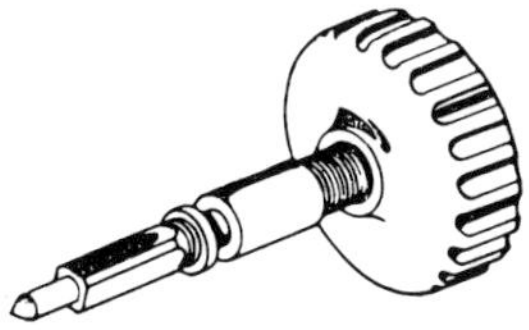

STEM AND CROWN

WINDING PINION

CLUTCH WHEEL

SETTING LEVER (detent)

CLUTCH LEVER

CLUTCH LEVER SPRING

SETTING WHEEL

YOKE (set bridge)

MINUTE WHEEL

TRAINING UNIT NUMBER 4

★

BURNISHING BALANCE PIVOTS

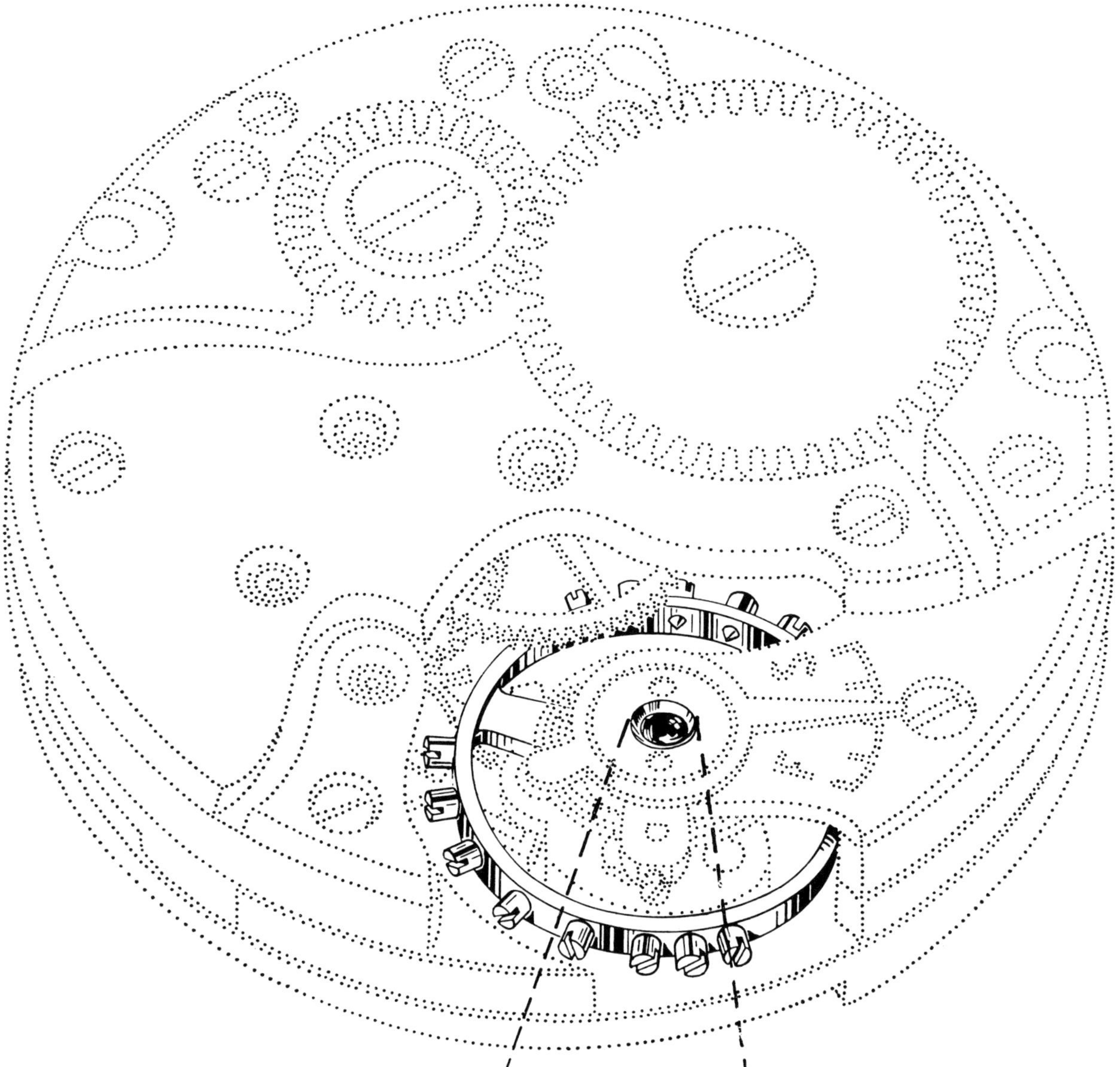

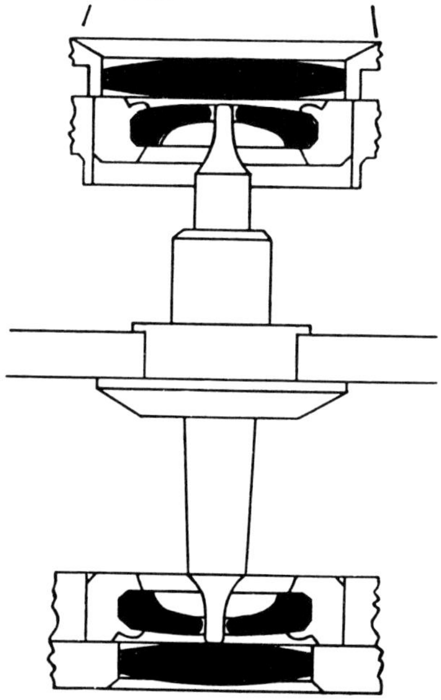

Since the balance wheel and hairspring constitute the timekeeping element in the watch, it follows that the condition of the balance staff pivots can add to or detract from its timekeeping qualities.

Therefore, it is important at this time to make a study of balance staff pivots and to acquire a skill necessary in preparing them to operate at maximum efficiency.

JOSEPH BULOVA SCHOOL *of* WATCHMAKING

TRAINING UNIT NUMBER 4

Subject: Burnishing Balance Staff Pivots.

Objectives:

1. To study the various types of pivot defects and the methods for eliminating them.
2. To develop a high degree of skill in burnishing balance pivots.

New Tools and Materials Provided:

1. Jacot Lathe.
2. Dorrington Pivot Polisher.
3. Balance Chuck.
4. Hardened steel burnisher.
5. Heavy brass or soft steel tweezers.
6. Watch oil.
7. 2/0 Emery cloth.
8. Fiddle bow.

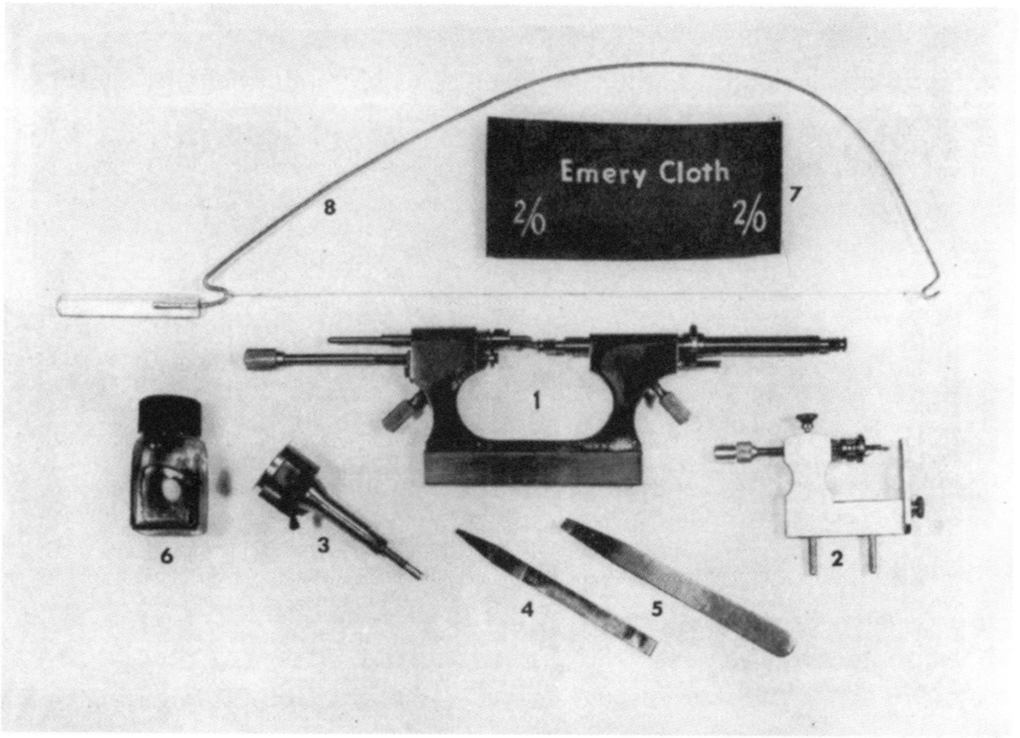

BALANCE STAFF PIVOTS AND JEWELS:

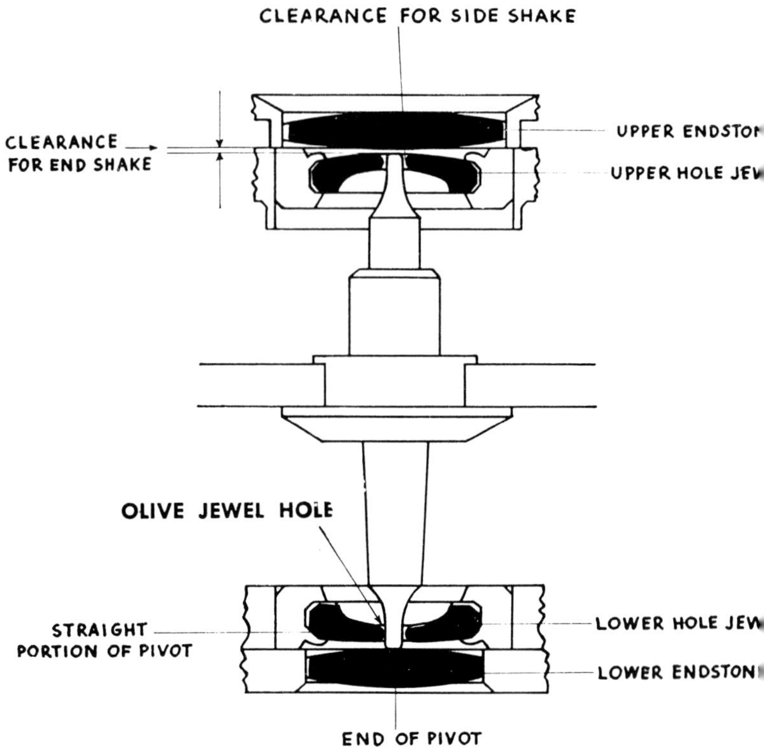

Fig. 1

The jewel bearings in which the balance staff pivots are mounted consist of a hole jewel and an endstone.

The pivot itself is equipped with two bearing surfaces:

a. The straight part of the pivot which rides on the hole jewel when the watch is in the "pendant up" position. See Fig. 2A.

b. The end of the pivot which rides on the endstone when the watch is in the "dial up" or "dial down" position. See Fig. 2B.

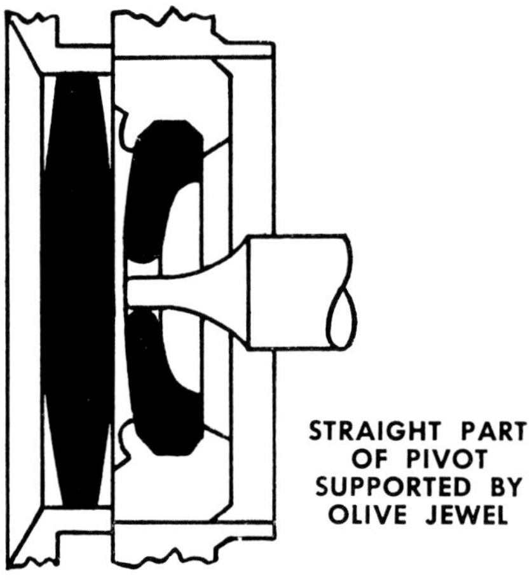

Fig. 2A

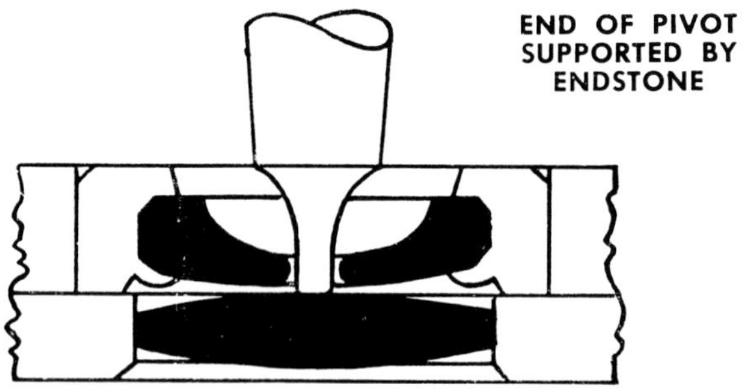

Fig. 2B

The clearance between the hole jewel and the straight part of the pivot is called "side shake" and the clearance between the endstone and the end of the pivot is called "end shake". The purpose of any bearing is to reduce friction. However, no bearing can completely eliminate friction. Therefore, this friction must be taken into consideration when designing or repairing a watch. Being a portable timepiece, a good watch must function with a high degree of accuracy in all positions. This means that the frictional losses between the pivots and jewels must not only be minimized, but equalized insofar as possible for any operating condition. More precisely, it means that the friction between the hole jewel and the straight portion of the pivot must be as close as possible to the friction between the endstone and the end of the pivot.

The solution of the problems involving the equalization of frictional loss in various positions lies in the shape given to the pivot. If the pivot is designed so that the end offers as much friction to the endstone as the straight portion does to the hole jewel, the problem is solved within reasonable limits. This is accomplished by making the end of the pivot almost perfectly flat. The edge is rounded just enough to insure an even surface without ridges or burrs.

After the shape of the pivots has been established, the frictional losses must be decreased as much as possible. This is best accomplished by burnishing the pivots with a steel burnisher. It is the function of the burnisher to remove tarnish, slight scratches, burrs, superficial rust and corrosion. The burnisher acts as a superfine file. The process of burnishing also tends to work-harden the outer surface of the pivot. It must be emphasized that in the event a pivot has other than superficial defects, the process of removing them may reduce the diameter of the pivot so that it will not fit correctly in its jeweled bearing. The excessive side shake that would result in such a case would interfere with the proper functioning of the balance wheel assembly. Where pivot defects are too great, the watchmaker would be well advised to replace the staff with a new one.

PIVOT DEFECTS

For comparison purposes a correctly shaped pivot is shown in Figure 3A.

In Figure 3B is shown a bent pivot. This type of pivot causes erratic timing and may even stop the watch.

In Figure 3C a scratched pivot is shown. This condition may be brought about by a broken hole jewel or dirt in the hole jewel. This type of fault may also be present on the very end of a pivot, which may be caused by a cracked or pitted cap jewel or dirt on the inner surface of the cap jewel.

In Figure 3D is illustrated a pivot with a burr at its edge. This may be brought about by burnishing the end of the pivot and neglecting to put the finishing touch to the whole procedure by drawing the burnisher across the edge of the pivot. This fault may also result from dropping the watch.

In Figure 3E are shown the pitted and surface-marked effects due to corrosion and rust.

In Figure 3F is shown a pivot with its end too rounded.

A B C D E F

Fig. 3

ELIMINATION OF PIVOT DEFECTS

The Burnisher and Its Preparation

The object in burnishing is to reduce the friction between the staff and its jewel bearing to a minimum. By use of a hardened steel burnisher the surface of a pivot will be made smoother. In Figure 4 is shown a portion of a pivot as it might appear under a high powered microscope.

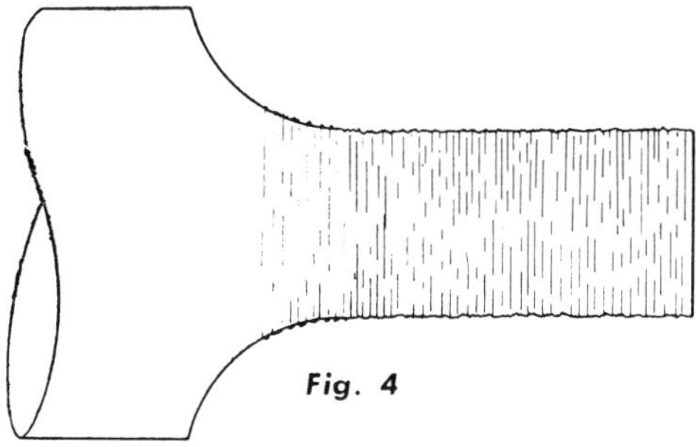

Fig. 4

The result of the burnishing is shown in Figure 5. The uneven and irregular surface (Figure 4), is now smoothed out.

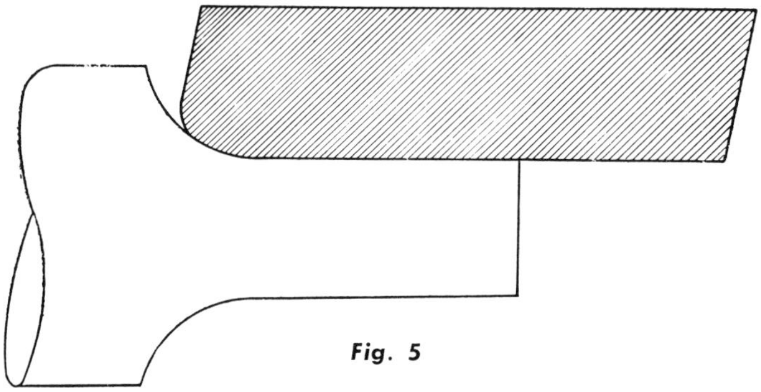

Fig. 5

Due to the fact that many steel burnishers, as they are received from the tool manufacturer, do not have the finish and flatness required for effective burnishing. It therefore, is necessary to discuss the method of refinishing them. Generally, the lines are too rough and the edges do not have the proper form.

Preparing the Burnisher

1. Place the burnisher flat on the oilstone which is covered with a thin film of light machine oil. Figure 6. It should be stroked at an angle of about 90° to the edge of the stone. While using the lengthwise stroke, an in and out stroke along the length of the burnisher should also be carried out. This technique will make sure that the entire surface of the burnisher will be affected. Stone all four sides.

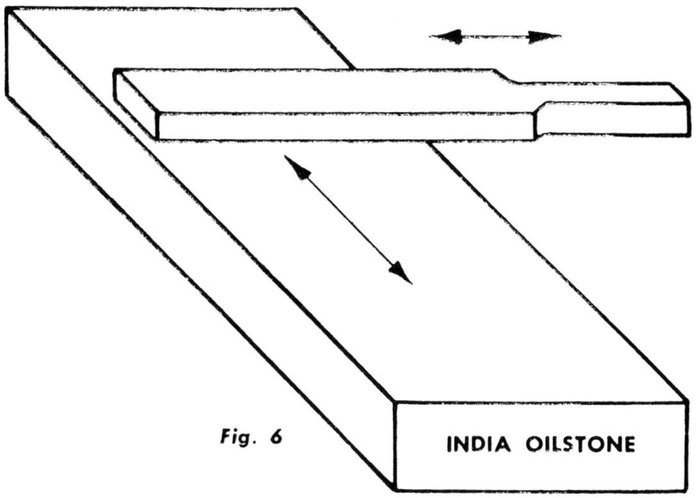

Fig. 6

2. In order that the sharp edge of the burnisher may not cut a groove at the base of the cone while burnishing, it will be necessary to round the edge of the burnisher. This may be done on the large oilstone. While stroking lengthwise on the stone the burnisher is rotated on its edge. Figure 7 shows the relationship that the burnisher will then have to a pivot. The diagonally opposite edge is left sharp for the purpose of burnishing pivots with square shoulders.

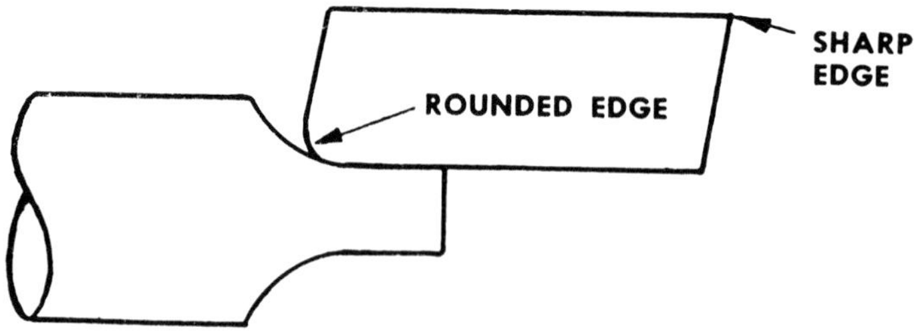

Fig. 7

3. The final stage in preparing the burnisher consists in putting on the lines. It is these lines which convert the burnisher into a superfine file. The burnisher will now be able to smooth out most defects and leave in their place a very even, mirror-like surface which will produce less friction.

   Hold a strip of 2/0 emery cloth taut on the surface of the oilstone, Figure 8. Lay the burnisher on the cloth right angles to its edge. Using a moderately heavy pressure on the burnisher, draw it along the cloth in one direction only. Lift the burnisher and repeat for a few more times. Lines should also be formed on the rounded edge. The flat working surface of the burnisher with its lines is shown in Figure 8A.

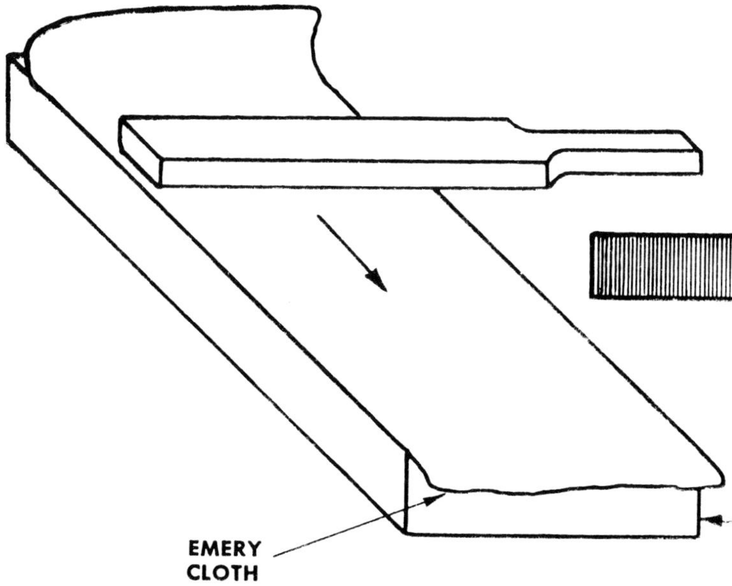

Fig. 8A

Fig. 8

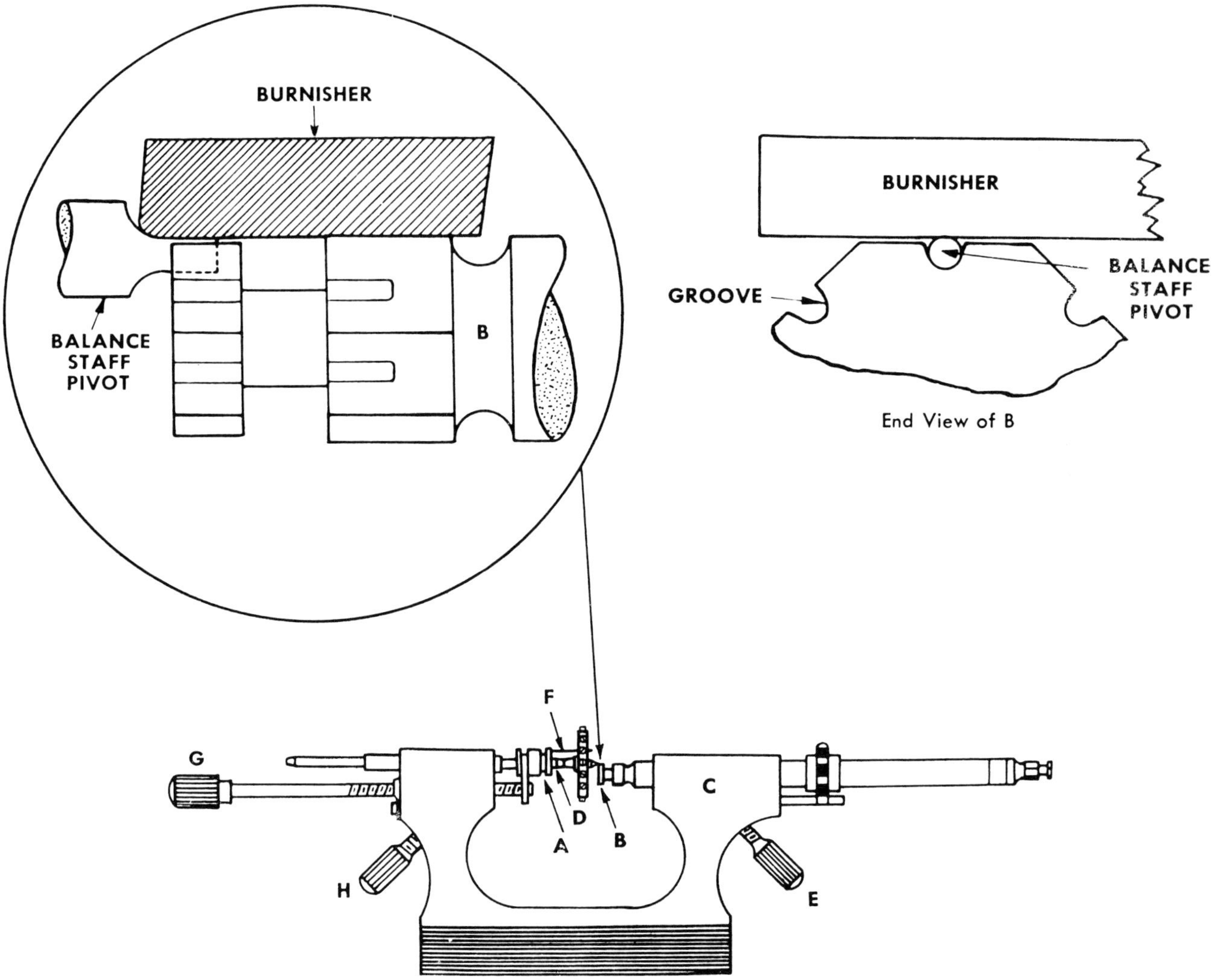
BURNISHER
BALANCE STAFF PIVOT
B
BURNISHER
GROOVE
BALANCE STAFF PIVOT
End View of B
F
G
C
D
A
B
H
E

Fig. 9

THE JACOT LATHE

The Jacot Lathe offers one of the very best means of burnishing pivots. Figure 9.

The pivot to be burnished is placed in a cylindrical groove of the spindle. This groove is on a precise line of centers with the pulley rod which supports the other pivot. The risk of breaking or bending a pivot while burnishing is reduced because it is solidly supported in the groove. The cone and the end of a pivot as well as its cylinder surface may be easily and efficiently burnished in the Jacot Lathe. The progressive steps listed below explain how this may be accomplished.

Burnishing the Pivot Cylinder

1. Mount the fiddle bow. To do this, place the silk thread on the bottom of the pulley A with the handle pointing away from you. Hold the thread in the pulley groove with one hand; with the other hand bring the handle forward so as to form a complete loop around the pulley.
2. Select the proper size cylindrical groove in the spindle B that will accommodate the pivot to be burnished. The requirements are that the pivot rests on the bottom of the groove and extends a few hundredths of a millimeter above the level of the flat top.
3. Insert the spindle in the main frame of the tool C with the selected groove uppermost.
4. Oil the burnisher with watch oil, also the pivot — supporting end of the centering arbor D.
5. Carefully insert a pivot in the centering arbor D. Tighten screw H.
6. Adjust the spindle containing the selected groove so that there is a bit of endshake. Not quite all of the cylinder is supported in the groove. This prevents the edge of the groove from cutting a ridge in the cone. Tighten screw E.
7. Move the wire dogs F toward the balance wheel until they straddle one of its arms. This is done by turning screw G.
8. Place the burnisher on the top of the pivot resting in the groove. Do not permit the wheel to turn until this step has been taken, otherwise damage to the wheel and pivot will result. While burnishing, keep the rounded edge of the burnisher in contact with the cone of the pivot and the flat part of the burnisher in solid contact with the cylinder of the pivot.
9. Move the bow downward and the burnisher forward; then the bow up and the burnisher toward you. The pivot and burnisher at the point of contact will thus move opposite to each other. This establishes the action necessary to burnish the pivots and should be continued until the pivots are restored to a high finish. This action is the same as is used in "roll filing."

Burnishing the End of a Pivot

One of the arbors or spindles of the Jacot Lathe set is provided with two discs which contain a series of graduated holes. Each of the holes is chamfered so that it will accommodate the cone of a pivot. This permits more of the pivot cylinder to extend beyond the outside of the disc and provides a stronger working support for the pivot while its end is being burnished.

1. Select a hole in the disc which will permit its chamfer to support the cone of the balance staff. Figure 10.
2. Rotate the balance with the fiddle bow and keep the oily burnisher held solidly against the end of the pivot while moving it back and forth as was done while burnishing the cylinder surface of the pivot.
3. To flatten or shorten the end of a pivot keep the burnisher moving on a line perpendicular to the axis of the pivot.
4. Burnishing either the cylinder surface or the end of a pivot may produce a sharp or burred edge around its end. This must be removed and is accomplished by presenting the oily burnisher on an angle to the axis of the pivot. Figure 11. The burnisher and wheel are moved in the same manner as before.

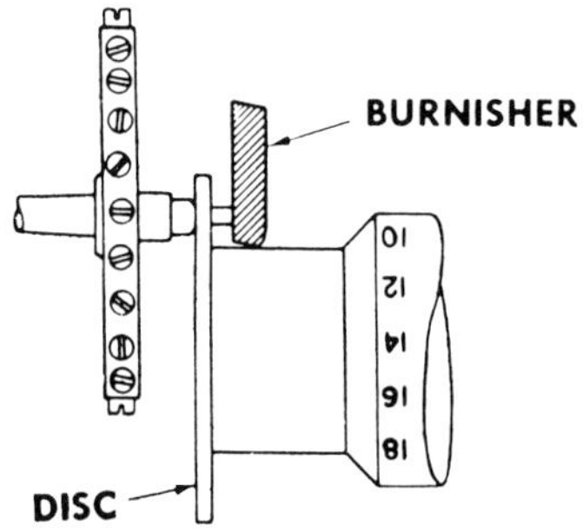

Fig. 10

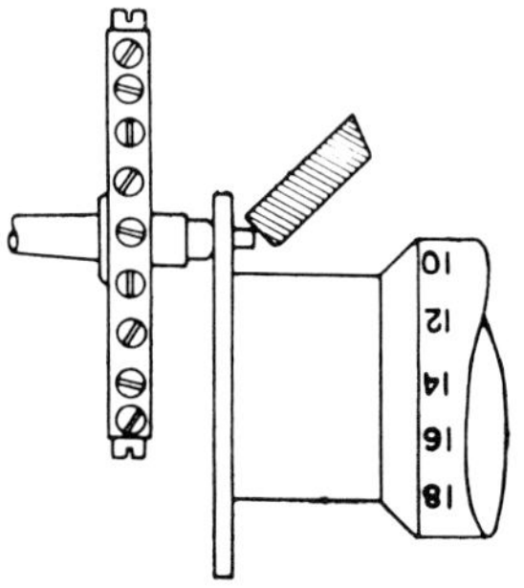

Fig. 11

Burnishing Train Wheel Pivots

The Jacot Tool offers a very effective and efficient means for burnishing train wheel pivots. Figure 12. The procedure for selecting the proper groove in the spindle is the same as that used when burnishing conical pivots. The one selected should permit the pivot to be cradled in the bottom with a few hundredths of a millimeter extending above the level of the two flat surfaces. The sharp edged side of the burnisher is used for square shouldered pivots.

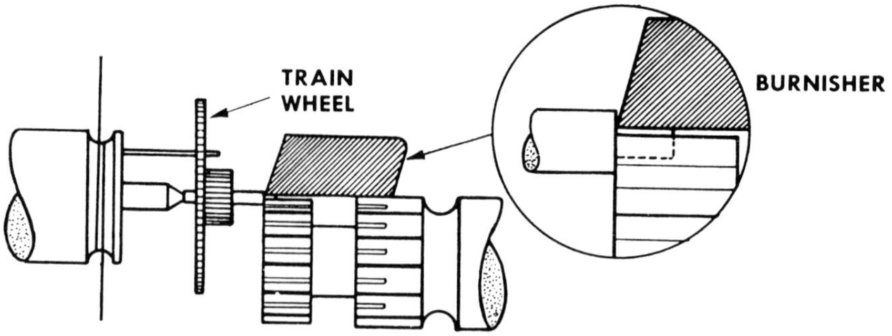

Fig. 12

THE DORRINGTON PIVOT LATHE

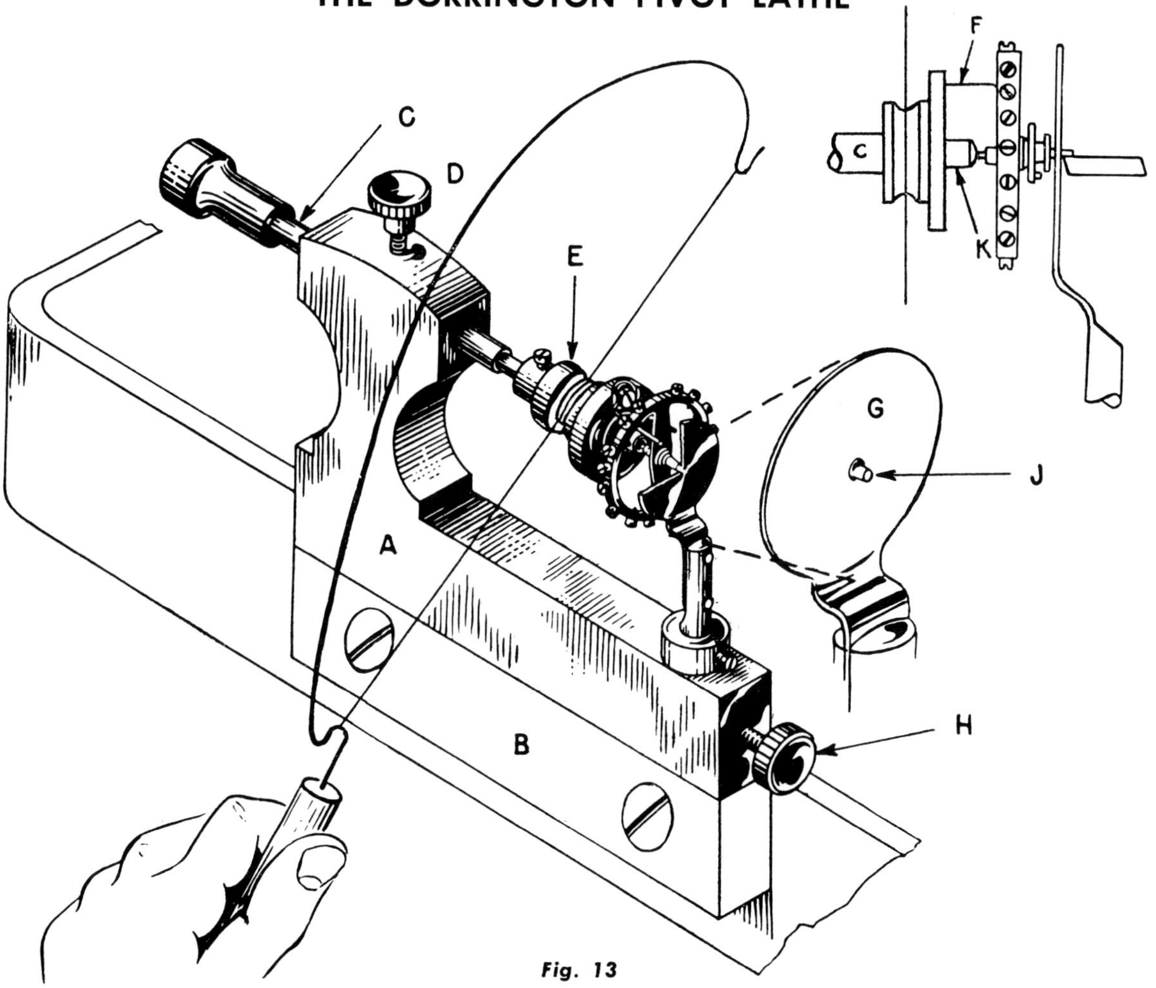

Fig. 13

Although not as precise as the Jacot Lathe, the Dorrington Pivot Lathe, Figure 13, offers a quick and efficient way to burnish pivots. It is a dead center lathe mounted on a base which is screwed to the watchmaker's bench.

The main frame A has two mounting pins which fit into the base B. When not in use, the main frame is pulled up and out of the base and is thus kept out of the way while the watchmaker is working at something else. Rod C fits into the headstock and may be locked in any position by screw D. One end, K, is drilled and countersunk in order to receive balance staff pivots. There is a pulley wheel E on the rod which is free to rotate when motivated by the fiddle bow. A pair of pins or "dogs" F are fastened to the pulley which are used to straddle the arm of the balance wheel. By moving the fiddle bow, the wheel is made to turn. Disc G which is fastened by screw H supports the pivot to be burnished. Some tools have three different size holes in one disc. A hole is selected that will permit all of the cylindrical part of the pivot J to extend out of the disc.

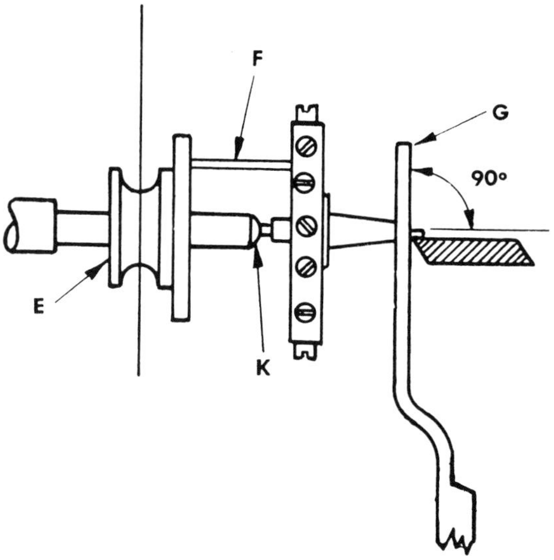

Fig. 13A

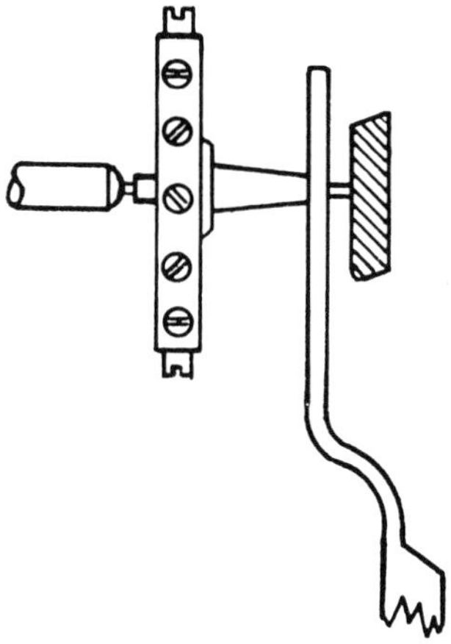

Fig. 13B

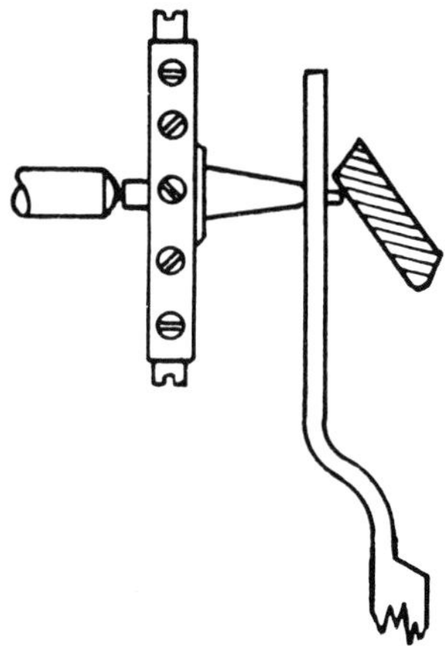

Fig. 13C

The fiddle bow is mounted as in the use of the Jacot Lathe. The pivot to be burnished is then inserted in the proper hole in the disc and the rod is brought forward until the other pivot is inserted in its end. There should be no end shake in the staff between the rod hole and the disc hole. The height of the disc must be adjusted so that the plane of the wheel is parallel to the disc. This will make certain that the pivot to be burnished will be perpendicular to the disc. Before burnishing, the burnisher and hole K in rod C should be moistened with watch oil. Figure 13A.

Care should be taken not to apply the burnisher against the pivot with too much pressure. As the fiddle bow is pulled downward the burnisher, which is held underneath the pivot, should be pushed forward. This will produce opposing motions between the burnisher and pivot. Failure to keep the burnisher flush against the pivot may develop a forward or a back taper.

To burnish the end of the pivot, the wheel is rotated with the fiddle bow while the burnisher is moved back and forth across the end of the pivot. Figure 13B.

Any burrs or sharp edges that have been formed after burnishing the cylinder or the end of the pivot should be removed by presenting the burnisher at an angle to it while turning the wheel with the bow. Figure 13C. A few very light strokes of the burnisher should successfully remove the sharp edge.

The burnisher used with this type of tool should be very slightly rounded on its edge in order to prevent wearing a groove in the pivot. However, the amount of roundedness used on the burnisher for the Jacot Lathe would be too great to do a successful job of burnishing with the Dorrington type of tool.

Use of the Balance Chuck

The balance chuck, which is an accessory for a regular jeweler's lathe, offers the watchmaker another means for burnishing pivots. Though it is advisable to remove the hairspring before attempting to burnish, it is not always necessary, especially where the hairspring is stronger and bigger as in pocket watches.

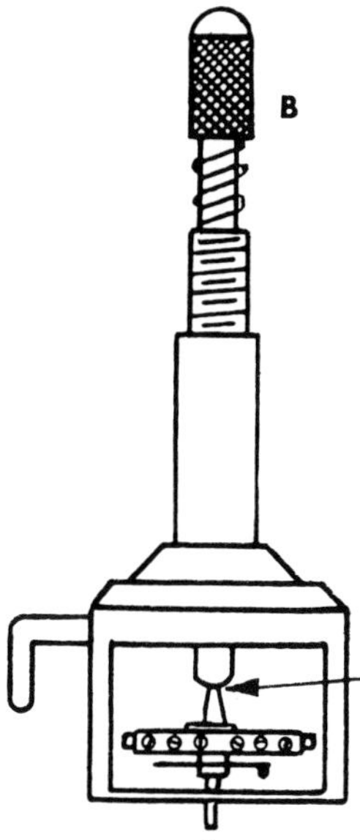

Fig. 14

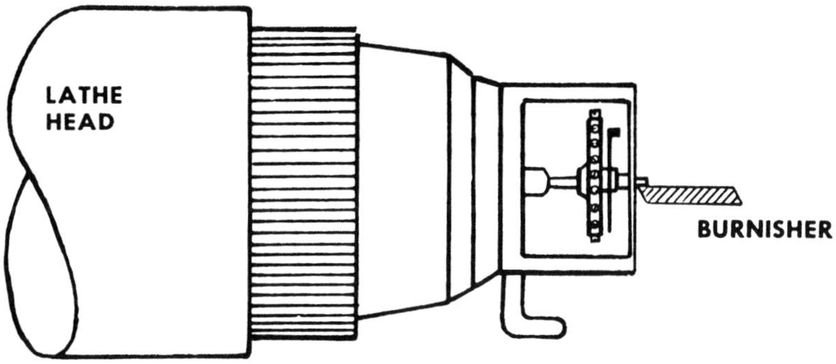

Fig. 15

After inserting the pivot A (Figure 14) through the front end of the chuck, the plunger B is brought down until pivot C can be inserted into the pivot hole of the plunger. The plunger is fixed in this position by tightening locking screw D.

The balance chuck and wheel may now be inserted into the jeweler's lathe head. Figure 15. A high speed may be used if the hairspring has been removed from the wheel. However, if the hairspring has been left on the wheel, then consideration must be given the possible harmful effect of centrifugal force. In such cases the speed should not be too high or else the hairspring may be damaged.

The steel burnisher used with the balance chuck should have a slightly rounded edge similar to that used with the Dorrington type of burnishing tool.

Care should be taken to present the burnisher to the pivot evenly along its surface, otherwise a forward or a back taper may develop in the pivot. While the chuck is turning in the lathe, a back and forth stroke is executed with the oily burnisher under the cylindrical side of the pivot.

Straightening Bent Pivots

If it is found that the bent pivot has a kink or double bend in it, Figure 16, any attempt to straighten it will be futile. In cases like this, the balance staff should be replaced.

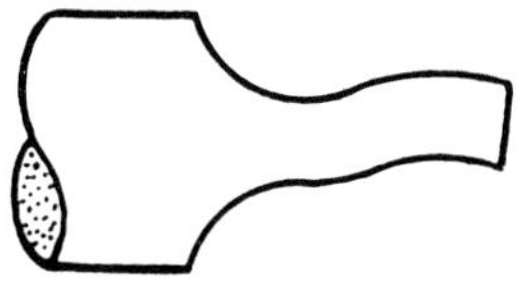

Fig. 16

However, if the bend is a smooth, single one, there is a good chance of straightening it. The tool used for this straightening job is a broad, heavy, snubnosed tweezer made of soft steel or brass. Figure 17. The tweezer is heated so that when it is applied to the bent pivot, the pivot will, upon absorbing heat, be less likely to break. The heat will temporarily increase the elasticity of the pivot during the application of the heated tweezer.

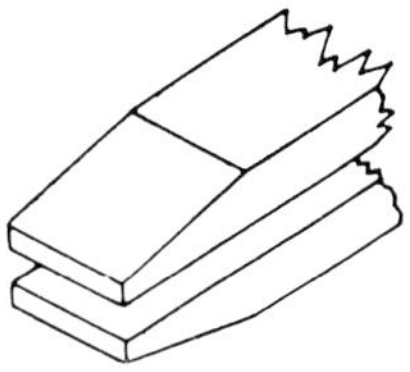

Fig. 17

The balance wheel and staff with the bent pivot to be straightened may be mounted in either the regular lathe head or the Dorrington Pivot Lathe. If mounted in the regular jeweler's lathe, care should be taken to see that the staff is accurately centered in the chuck before finally tightening up on the chuck. It is advisable to remove the roller before attempting to straighten the upper pivot because of the uncertainty of the roller being concentric with respect to the staff.

When the staff is centered, it is possible to determine how true or bent it is by using the tee rest as a fixed reference. By adjusting the left side edge of the rest so that a very narrow space of light appears between the pivot and the tee rest, the condition of the pivot may be seen by rotating the lathe head by hand.

After the error has been located, hold the pivot between the jaws of the tweezer and apply a pulling, bending action to it. Figure 18.

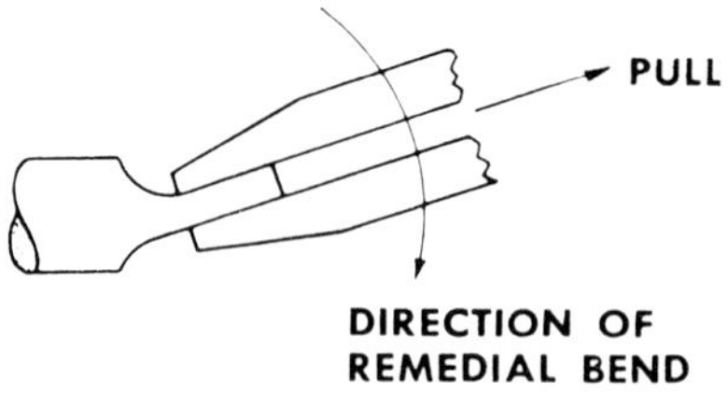

Fig. 18

Pivot straightening may also be accomplished by holding the balance wheel in the hand and using the straightening tweezer in the manner described before. The results should be checked for trueness by use of the Dorrington pivot lathe or the regular jeweler's lathe.

Every pivot straightening job should be followed by a burnishing operation to assure a proper finish to the pivot.

TRAINING UNIT
NUMBER 5

★

POISING BALANCE WHEELS

POISING BALANCE WHEELS

Because the balance wheel and hairspring assembly is the time keeping element of a watch, the balance wheel must be properly poised.

In the poising operation, the mass of the balance wheel is distributed equally around the axis of rotation so that the effect of the force of gravity is eliminated. Only the forces inherent in the balance and hairspring should govern the accuracy of the balance wheel movement.

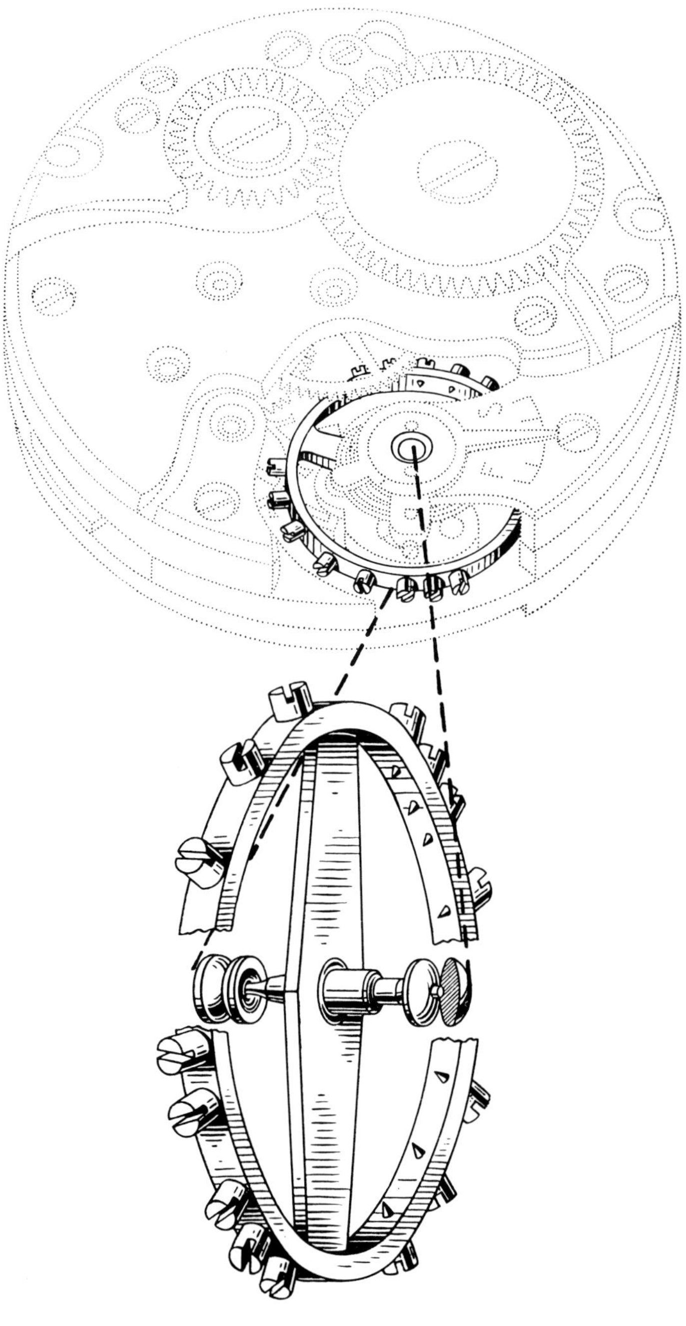

BULOVA SCHOOL *of* WATCHMAKING

TRAINING UNIT NUMBER 5

SUBJECT: Balance Wheel Poising.

OBJECTIVE: To establish the Center of gravity of a Balance at the intersection of the vertical and horizontal axis.

or

To establish equal distribution of the mass around the Balance Wheel Staff.

Practical work — 40 hours.

Related theory — 8 hours.

TOOLS REQUIRED:
1. Poising Calipers or Poising Tool.
2. Balance screw undercutter.
3. Balance screw saw.

Balance screw driver.

Slotted block.

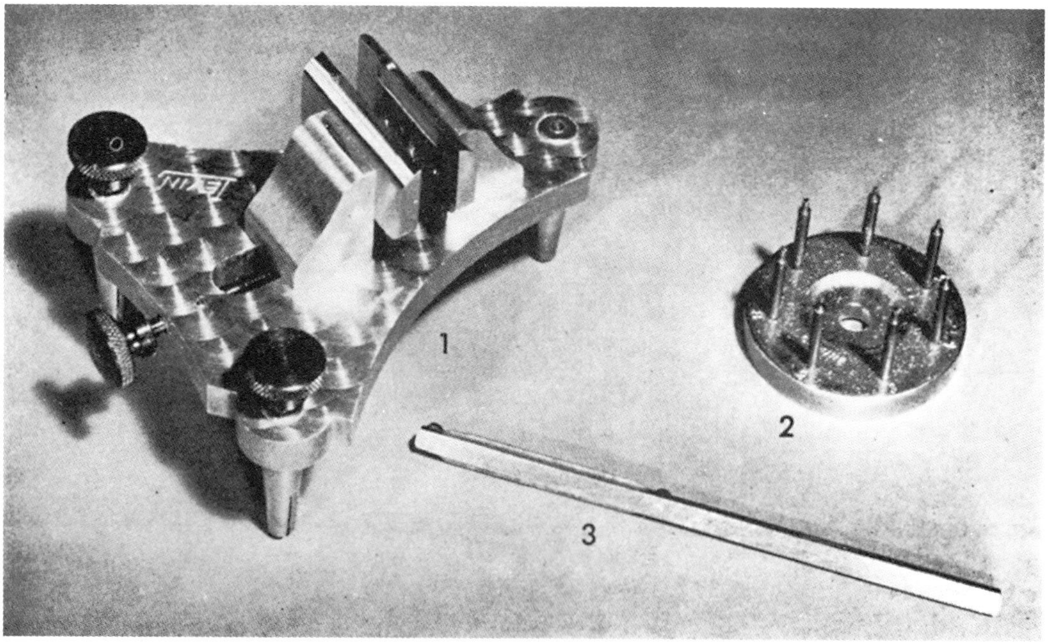

Fig. 1

POISING TOOLS AND POISING CALIPERS

The proper use of the poising tool and poising caliper is shown in Fig. 2A and 2B. The poising tool utilizes jeweled knife edges to support the balance while the calipers are equipped with jeweled stumps that suspend the balance in such a manner that all forces acting on the balance are eliminated except the force of gravity.

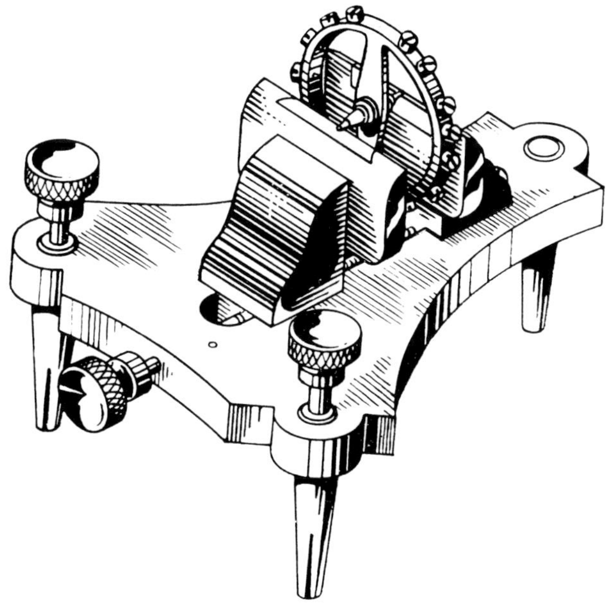

Fig. 2A

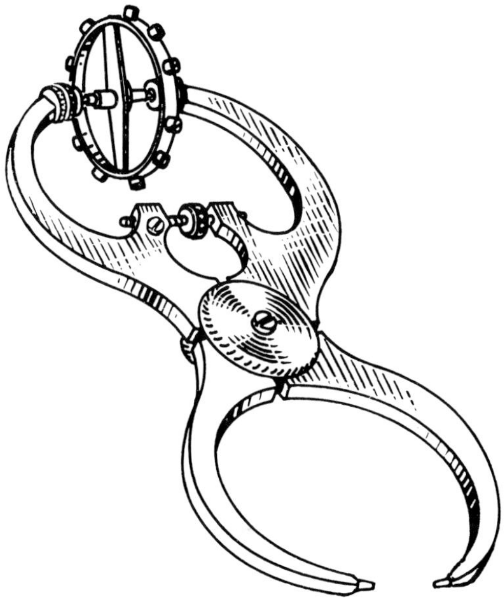

Fig. 2B

Under these conditions it is reasonable to assume that any movement of the balance around its staff axis can be attributed to the force of gravity acting upon some portion of the balance to a greater extent than upon other portions. In other words, the weight of the balance is not equally distributed around the staff axis. This condition places the center of gravity away from the intersection of the two axis of the balance and the piece is said to be "out of poise".

It would be well to recall the definition of "weight" at this time, in order to clarify this problem of poise. Weight, by definition, is merely the amount of force exerted on an object by gravity. Therefore, in the balance wheel under discussion, it is obvious that if there is a point on the rim that is heavier than other points then that point, under the pull of gravity, will be pulled downward. In short, the balance wheel will rotate until the heaviest part finally settles at the lowest point.

Once this heavy point has been determined, the whole problem resolves itself into one of redistribution of the weight in such a manner that the center of gravity will be relocated exactly at the intersection of the staff and rim axis.

Particular attention is called to the phrase, "redistribution of weight", the relationship between the weight of the balance wheel and the strength of the hairspring is an important factor controlling the accuracy of a watch. There is a definite cooperation of physical laws in which a balance wheel that is too heavy or too light will cause a watch to lose or gain time. In view of these facts, it can be seen that it is of the utmost importance that the weight of the balance wheel is not appreciably increased or decreased during the poising operation.

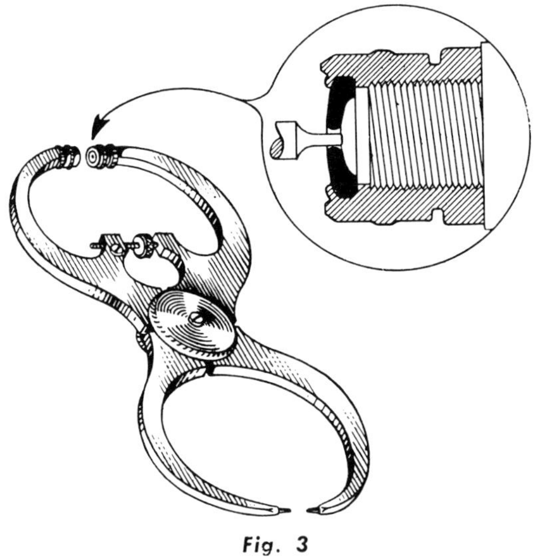

Fig. 3

The actual work involved in poising a balance wheel starts with placing the balance in the poising caliper in such a manner that only the extreme end of the pivots make contact with the stumps, See Fig. 3. In this position the calipers duplicate the conditions under which the balance must function in a watch.

It must be remembered that in actual practice a balance wheel must be poised with the roller in place. However, in this training unit the principles of balance poising will be demonstrated without the roller on the staff.

A perfectly poised balance wheel will rotate slowly as the calipers are agitated. See Fig. 4.

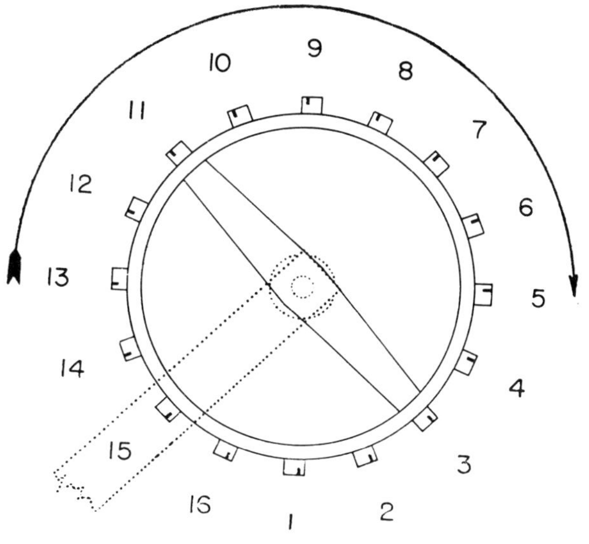

Fig. 4

In an unpoised balance the heaviest portion of the rim will be drawn to the bottom of the wheel. In this case screw No. 13 is the indicated heavy point and will be drawn down to the lowest part of the wheel as shown by the arrow. See Fig. 5.

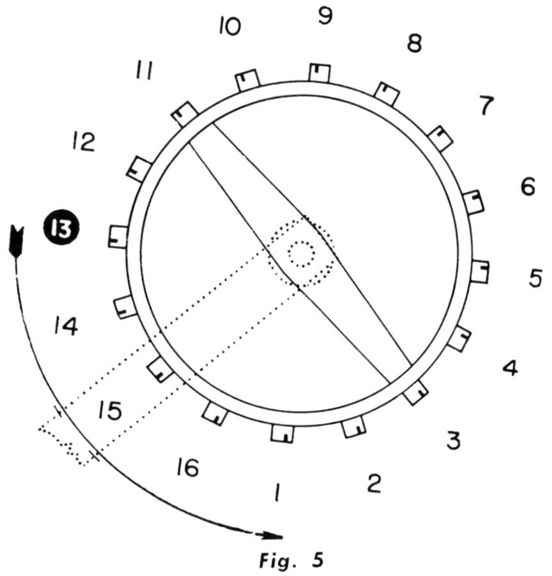

Fig. 5

Likewise, if screw No. 8 was the heaviest it would cause the wheel to turn until this heavy point was at the bottom of the rim. See Fig. 6.

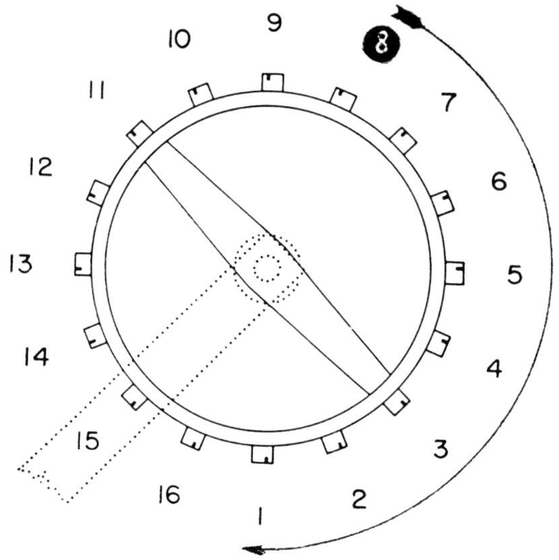

Fig. 6

By this method it can be determined when and where the balance wheel is out of poise. See Fig. 7.

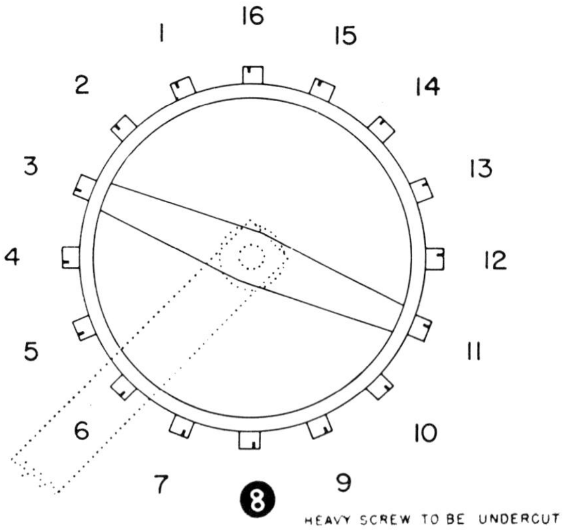

Fig. 7

After having determined the location of the heavy screw it can be removed and replaced by the use of the balance screw remover as shown in Fig. 8A. Another method of removing the balance screw is by the use of a screw driver. After undercutting the screw it can be replaced by wedging the blade of the screw driver into the slot by the use of a block, See Fig. 8B. This wedge will support the screw so that it can be replaced into the balance wheel as shown in Fig. 8C. Care must be used in grinding the blade so that the wedge will not spread the screw slot.

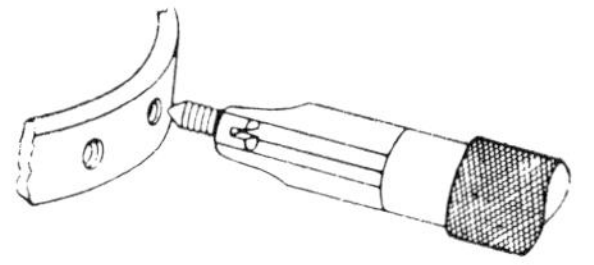

Fig. 8A

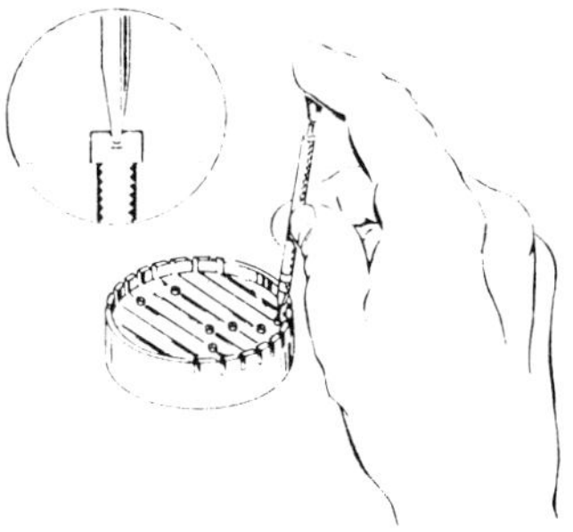

Fig. 8B

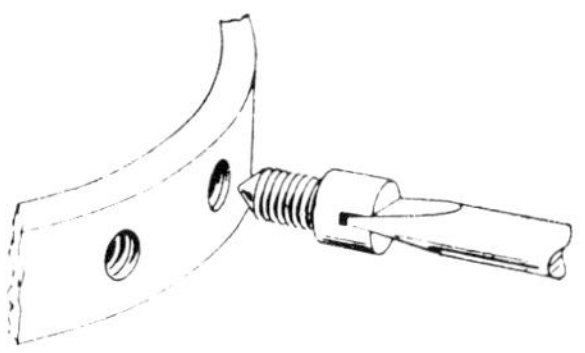

Fig. 8C

After removing the heavy screw it will be necessary to under-cut it in order to eliminate the excess weight, as shown in Fig. 9.

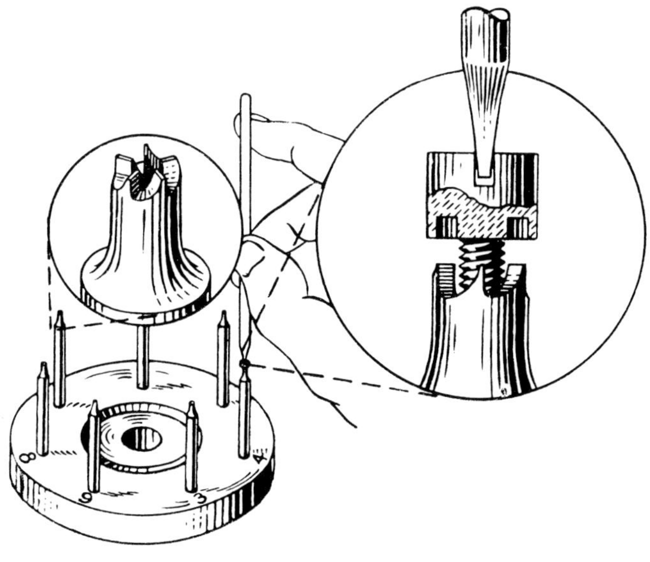

Fig. 9

If, after replacing the under-cut screw in the same hole from which it was taken, the opposite screw shows heavy, (in this case screw No. 16) the indication is that too much metal was removed in the under-cutting process on the original screw. See Fig. 10. This condition is treated as if it was the first operation of the poising procedure.

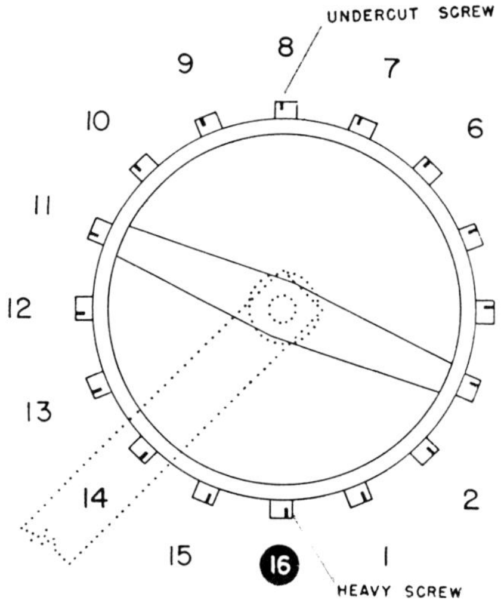

Fig. 10

Under-cut screw No. 16 using great care not to remove too much weight with the under-cutter. Insert the screw in its proper hole and check the balance for poise.

In case too much weight has again been removed it will be found that the heavy point has shifted back to the opposite side of the balance showing the rim heavy at screw No. 8, See Fig. 11. This screw must be under-cut again before the wheel can be brought into poise.

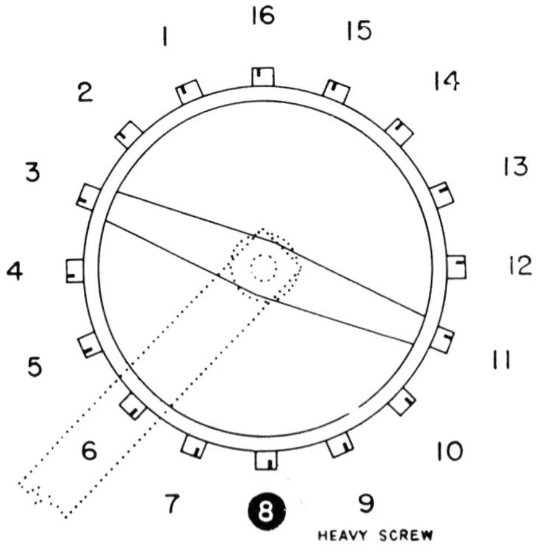

Fig. 11

During these successive under-cutting operations, a certain amount of weight has been removed from the balance wheel. Bearing in mind that any appreciable change in the overall weight of the balance will affect the rate of the watch, it might be well to consider this rate problem before attempting further under-cutting of the balance screws.

An indication of the change in the rate of the watch caused by the preceding under-cutting can be had by comparing the weight of the chips removed from the screws with the weight of a 10AK, 1 minute timing washer. A pair of these washers added to a 10AK balance wheel will cause the watch to run one minute slow in twenty four hours. Likewise, if the weight removed from a balance during the poising operation approximates the weight of one of these timing washers, the rate of the watch will be increased about thirty seconds per twenty-four hours.

With this in mind, it is well to be constantly aware of the amount of weight being removed while poising a balance when it is apparent that a sizeable reduction in weight must be made at a point, the overall weight of the balance could be maintained by adding a timing washer to the light side and then proceed with the poising. Fig. 12 and 13.

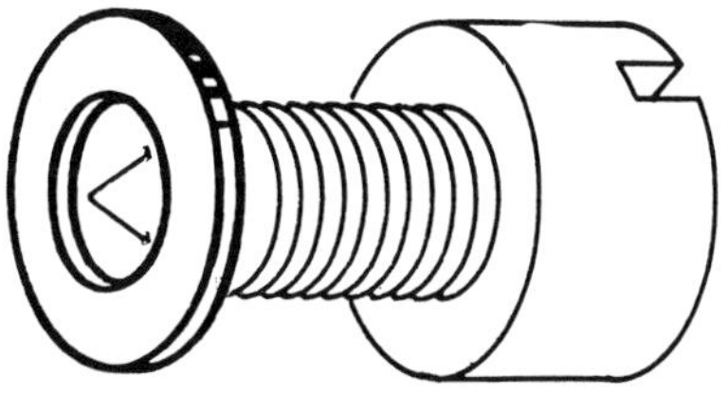

Fig. 12

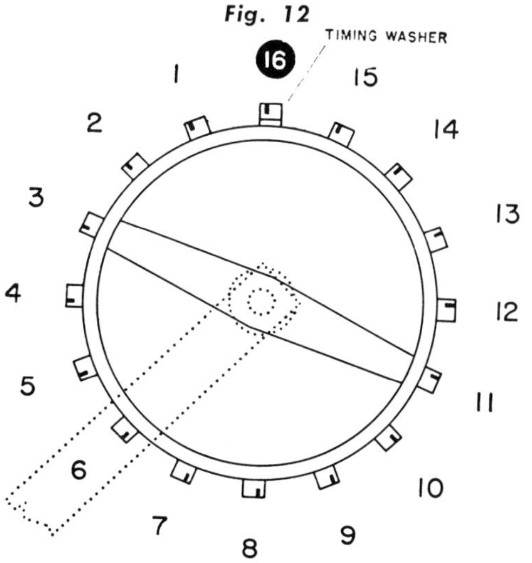

Fig. 13

If considerable weight has been removed from a monometalic balance the former weight may be restored by the addition of a pair of timing washers to any two opposite screws.

In the case of a bi-metallic balance wheel, temperature compensation would be less affected by adding the timing washers to the pair of screws nearest the end of the balance arm.

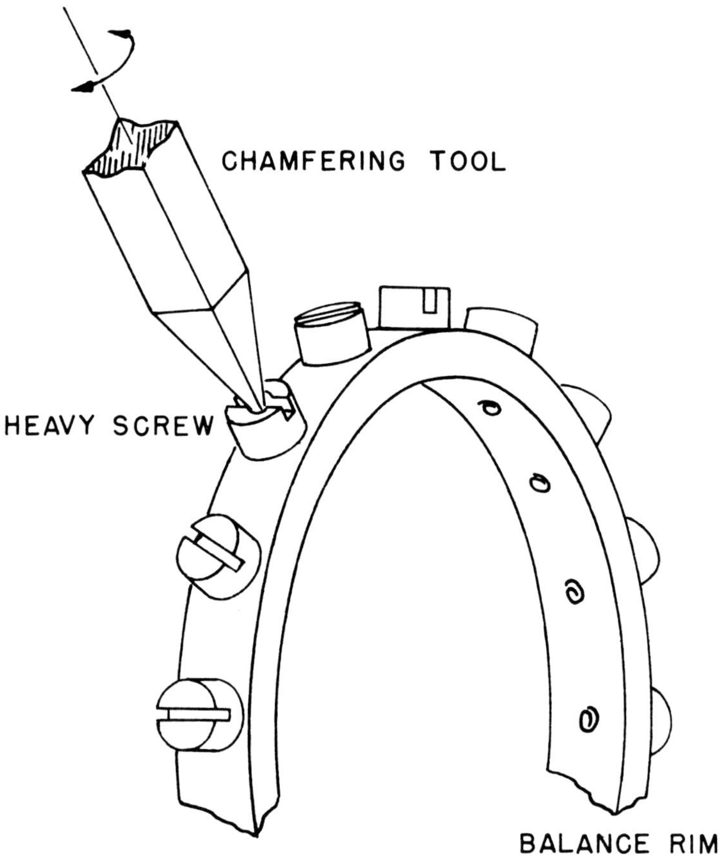

Fig. 14

Some balance wheels are equipped with two or four meantime screws. When two screws are used they are usually located at the ends of the arm. When 4 screws are used the extra pair are located 90° from these screws. It should be remembered that meantime screws are never undercut during the poising operation.

These screws are used in the timing operation to establish the proper rate of vibration of the balance wheel. They are usually of a different color than the balance screws and slightly different in size.

Before starting to poise a wheel equipped with meantime screws it is well to adjust these screws so that an even amount of thread is visible on either side of the rim. This permits a greater range of adjustment during the timing operation.

POISING SUPPLEMENT

When poising large balance wheels such as the 21AE model, the instructions given in the manual should be followed with respect to the use of the balance screw undercutter. However, filing screw slots with the balance screw saw should be confined to only very fine weight adjustments to avoid marring the appearance of the balance screws. Under no circumstances, should the screw head itself be ground or filed.

When poising smaller balance wheels such as 10AK, it is not possible to use the balance screw undercutter. If weight must be removed from this size wheel, this can be accomplished very nicely without spoiling the appearance of the screw head by using a home made type of reamer or chamfer specially prepared from a rat tail file. Four sharp cornered sides are taper ground towards the point of an old rat tail file so that the chamfering tool becomes suitable for removing metal from the screw head center. The tool can be turned back and forth and the screw need not be removed from the balance wheel. See Fig. 14.

On the subject of poising, one important caution should be followed. Whereever possible, the original weight of the balance wheel should not be changed or at worst, only very minutely in order to avoid the necessity of altering or replacing the hairspring. Therefore, the addition of timing washers is equally as necessary as the removal of weight when putting a balance wheel in poise.

If checking a balance wheel on the poising tool indicates a heavy screw, the error frequently may be corrected by adding a timing washer to the opposite screw. This can first be tested by trying one, then perhaps a slightly heavier timing washer by causing it to stick to the opposite lighter balance screw by means of a dab of moisture or watch oil. If the trial poise tests out satisfactorily, then the selected timing washer can be screwed into place.

Now clean the balance wheel assembly and if a final adjustment should be necessary, it can be made by applying the chamfer tool described above to the appropriate screw.

Following is another problem that may manifest itself during the poising operation.

At the start of the operation a certain screw appears to be heavy, in this case, screw No. 8. (Fig. 15) This screw is carefully undercut and returned to its proper hole.

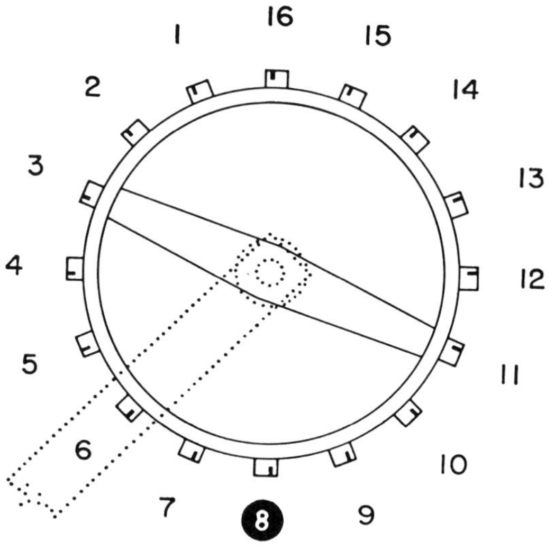

Fig. 15

Upon testing for poise, it is found that the heavy point has shifted to another screw, No. 6, on the same side of the rim. Fig. 16.

This condition is brought about by unequal distribution of weight at two or more points within the balance wheel.

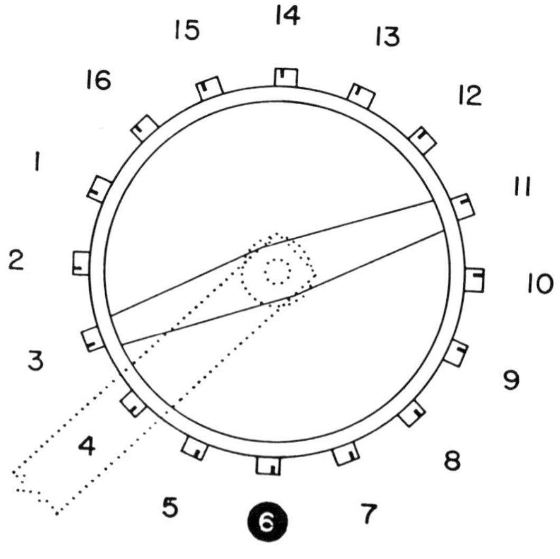

Fig. 16

Suppose for example that screw No. 12 was slightly heavy and that screw No. 6 was even heavier.

When tested for poise this balance would settle somewhere in between these two screws, See Fig. 17.

Let's say, for convenience, that screw No. 8 is low. According to the rules governing poising, number 8 screw is under-cut and the wheel again tested for poise.

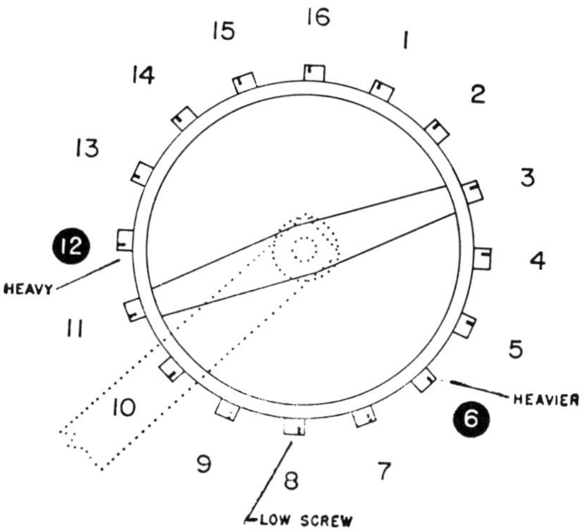

Fig. 17

It is entirely possible that under these conditions screw No. 10 will now show heavy. See Fig. 18. This shifting of the heavy point within a certain segment of the balance rim can continue through several under-cuttings. Each successive heavy screw should be under-cut until the wheel is poised. However, if, after any one of these undercutting operations the heavy point shifts to the opposite side of the balance, it is a clear indication that too much weight was removed by the last under-cut. In any event the successive heavy screws are under-cut until the balance is in poise.

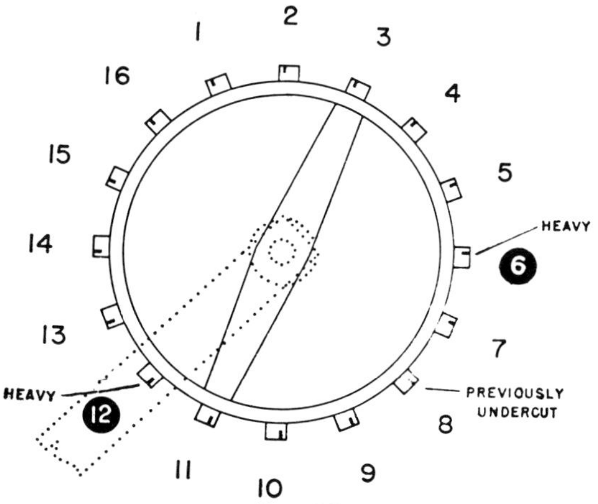

Fig. 18

With the wheel now poised, it will be necessary to bring its weight back to the original amount, so that the rate of the watch will not have been affected by the poising.

To accomplish this, timing washers must be added as explained before.

The wheel must be checked for poise after the addition of these timing washers in order to detect any unpoised condition due to any variation in the weight of the washers.

TRAINING UNIT NUMBER 6

★

HAIRSPRING TRUING

THE ART *of* HAIRSPRING TRUING

Hairspring technique is most important in the study of horology. The high degree of manipulative skill it requires commands the respect and admiration of all experienced horologists.

The general subject of hairsprings is many-sided. Besides attracting the talents of the mathematician, the physicist, and the metallurgist, it affords an exacting field for those who would develop their manual dexterity. The repairman who has knowledge of the physical characteristics of hairsprings and the skill to put such knowledge to work possesses an advantage over a less equipped fellow-worker; his important knowledge and skill wins the appreciation of the customer who observes him make a deft touch on a hairspring that brings his watch back into serviceability.

The tools used are simple ones. The work of manipulating the watch hairsprings is primarily a matter of obtaining an appreciation of the delicacy of the spring, together with a knowledge of the various bending and breaking techniques and how the spring reacts to them. Some hand and finger movements are possessed in part by the learner; the others, acquired by instruction and practice. Mistakes made may be used as stepping stones to success during the learning process. Eventually the spring begins to obey. Error-making gradually is eliminated and correct obedience of the hands to the mind show a progressive change toward success each time the proper manipulative movements are repeated. Finally, after diligent application of correct movements of the hands and fingers, these movements become organized into a pattern, a chain of movements that fit together without delays or confusion and with meaningful relationships to one another and to the job at hand. In short, a skill is born.

This particular unit is set up to serve as a guide and to lay a broad foundation of the proper manipulative operations required to become proficient in the art of hairspring truing. It can be used to great advantage by the watchmaker who wishes to advance himself, as well as the student who requires a well organized schedule of operations.

In action, the hairspring reveals grace and beauty. It appears to have life. Its expansion and contraction give the semblance of breathing. Its dynamic relationship to the balance wheel, to which it is attached, is so established that it causes the balance to oscillate with a definite period. The spring is like a brain. Through its elastic properties it is the controlling factor in the time keeping function of the watch, being the governor or regulator, over the movement of the balance wheel.

The hairspring of a watch takes the form of a spiral and is rectangular in cross section. It is generally made of steel, but in recent times metallurgical research has brought forth a spring that is practically insensitive to temperature and magnetic changes.

Hairsprings are made by inserting several lengths of ribbon-like wire metal into a hollow, steel, barrel-like forming device. This tool is designed so that several pieces of wire are wound each on top of the other, each taking a flat spiral form. When a sufficient number of turns or coils are wound in a tool, the coils are subjected to a heat treating process which establishes the required degree of elasticity, thus preparing them for the watchmaker whose work begins at this point.

The Hairspring

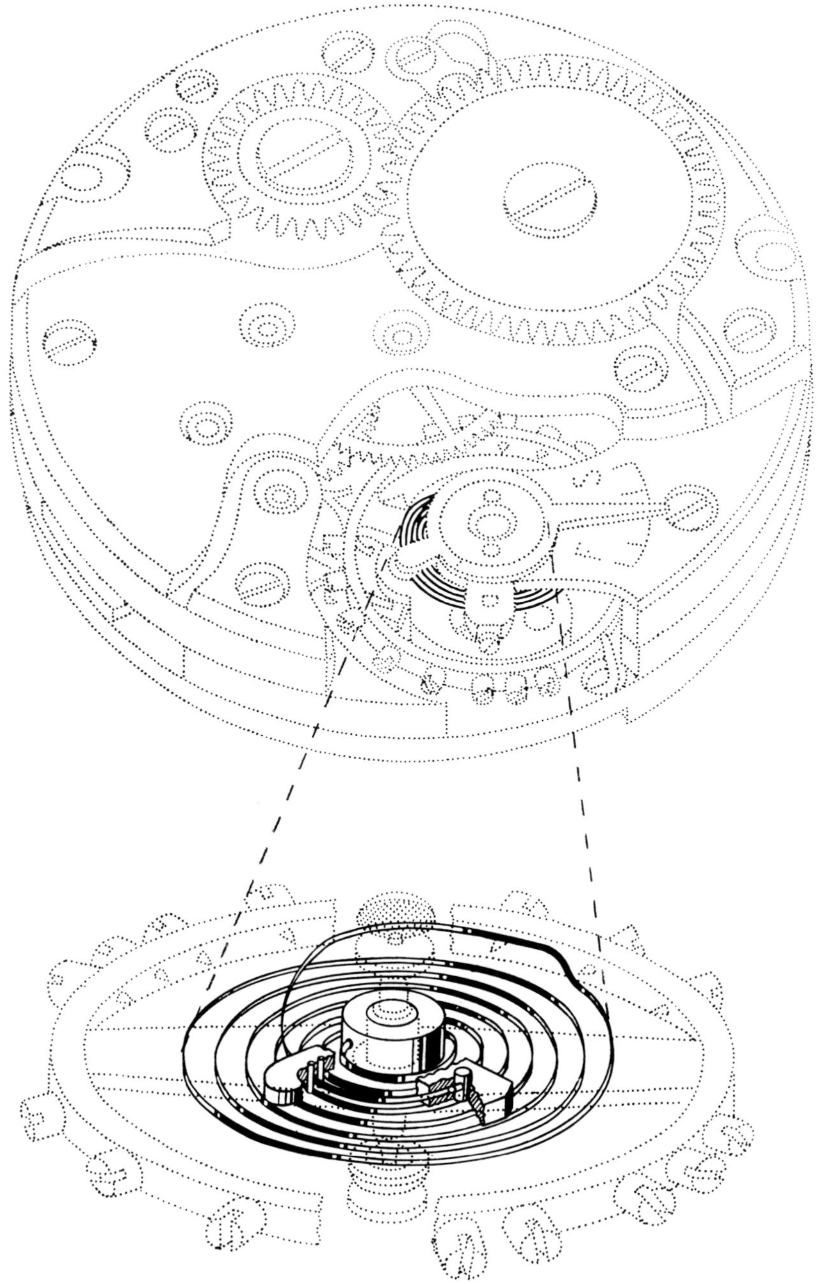

BULOVA SCHOOL *of* WATCHMAKING

TRAINING UNIT No. 6

Subject:

Practical Manipulation of Watch Hairspring.

Objective:

To develop skill in hairspring work.
Practical work 100 hours.
Related theory 20 hours

Tools Needed:

1. 1 Colleting arbor.
2. 2 Pair tweezers.
3. 1 Pair snipe nose pliers.
4. 1 Pair hairspring truing calipers.
5. 1 Pair side cutting pliers.
6. 1 Loupe.

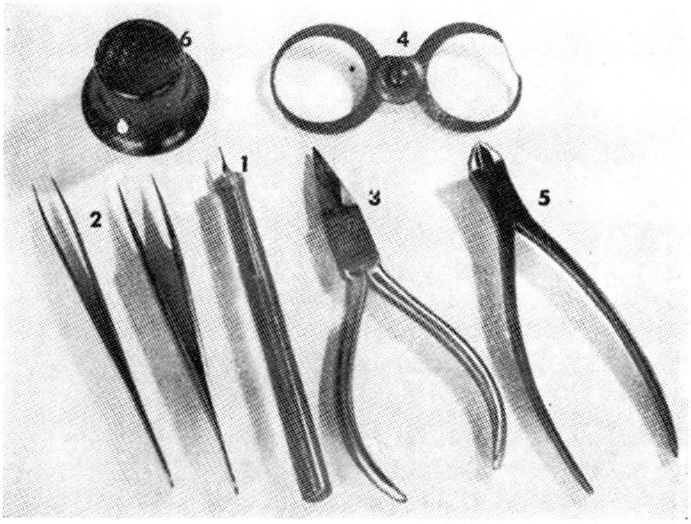

Material Needed:

1. Balance wheels with staff fitted.
2. Supply of hairsprings.
3. Supply of hairspring collets.
4. Supply of hairspring colleting pins.

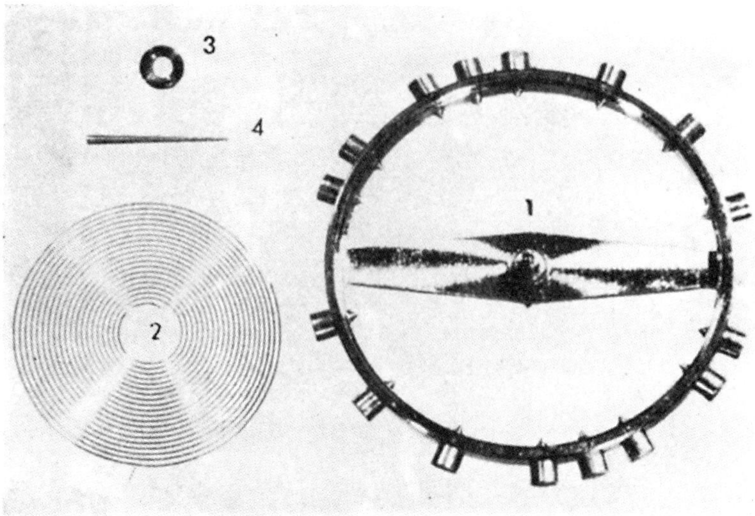

HAIRSPRING TRUING

Practical Manipulation of Watch Hairsprings will be presented in four main sub divisions:

I. Special hand manipulative operations on hairsprings.

II. Colleting.

III. Staking the hairspring and collet to the balance staff.

IV. Truing in the round and flat.

1. SPECIAL HAND MANIPULATIVE OPERATION ON HAIRSPRING

Manipulative Breaking Exercises

Method No. 1, Fig. 2

To break off outer coil, grasp it with a tweezer at the desired point of break. (See point A.) Grasp the coil at point B with another tweezer and bend the coil outward and away from the center of the spring and back again to the original position. Repeat this bending procedure until the break occurs.

Break off one piece from the outer coil of five hairsprings at 45°, five at 90°, five at 135°, five at 180° and five at 225°.

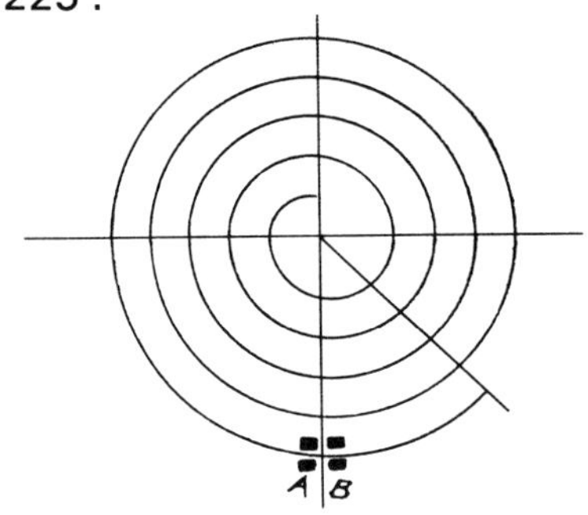

FIG. 2 45°

Method No. 2, Fig. 3

Grasp the coil at the desired point of break. (See point A.) Use another tweezer, placed at B, next to A on the inside of outer coil, break coil away by pulling end with tweezer B as far as necessary for break to occur.

Break off one piece from the outer coil of five hairsprings at 45°, five at 90°, five at 135°, five at 180° and five at 225°.

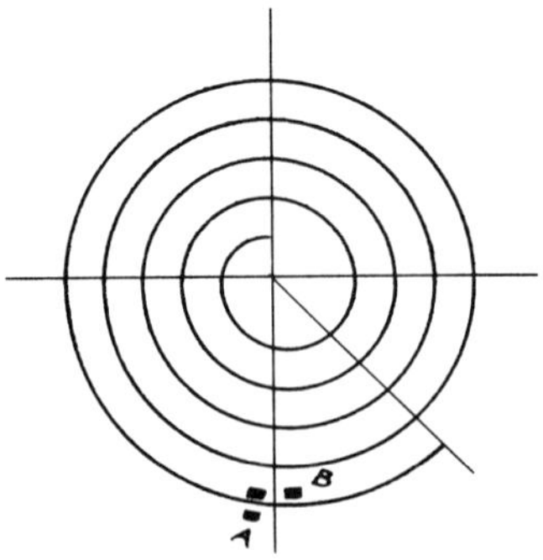

FIG. 3 45°

Manipulative Bending Exercises

Method 1, Fig. 4

First, grasp the outside coil of the spring with a pair of tweezers at point A, 90 degrees from the end. Second, grasp the outside coil at point B with another tweezer and bend the coil inward until it just touches the adjacent coil.

Note: By reversing the direction of the bend, the end of the coil may be bent outward.

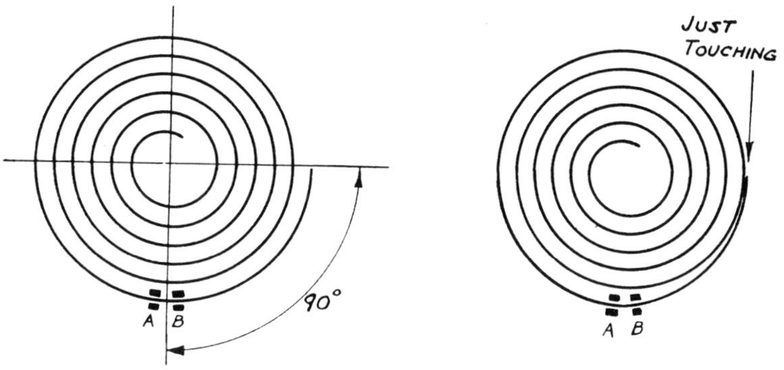

Fig. 4

Method 2, Fig. 5

Grasp the outside coil of the spring with a tweezer at point A, 90 degrees away from the end. Place another tweezer at point B and push on outer coil until it just touches the adjacent coil.

Note: To reverse the direction of the bend, place tweezer at point C, and pull coil outward.

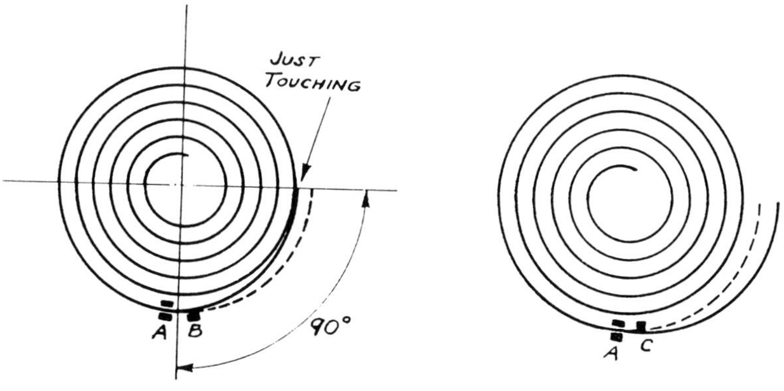

Fig. 5

Fig. 6

In this exercise, follow bending Method 1 with the exception that in this instance the bend is to be made 180 degrees from the outside end of the spring.

Note: Bending method 2 can also be used to accomplish the same result in this and the following exercises.

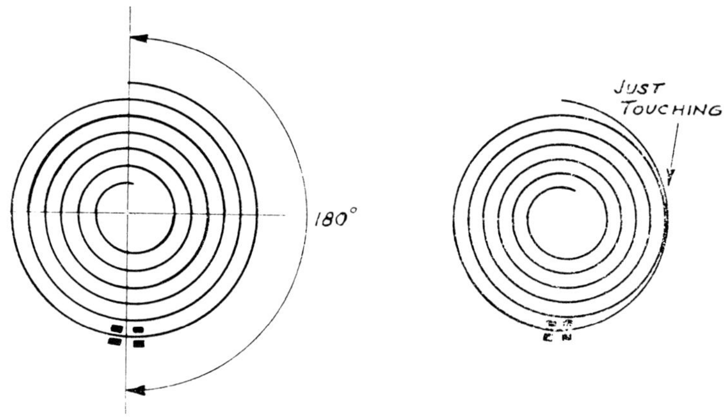

Fig. 6

Fig. 7

Again, this exercise is a duplication of the preceding one with the exception that the bend is made 270 degrees from the outside end of the spring.

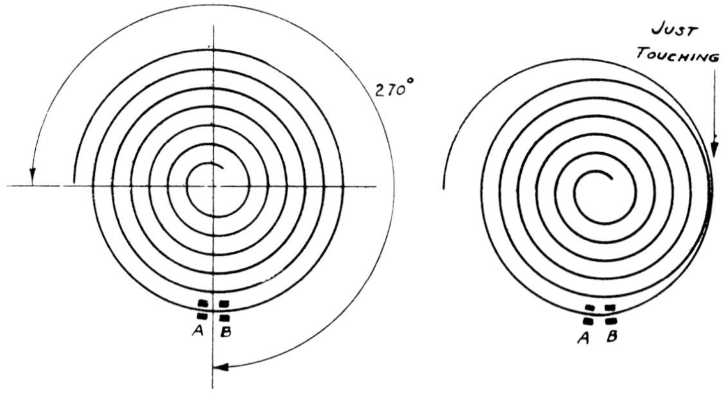

Fig. 7

Fig. 8

Duplicate the procedure already established, but this time the bend is to be made 360 degrees from the outside end of the spring.

After completing the above exercises, bend each coil, regardless of its place of bend, back to its original position.

Having performed this operation, break off the outer coil of each hairspring at the point where the bend was made. Use Method 2 under manipulative breaking exercises.

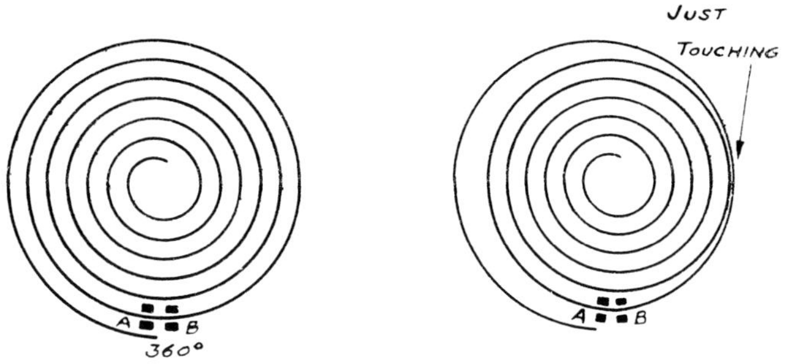

Fig. 8

Fig. 9

To straighten a coil, grasp the coil with a tweezer at desired point (See point A). Use flat pointed tweezer starting close to A and while maintaining gentle pressure between the flat points, pull and pinch coil straight outward toward the end.

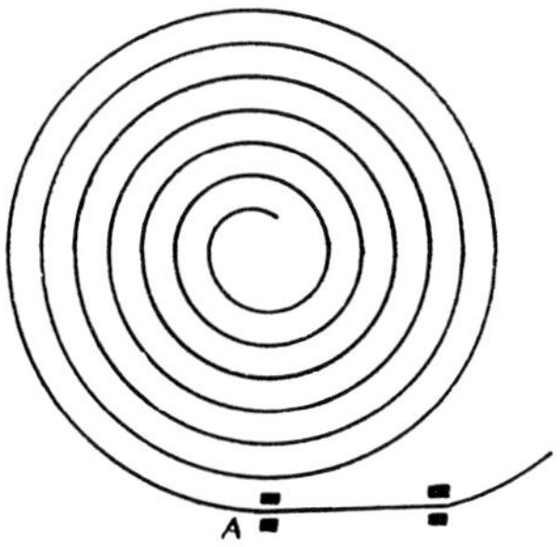

Fig. 9

2. COLLETING

Attaching Hairspring to Collet

Colleting is the procedure of attaching the inside coil of the hairspring to the collet by means of performing the following operations in the order listed:

Breaking out inner coil.
Forming the tongue.
Pinning-in.

When the inner coil is broken out preparatory to attaching the spring to the collet, it will amount to distorting the spring for the first time. Therefore, take utmost care so as not to damage the remaining portion of the spring while performing this operation.

Breaking Out Inner Coil

To break out the inner coil, grasp the coil with a tweezer at the desired point of break. See point A, Fig. 10 and follow with the procedure in Method 2, Fig. 3 in the breaking exercises.

In the breaking out operation, the inner coils are eliminated to an extent necessary to leave room for the collet. There should be enough broken out so that when the collet is placed in the center of the spring the space from the edge of the collet to the first inner coil should be one and one half times as large as the space between any two coils of the spring, Fig. 11.

Due to the fact that these tiny dimensions cannot be measured, the exact space between the edge of the collet and the first coil may be a little more or less than this stated amount.

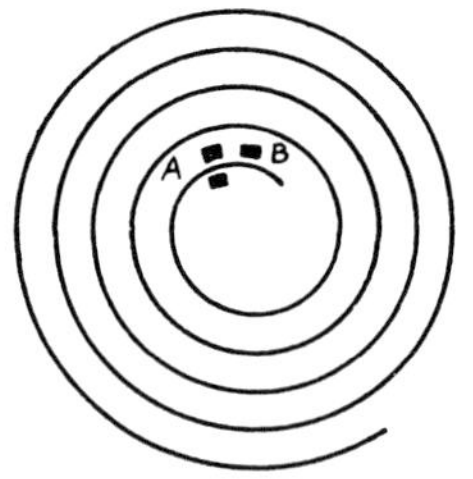

Fig. 10

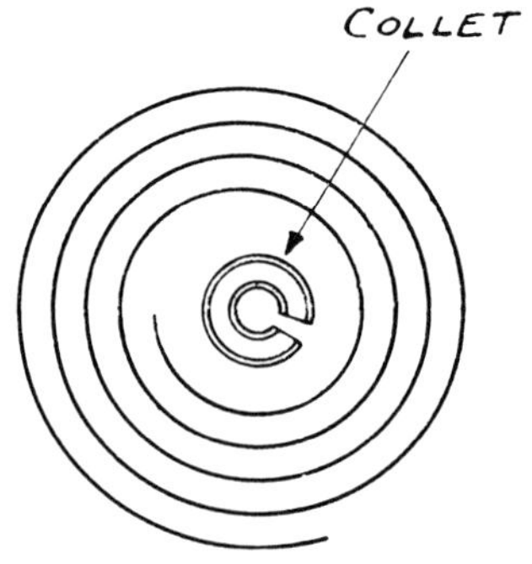

Fig. 11

Forming the Tongue

The tongue consists of a straight portion and an elbow which joins it with the first inner coil. The length of the tongue from point B to point A, part of which is fastened to the collet, should be as long as the collet pin hole plus the distance from the pin hole to the first coil. See Fig. 12.

To accomplish this operation, lay the hairspring flat on the bench. Grasp the inner coil at point A, Fig. 13, with a tweezer. This point will be about 90 degrees away from the inner end of the coil.

Use another pair of tweezers at B, Fig. 14, and bend the coil in the direction of the arrow. The amount of bend can be determined only through experience. Avoid a *sharp* bend at this point.

The curved tongue thus formed must now be straightened. See Fig. 15. The procedure for straightening the coil is shown in Fig. 9.

After completing that operation, the spring is ready to be pinned to the collet. This operation is called "Pinning-in".

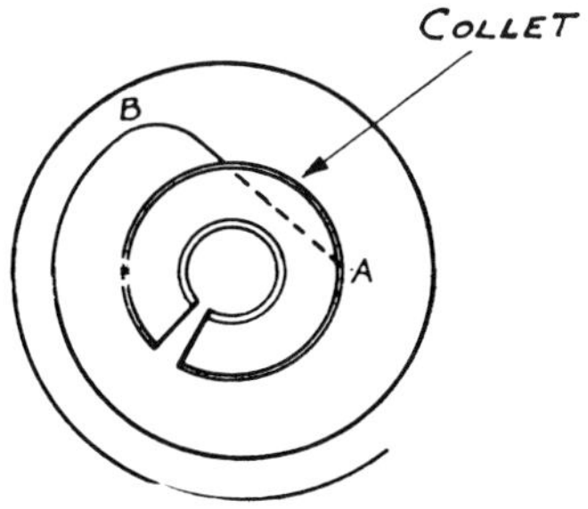

Fig. 12

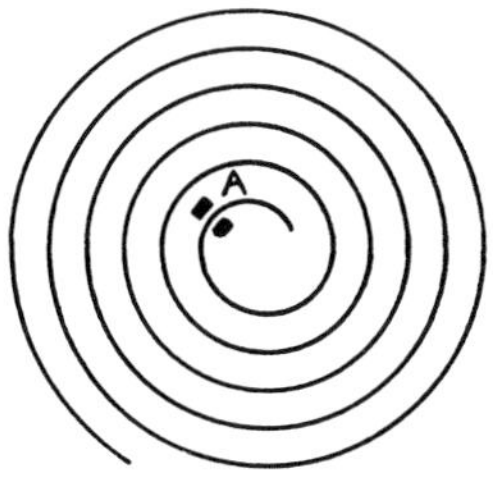

Fig. 13

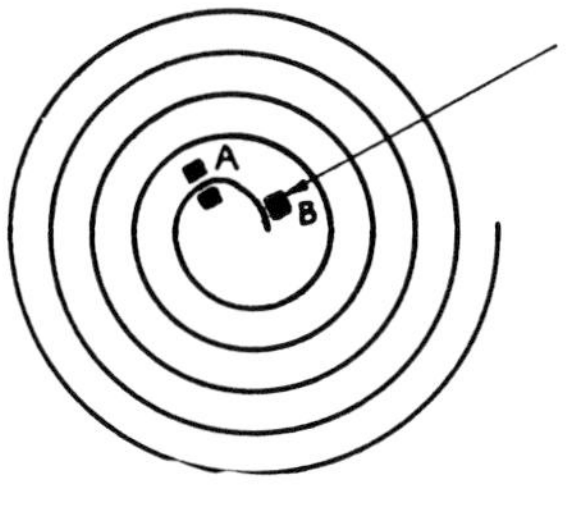

Fig. 14

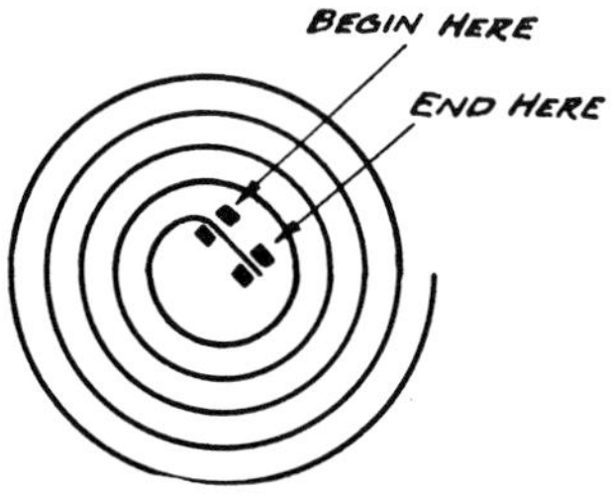

Fig. 15

PINNING-IN

The mechanics of actually pinning-in the spring to the collet will be explained and shown first, followed with a study of the usual errors that are the results of colleting. This is an operation that requires great care and skill. Take extreme precaution not to bend or distort any inside coils.

To Start the Actual Pinning-in Operation

Place the collet on a colleting arbor with top of collet upward and point D, Fig. 16 of pinhole toward you. Hold arbor in left hand.

To insert tongue in collet hole, grasp coil with tweezers at point B, Fig. 17. Bring spring over and down colleting arbor with end of tongue A, Fig. 17 in direct line with point D, Fig. 16 of the pinhole in collet.

While holding tongue A in proper position, turn colleting arbor in a counter-clockwise direction gently enough to bring tongue into pin hole D of collet. Fig. 18. After placing tongue in pin hole, use thumb and first finger to support the spring and prevent it from tipping.

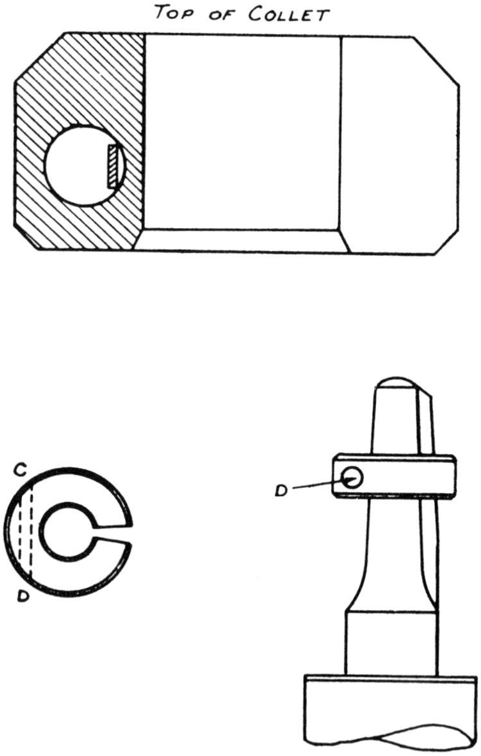

Fig. 16

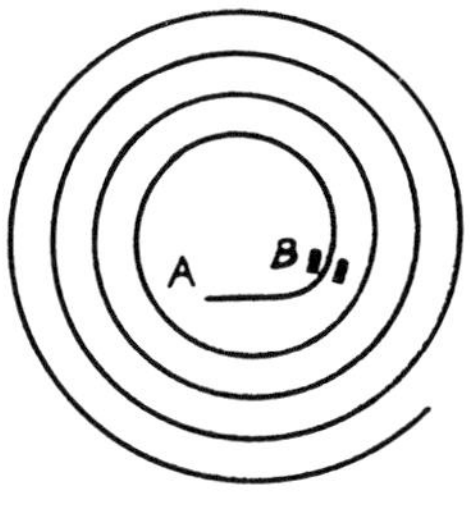

Fig. 17

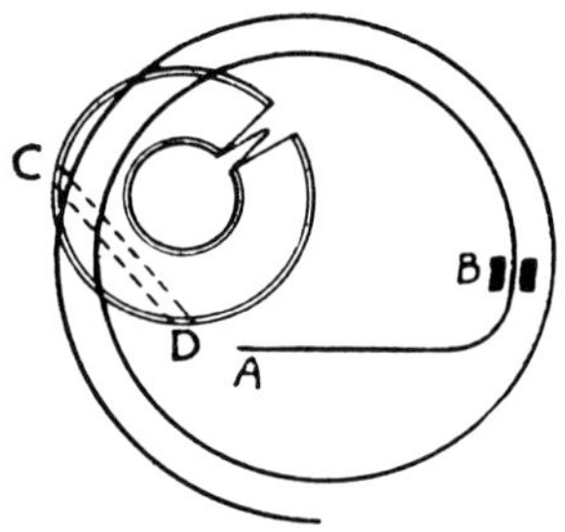

Fig. 18

From the same end D of the pin hole that the tongue entered, insert a tapered brass pin against the side of the tongue by grasping the pin at the thickest part with a tweezer. The pin must extend beyond the collet pin hole above the main body of the spring as shown at E, Fig. 19.

Push the pin just tight enough to hold the spring in place. It will now be necessary to make a preliminary inspection of the spring to determine the position of the inside coil in relation to the collet. Listed below are two of the outstanding errors that may have been established in the pinning in process. Eliminate these errors now, to avoid corrective manipulations in the future.

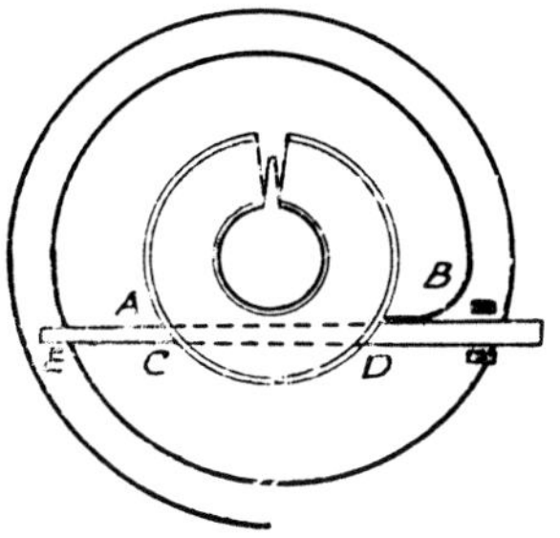

Fig. 19

Preliminary Pinning-in Errors

Fig. 20 shows tongue pinned too far in pin hole.

Fig. 21 shows tongue pinned too far out of pin hole.

If one of the two errors shown above is present, loosen the tapered pin sufficiently so that point B can be pulled or pushed until it is the proper distance from the collet. Inspect position of tongue again. Repeat this procedure until the tongue is pinned-in similar to Fig. 22 which shows tongue pinned in correctly.

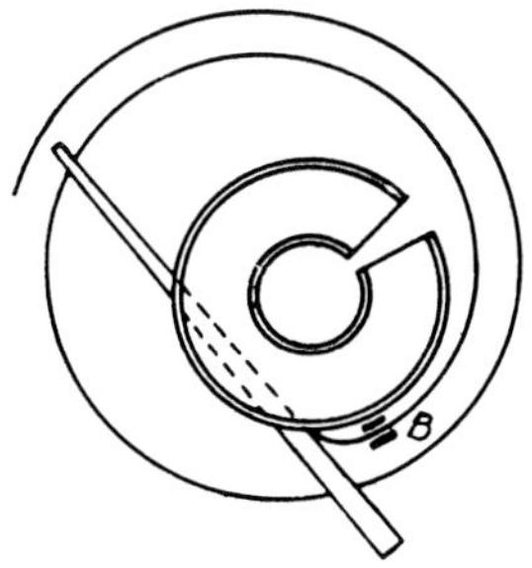

Fig. 20

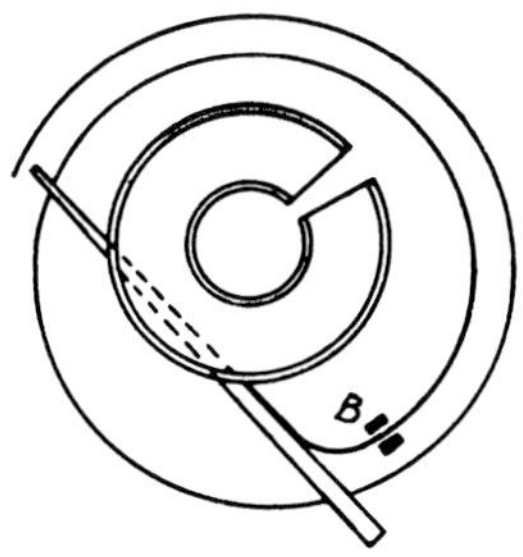

Fig. 21

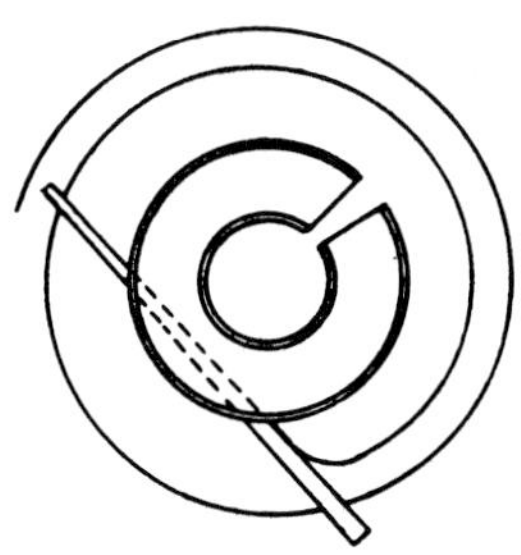

Fig. 22

After this has been accomplished, inspect for errors in the flat. To do this, hold the colleting arbor at such an angle, convenient to the observer, that the plane of the spring is on the line of sight. Use a loupe and observe whether or not the plane of the hairspring is parallel with the plane of the collet as shown in Fig. 23. If the two planes are not parallel, point A of the tongue, Fig. 24 was fixed too low or too high in the collet hole and must be corrected.

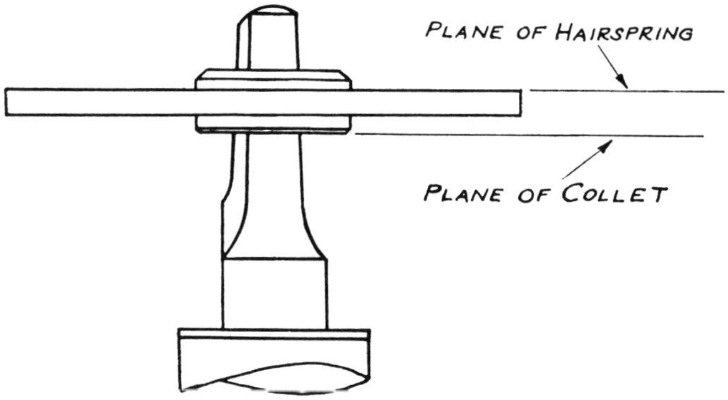

Fig. 23

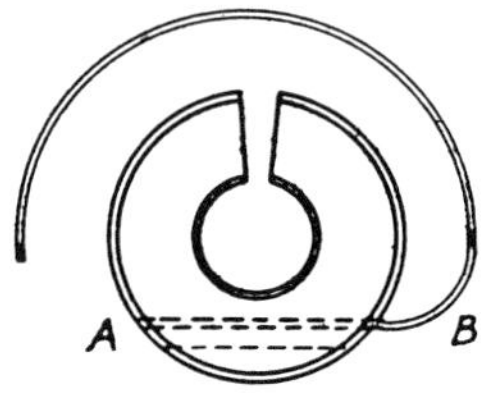

Fig. 24

Preliminary Errors in the Flat

When point A of the tongue becomes fixed too high in collet hole, it will cause all of the coils to be high opposite the pinning point D. See Fig. 25.

When point A of the tongue becomes fixed too low in collet hole, it will cause all of the coils to be low opposite the pinning point D. See Fig. 26.

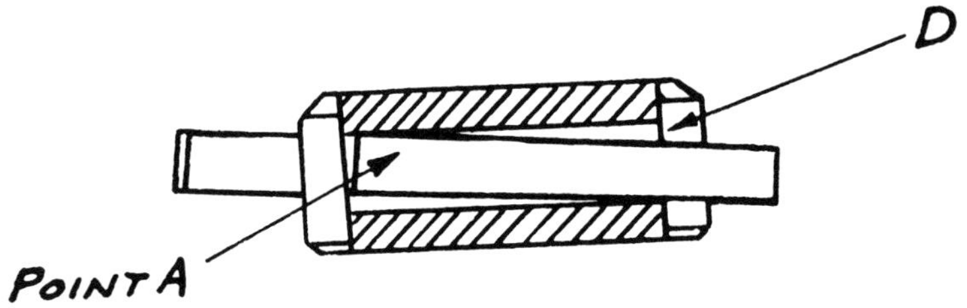

Fig. 25

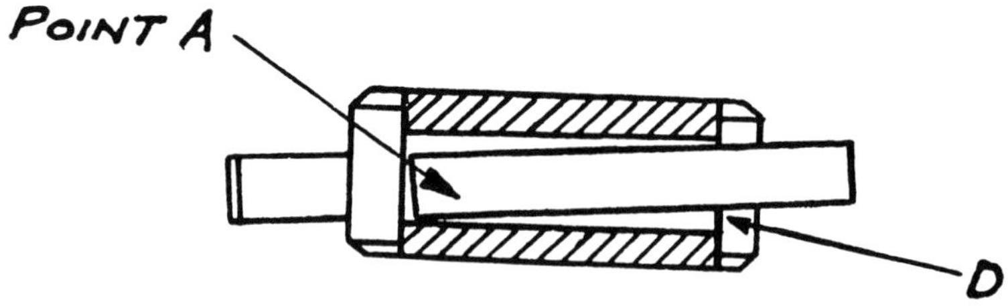

Fig. 26

Correct errors in the flat by holding the arbor in a vertical position and gently raise or lower the inner coil with a tweezer until the spring is reasonably parallel with the collet and perpendicular to the axis of the arbor.

TIGHTENING THE PIN IN THE COLLET

Method No. 1

Grasp the small end of the tapered pin with a pair of snipe nose pliers, and with a pulling and twisting motion, pull the pin tight. See Fig. 27.

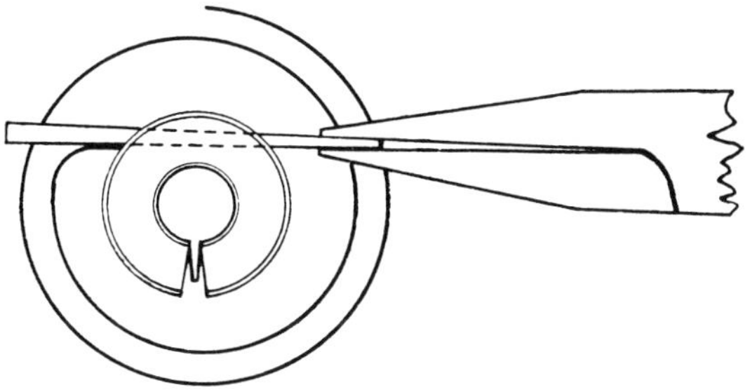

Fig. 27

Break off small end of the brass pin projecting from the hole. To accomplish this, grasp the small end of the pin with the pliers, and maintain a firm pull on the pin. Turn the arbor in a clockwise direction, thereby bending the pin at a sharp 90 degree angle close to the collet. Fig. 28.

Now turn the arbor in the opposite direction and the pin will break off clean. Fig. 29.

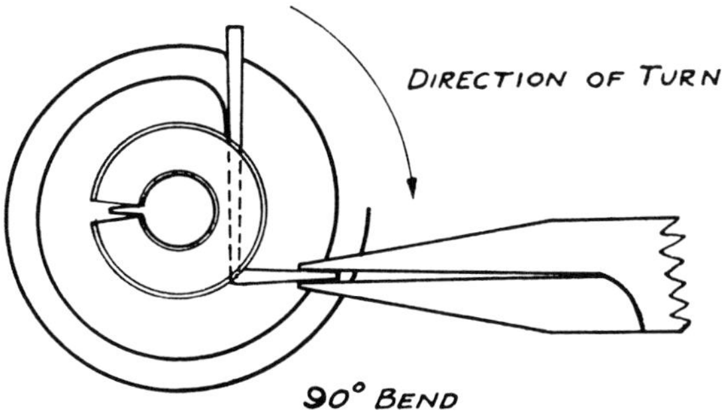

Fig. 28

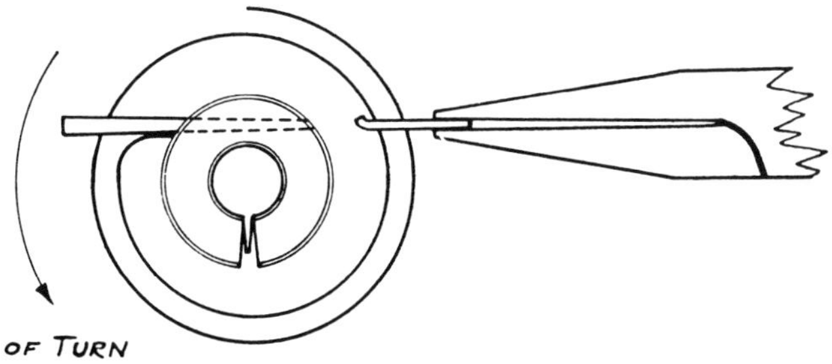

Fig. 29

The large end of the taper pin is broken off in much the same way. Grasp the pin with the snipe nose pliers and push on it while turning the arbor in counter-clockwise direction, thereby putting a sharp 90 degree bend in the pin close to the collet. Be careful while turning the arbor not to loosen the taper pin. See Fig. 30. By turning the arbor in a clock-wise direction the pin will break off close to the collet. Fig. 31.

After pinning-in, examine the position of the collet in relation to the first inner coil. The spring should not touch any part of the collet except at the pinning point.

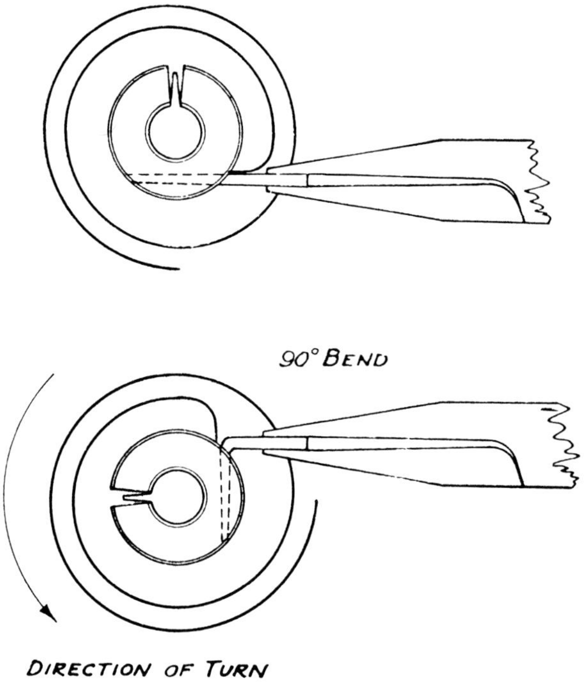

Fig. 30

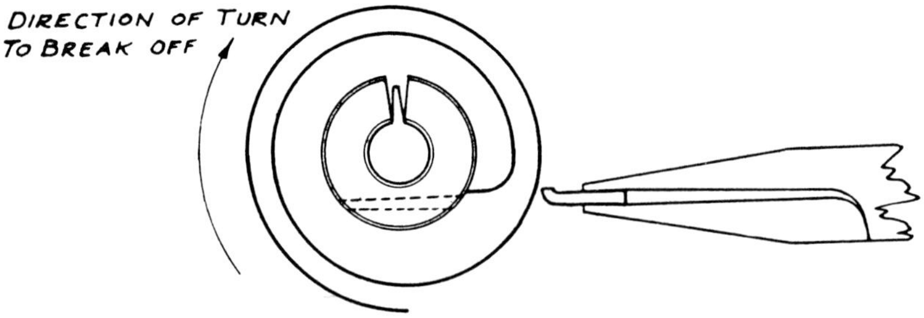

Fig. 31

TIGHTENING THE PIN IN THE COLLET

Method No. 2

Secure the taper pin firmly in place and remove the projecting ends by the following method. By use of the side cutting pliers cut off the large end of the taper pin at a distance from the collet equal to two to two and one half times the diameter of the large end of the pin. Fig. 27A.

Then use the snipe nose pliers to push the taper pin tightly in place by straddling the collet as shown in Fig. 27B. Place one leg of the pliers at A and the other on the large end of the pin at B. When it is pushed in tight, the large end of the taper pin should be flush with the collet.

To remove the small end of the taper pin, use the end cutting pliers. Hold the cutting edges of the pliers close up against the collet and snip the pin off. Fig. 27C.

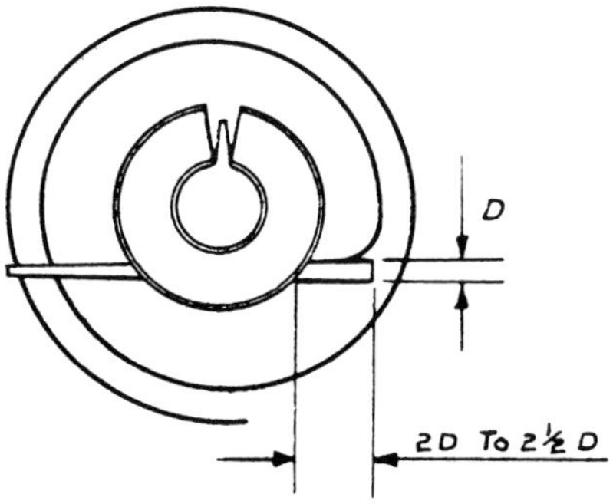

Fig. 27A

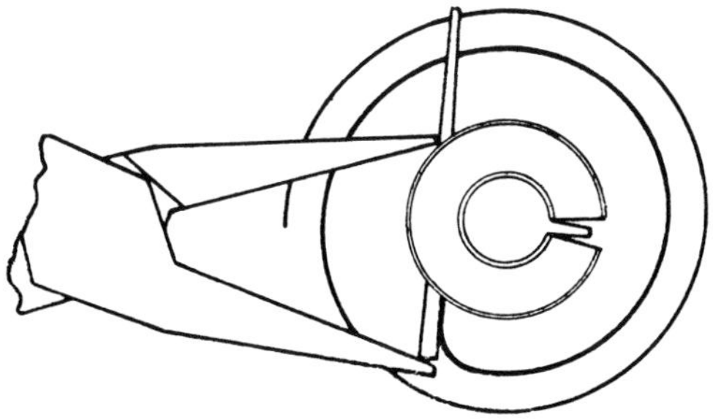

Fig. 27B

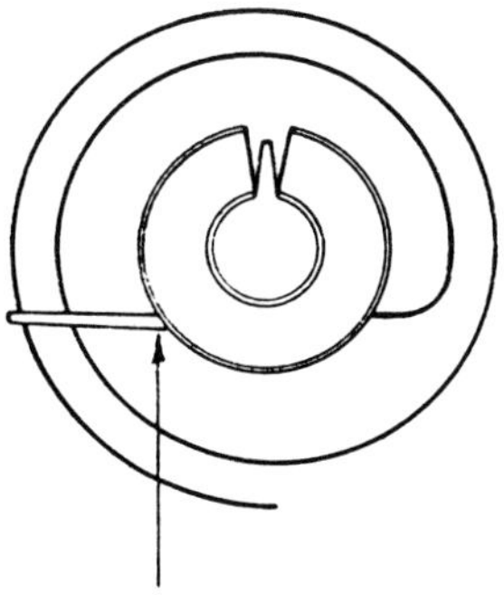

Fig. 27C

Preliminary Errors in the Round

Fig. 32 and Fig. 33 show two basic errors in the round that may have been made by improper forming of the tongue. Because of this, the collet is off center relative to the spring. Fig. 32 shows the result of bending the tongue too much at its elbow. Fig. 33 indicates the case wherein the tongue is not bent enough at the elbow. Fig. 34 shows how these errors compare with each other and with a tongue that is properly formed.

Try to bring the center of the spring to the center of the collet, as closely as possible. The bending is performed on the first one-eighth of the inside coil with a tweezer. Concentrate on centering only.

When preliminary errors in the flat and the round have been corrected, check the collet and spring for the following possible faults before removing from the arbor.

1. Both ends of the pin should be broken off clean at the collet.
2. Taper pin should be tight in collet hole.
3. If any of the inside coils are bent or distorted in any manner, it is best to discard the spring before going any further.

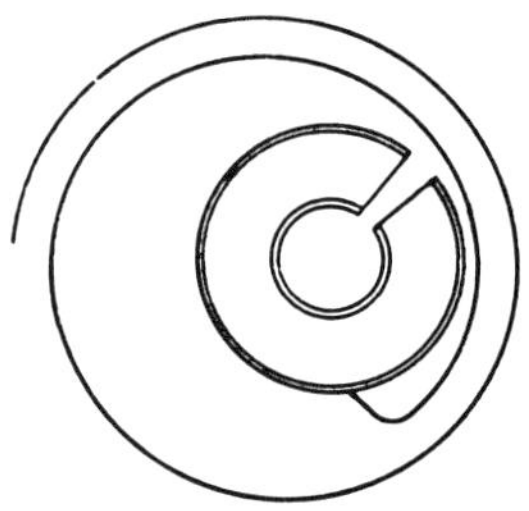

Fig. 32

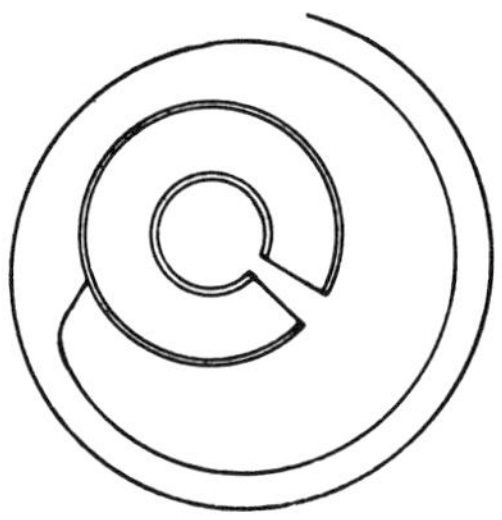

Fig. 33

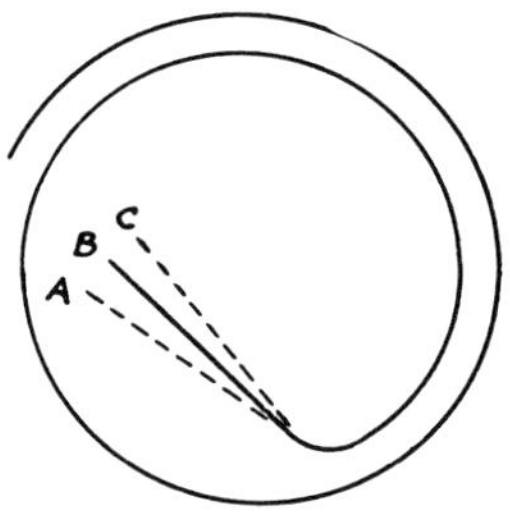

A Tongue Not Bent Enough
B Tongue Bent Right
C Tongue Bent Too Much

Fig. 34

III. Staking Hairspring and Collet on the Balance Staff

The hairspring and collet are now ready to be fastened to the balance staff. A staking stand is used to facilitate the operation.

Select a stump with a hole slightly larger than the roller shoulder of the staff, but smaller than the roller table seat. Fig. 35. Place the selected stump in the staking table and insert the balance staff into the stump as shown in Fig. 36.

Now, select a staking punch having a hole slightly larger than the collet shoulder of the balance staff, but smaller than the outside diameter of the collet. Fig. 36.

Lay the hairspring and collet on balance staff with top side of collet upward. Position spiral of spring in a counter-clockwise direction from pinning point. Bring the staking punch down on the collet and gently force the collet onto the collet shoulder until it is firmly seated on the collet seat. Fig. 36. The friction between a properly fitted collet and its collet shoulder is sufficient to hold the spring in place under all conditions.

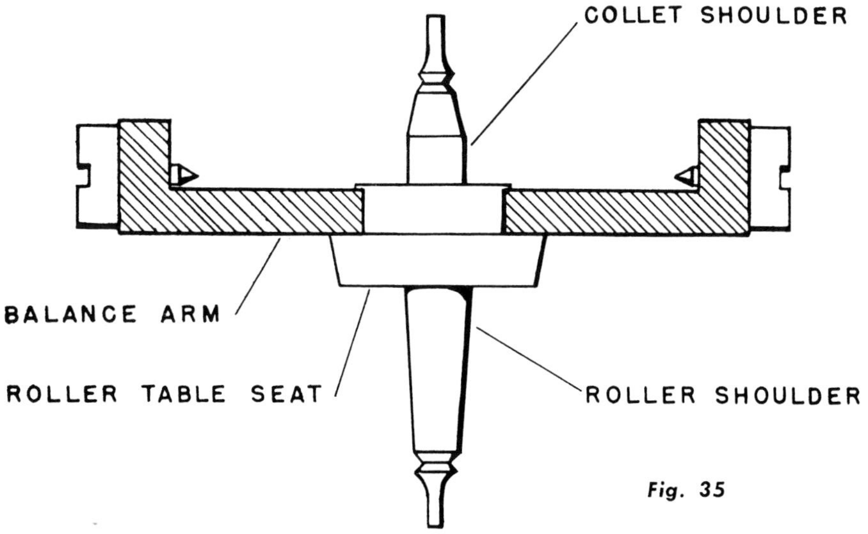

Fig. 35

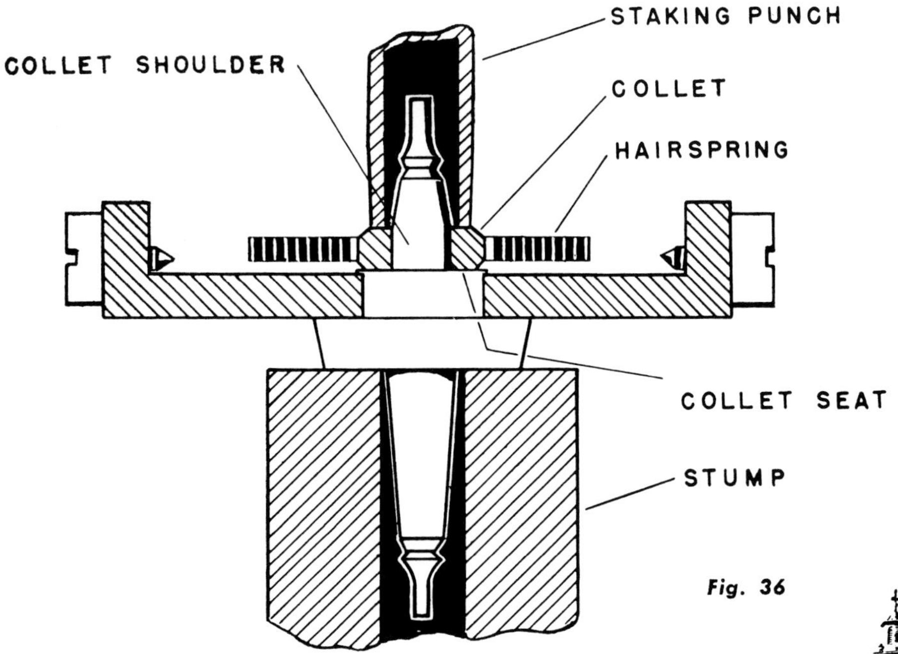

Fig. 36

After staking the springs on the staff, dip the entire assembly in a good cleaning solution, rinse properly, dip in alcohol and dry in box wood sawdust. When this is completed, place each balance wheel separately on the calipers as shown in Fig. 37. Use a camel's hair brush to remove the small particles of sawdust from the hairspring and balance wheel.

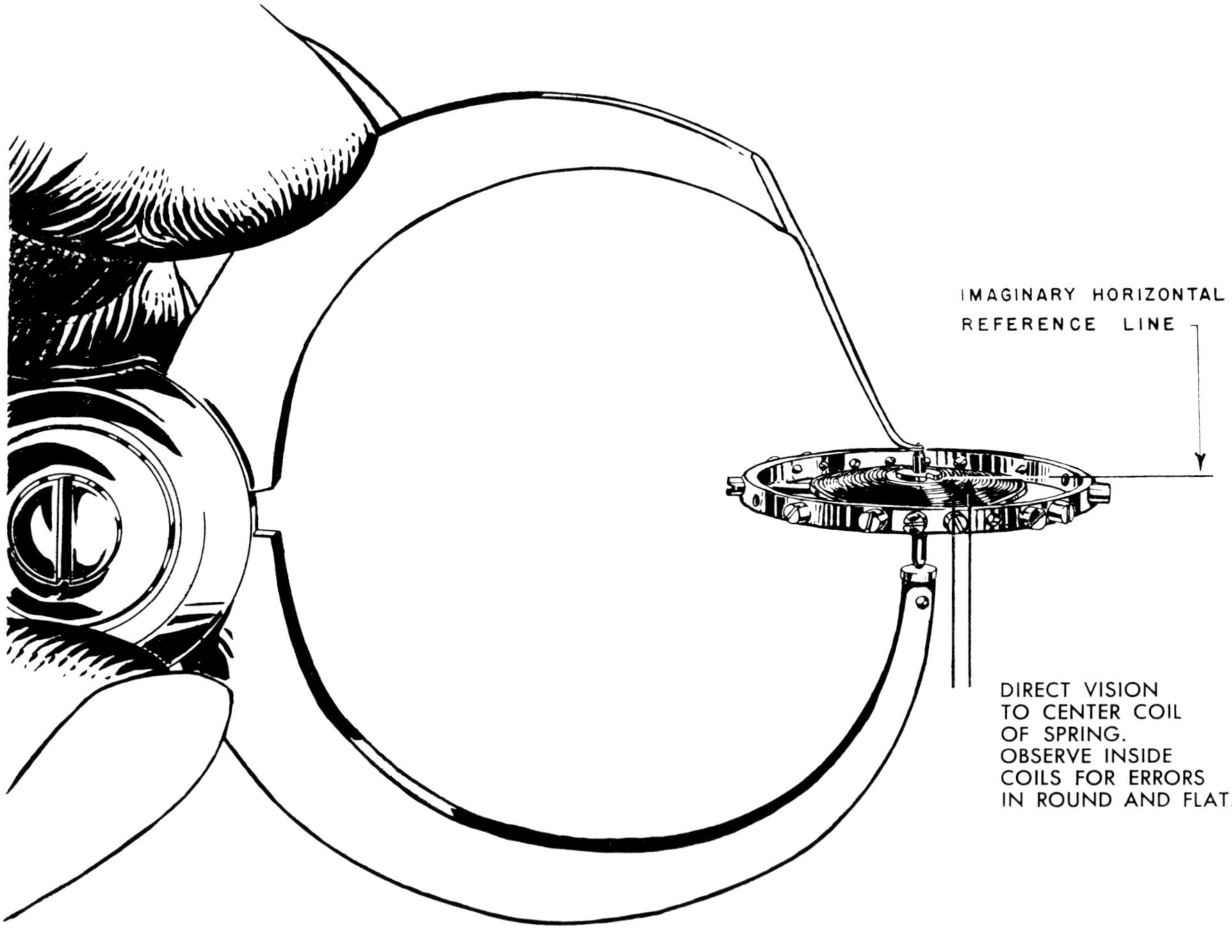

Fig. 37

IV. Truing in the Round and the Flat

Hairspring truing is the procedure of locating the collet so that it will be flat and central with the coils of the spring. It is quite obvious that the operation of attaching the hairspring to the collet requires a considerable amount of skill and it would be assuming too much on the part of the manipulator to perform this operation so that the collet would be perfectly central and flat with the spring. The operating of hairspring truing is necessary to correct the errors caused by the "breaking out", "forming the tongue" and "pinning-in" operations. With this in mind, proceed to locate the errors resulting from colleting.

With the use of calipers, examine the hairspring for the more refined errors and for the final truing in the round and flat.

Fig. 37 and Fig. 38 shows the proper position of a hairspring and balance wheel in the calipers for inspection in the flat and the round.

When making the examination, rotate the balance wheel in the caliper. Rotate the balance wheel slowly enough to observe the inner coil in relation to the collet by applying a light brushing motion to the balance wheel with the side of the forefinger.

While the balance wheel is rotating in the caliper, the spring can be considered true in the round if the first three or four inside coils appear as perfect circles. If the coils seem to jump or describe irregular circles, then there is an error in the round.

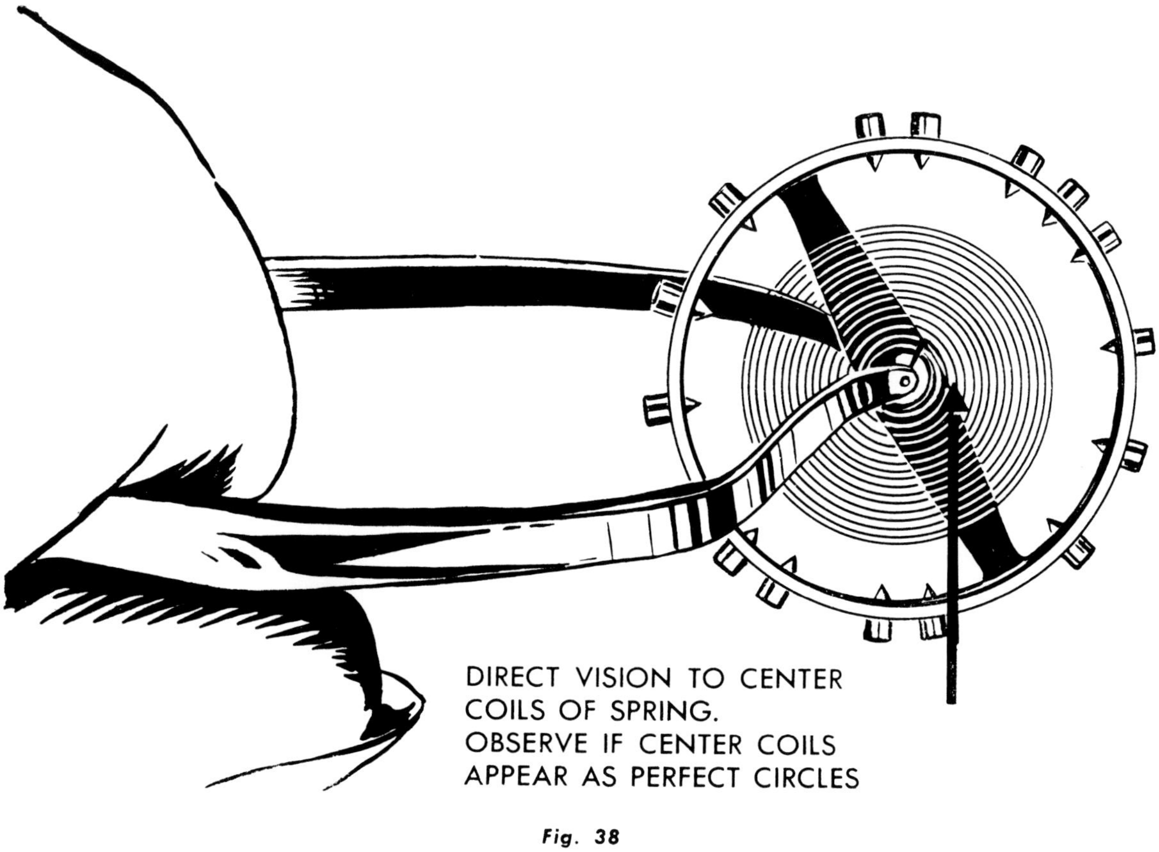

Fig. 38

Errors in the Round

When the collet is placed properly in center of spiral and the spring is correctly pinned to the collet, the spring will look like Fig. 39.

In Figs. 40 to 47 are a combination of eight errors that show the spring is out of true in the round and represents usual errors introduced by colleting. Each diagram has been greatly enlarged and exaggerated to help detect the errors that exist. Each figure describes the position of the collet which is off center in relation to the spring when the tongue is not "formed properly" and "pinned-in" correctly. The direction in which it is off center is indicated by the arrow.

The presence of any of the errors shown below denotes that corrections need to be applied to the spring. Before attempting to correct the errors, first study each spring and compare the errors with those shown in Fig. 40 to Fig. 47.

PERFECT

FIG. 39

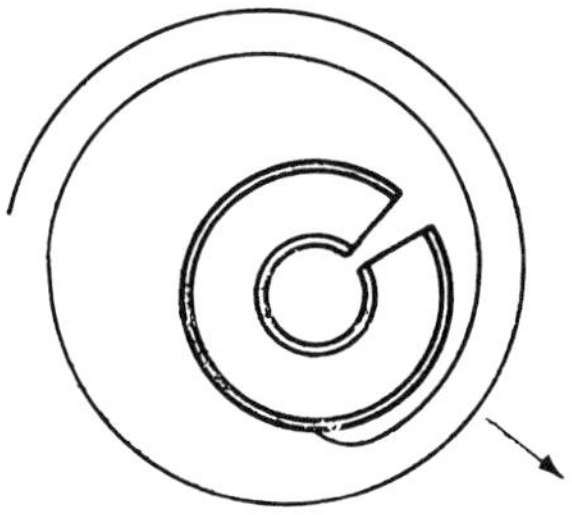

TONGUE IN TOO FAR

FIG. 40

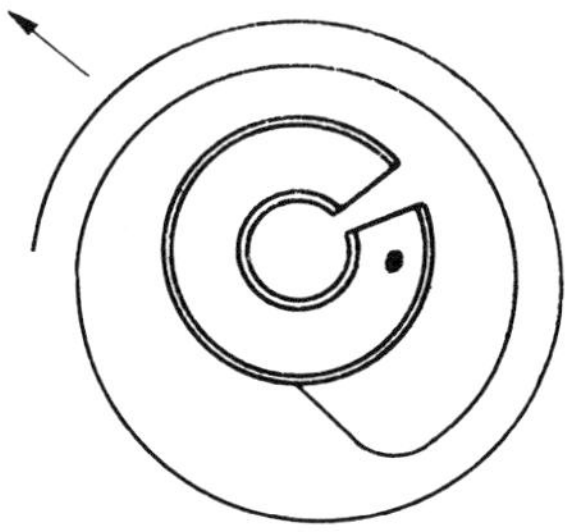

TONGUE OUT TOO FAR

FIG. 41

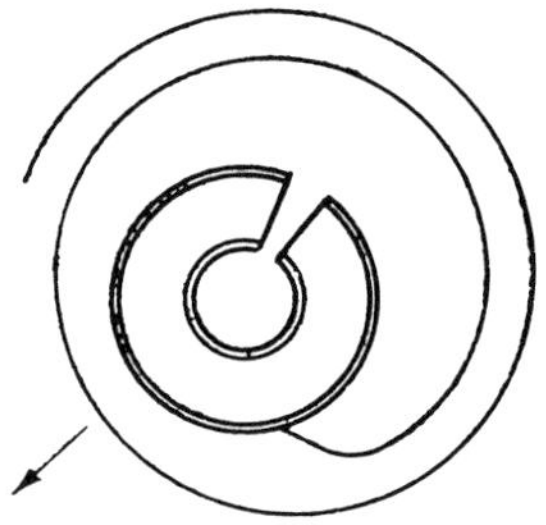

TONGUE NOT BENT ENOUGH

FIG. 42

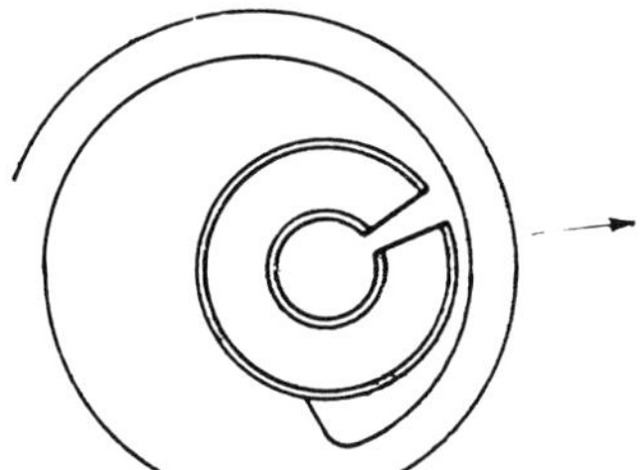

TONGUE BENT TOO MUCH

FIG. 43

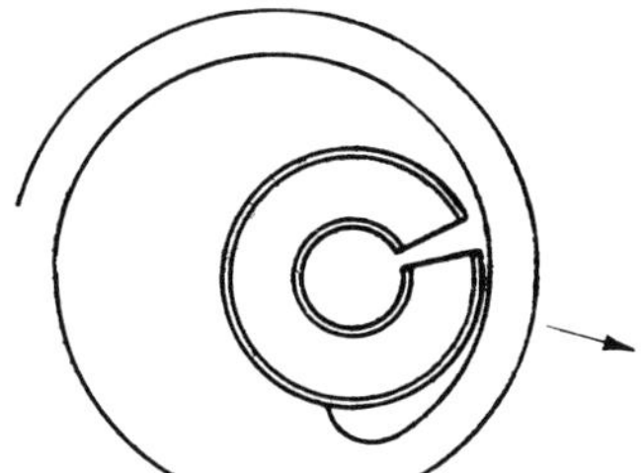

TONGUE BENT TOO MUCH & IN TOO FAR

FIG. 44

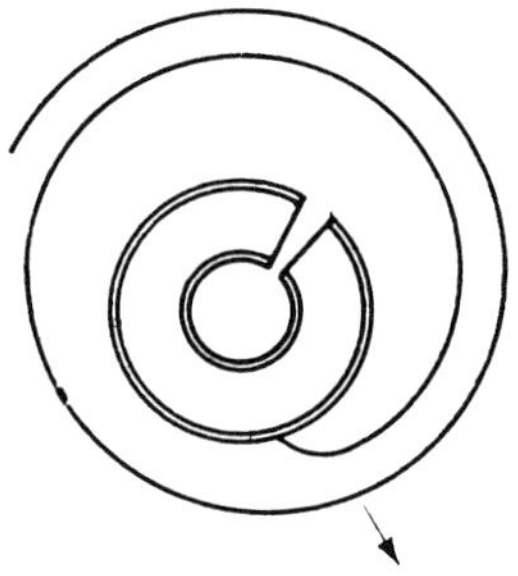

TONGUE NOT BENT ENOUGH & IN TOO FAR

FIG. 45

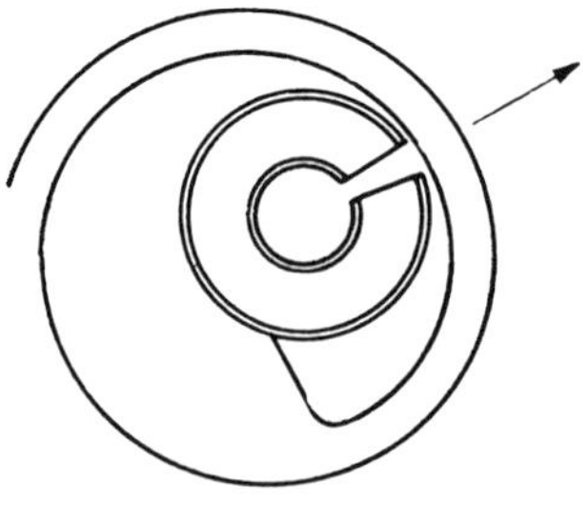

TONGUE BENT TOO MUCH & OUT TOO FAR

FIG. 46

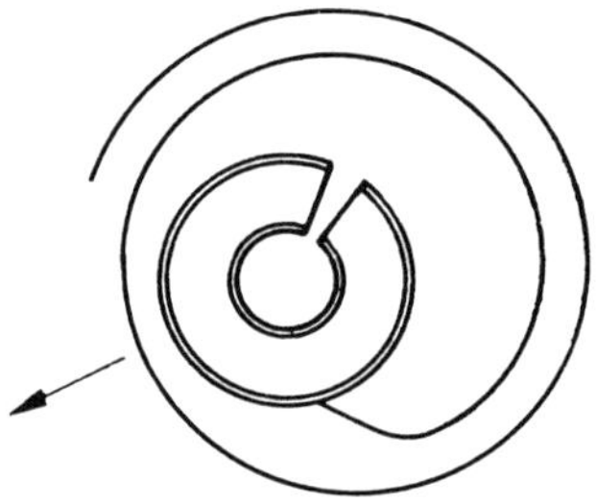

TONGUE NOT BENT ENOUGH & OUT TOO FAR

FIG. 47

After arriving at a decision concerning the errors of each spring, now turn attention to the corrective bending operations. Methods that are applied individually to correct the errors previously shown will be explained in the following figures. All bends necessary to eliminate the errors can readily be made with the tweezers while the balance wheel is supported in the calipers. Direct all corrective bendings toward locating the center of the collet in the center of the spiral by employing the bend on the first one-eighth coil.

A detailed breakdown of the bending operations used to correct the errors is given as follows:

ERROR shown in Fig. 48-A-B-C

CAUSE: Tongue pinned too far in.

To Correct This:

1. Grasp inside coil with tweezers at point B, Fig. 48A, and bend the coil away from collet as shown by arrow.
2. Then grasp the inside coil at point C, Fig. 48B, and bend the coil toward the collet as shown by the arrow. Fig. 48C shows the net results of the corrective bendings with the collet centered with the spring.

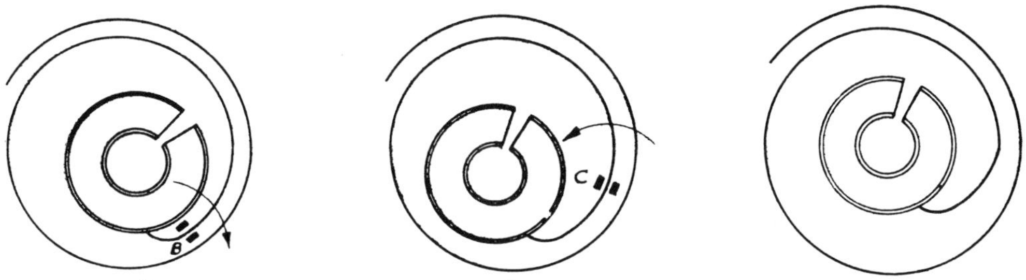

A B C

Fig. 48

ERROR shown in Fig. 49-A-B-C

CAUSE: Tongue pinned-to far out.

To Correct This:

1. Place tweezer points in the position shown as A and B, Fig. 49A and gently squeeze point B towards collet as indicated by the arrow.
2. Then grasp inside coil at point C, Fig. 49B, and bend the coil away from collet as shown by the arrow.

Fig. 49C shows the results of the above bendings with the collet centered with the spring.

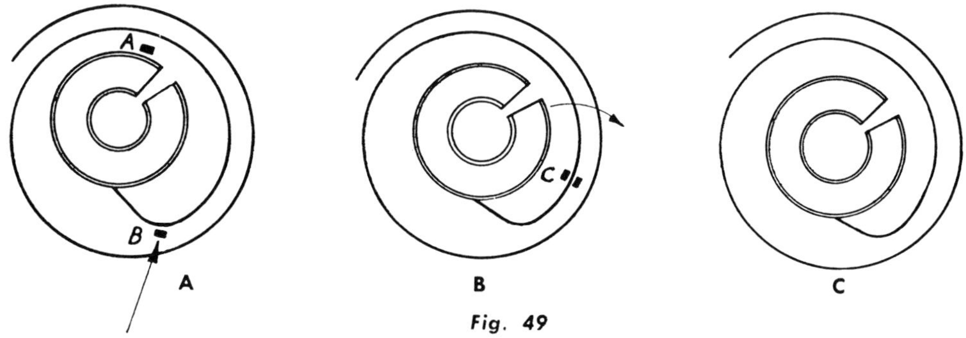

A B C

Fig. 49

ERROR shown in Fig. 50-A-B.

CAUSE: Tongue not bent enough.

To Correct This:

1. Grasp inside coil with tweezer at point B, Fig. 50A and bend the coil toward collet as shown by arrow. The collet is now centered with the spring as shown in Fig. 50B.

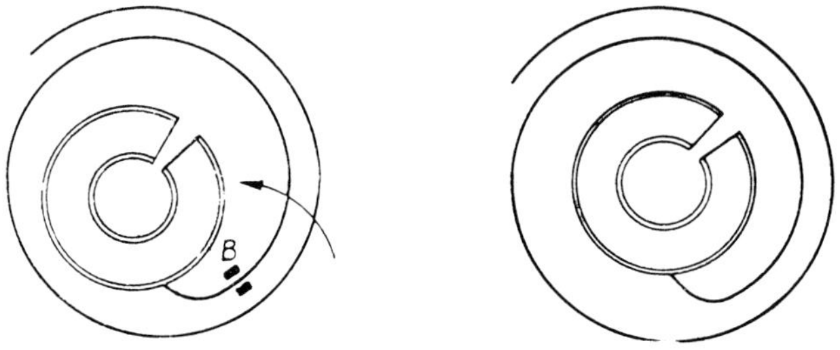

A B

Fig. 50

ERROR shown in Fig. 51-A-B.

CAUSE: Tongue bent too much.

To Correct This:

1. Grasp inside coil with tweezer at point B, Fig. 51A, and bend the coil away from collet as shown by arrow until collet and spring are centered as shown in Fig. 51B.

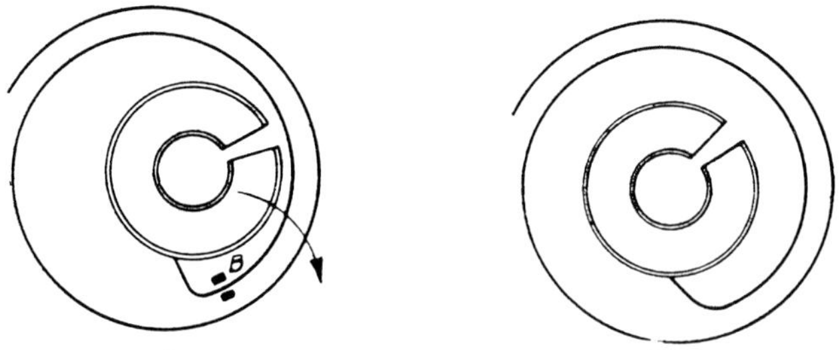

A B

Fig. 51

ERROR shown in Fig. 52-A-B-C.

CAUSE: 1. Tongue bent too much.
2. Pinned in too far.

To Correct This:

1. Grasp inside coil with tweezer at point B, Fig. 52A, and bend the coil away from collet as shown by arrow.

2. Then grasp the inside coil at point C, Fig. 52B, and bend the coil toward the collet as shown by arrow. Fig. 52C shows the collet and spring centered after proper manipulation.

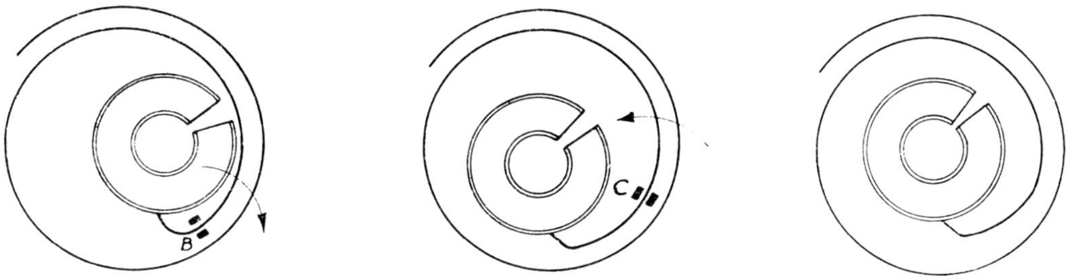

A B C

Fig. 52

ERROR shown in Fig. 53-A-B-C-D.

CAUSE: 1. Tongue not bent enough.
2. Pinned in too far.

To Correct This:

1. Grasp inside coil with tweezer at point B, Fig. 53A and bend the coil away from collet as shown by arrow.

2. Then grasp the inside coil at point C, Fig. 53B, and bend the coil toward the collet as shown by arrow.

3. Now grasp the inside coil at point D, Fig. 53C and again bend the coil toward the collet as shown by arrow.

Fig. 53D shows the results after the above bendings have eliminated the errors.

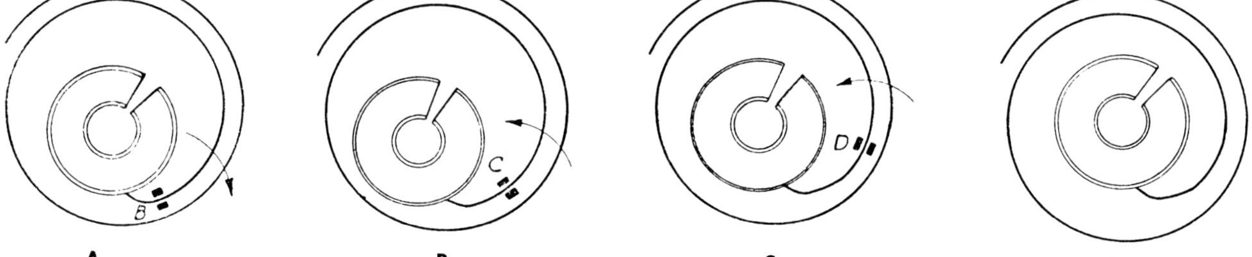

A B C D

Fig. 53

ERROR shown in Fig. 54-A-B-C.

CAUSE: 1. Tongue bent too much.

2. Tongue out too far.

To Correct This:

1. Place tweezer points in the position shown as A and B, Fig. 54A, and gently squeeze point B toward collet as indicated by arrow.

2. Then grasp inside coil at point C, Fig. 54B, and bend the coil away from collet as shown by arrow.

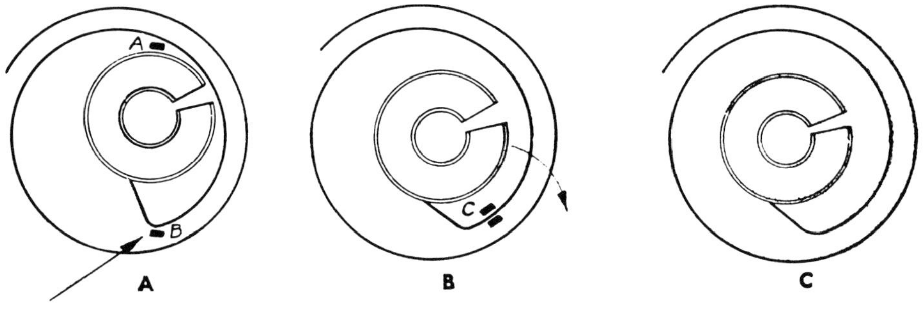

Fig. 54

ERROR shown in Fig. 55-A-B-C.

CAUSE: 1. Tongue not bent enough.

2. Tongue out too far.

To Correct This:

1. Place tweezer points in the position shown as A and B, Fig. 55A, and gently squeeze point B toward collet as indicated by arrow.

2. Then grasp inside coil at point C, Fig. 55B, and bend the coil away from collet as shown by arrow.

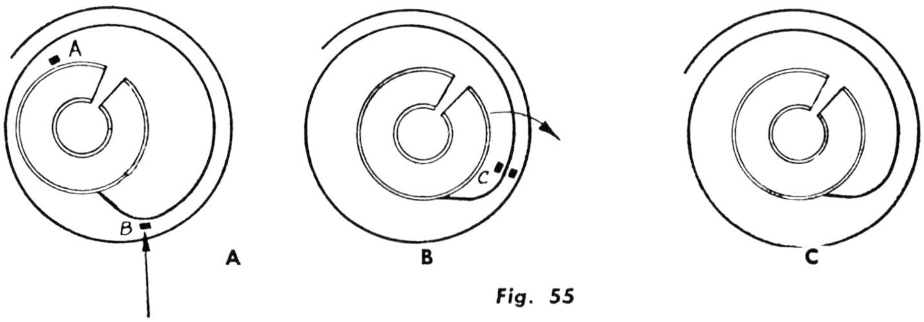

Fig. 55

Errors in the Flat

Here are four conditions that will throw the spring off true in the flat as a result of the spring being bent during the colleting procedure, or fixed in the collet hole other than in a vertical position.

When point A, Fig. 56 of the tongue is bent downward in relation to the spiral portion of the spring, it will cause all of the coils to be high opposite the pinning point.

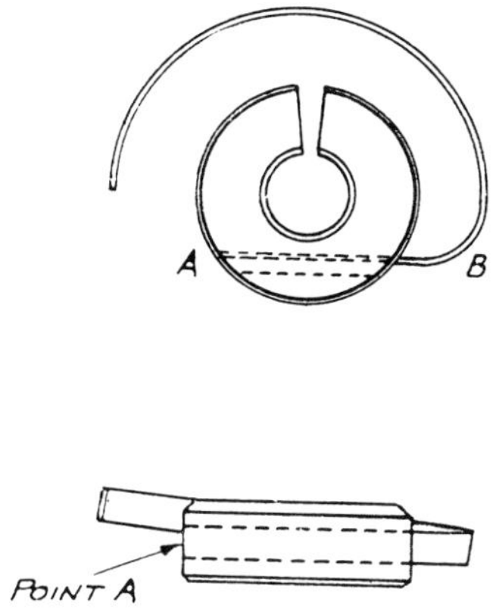

Fig. 56

When point A, Fig. 57 of the tongue is bent upward in relation to the spiral portion of the spring, it will cause all of the coils to be low opposite the pinning point.

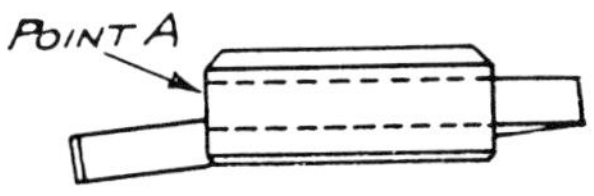

Fig. 57

When the flat face of the tongue is not pinned in a vertical position, but becomes fixed at some angle, as shown in Figs. 58 and 59, the spiral portion of the spring will be high or low, 90 degrees from the pinning point.

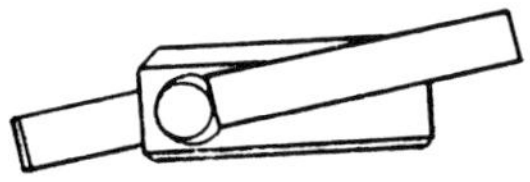

Fig. 58

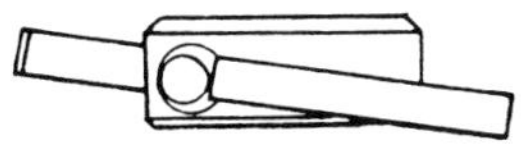

Fig. 59

Since all manipulations during the colleting procedure have been performed on the inside coil, the source of these errors can be corrected by applying a gentle force with a tweezer on the inner coil in the proper direction. To eliminate the error, raise the spring up on the low side or bend it down on the high side, thereby placing the spring level in relation to the collet.

When the spring is true in the flat, check it for trueness in the round. Very often the bending operations used to true the spring in the flat will introduce an error in the round and vice-versa. Therefore, check the spring alternately in the flat and round until it is as true as possible in both the flat and round.

TRAINING UNIT NUMBER 7

★

HAIRSPRING VIBRATING

THE ART *of* VIBRATING HAIRSPRINGS

Vibrating is the process of establishing the elastic reaction of a hairspring to the weight of a balance wheel so that the wheel will perform the required number of vibrations per hour. Vibrating is an art by itself; so much so, that in most of the large cities, a large amount of this work is done today by hairspring vibrating specialists.

However, an expert watch repairer should still be able to accomplish this very necessary procedure when the emergency arises. Very often, delays of three months and more in this service are required before the customer's watch can be returned to him. Therefore, the watch repairer who can perform this delicate operation is in an advantageous position.

Moreover, a thorough knowledge and skill in vibrating have a real value in making routine adjustments or repairs on the hairspring when performing a complete watch overhaul. Unless the watch repairer is thoroughly trained in the manipulative skill requisite to the fitting of a hairspring to a balance assembly, he cannot honestly turn out a finished or satisfactory repair job in every case.

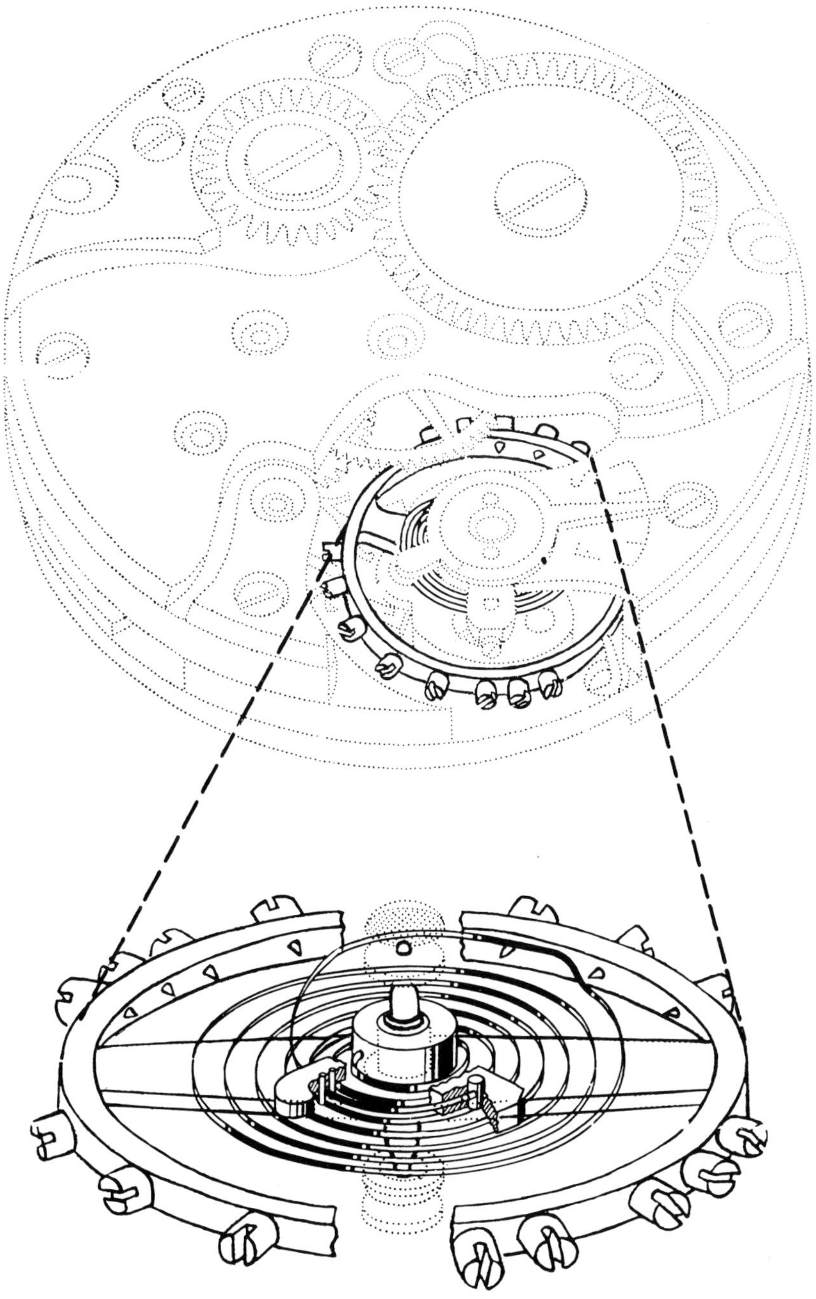

BULOVA SCHOOL *of* WATCHMAKING

Subject: Vibrating Hairsprings.

Objectives:

1. To learn the principles and relationships underlying hairspring vibrating.
2. To develop the skill necessary to vibrate a hairspring accurately.

New Tools Required:

Master Hairspring Vibrator.*

*Only used in production work.

Before entering into a detailed discussion on hairspring vibrating, some of the terms relative to the subject should be set forth and defined.

VIBRATION— A vibration is a single swing from one extreme limit to the other of an oscillating body. With regard to the balance wheel and hairspring, one vibration is the motion of the balance wheel from its position at which the hairspring is unstressed to its maximum limit and back again to the unstressed position of the hairspring. Or, stated in another way, the balance and hairspring make one vibration when the balance moves from an extreme position at the end of its swing back to the opposite extreme position at the other end of its swing.

OSCILLATION— An oscillation consists of two vibrations.

If an elastic rod is fixed at one end and deflected at the other, it will, upon being released, vibrate back and forth until the energy stored in the rod by the deflection is used up. (Fig. 1.) The outstanding feature about a vibrating rod is in the constancy of its rate of vibration. Because of the fact that it will continue to vibrate at the same rate until it comes to rest, we have a basis for a time keeping element in a watch or clock. It was this property of elasticity that led to the development of the hairspring. An early step in this development was the employment of a pig's bristle together with a balance wheel as the time keeping element of the watch. (Fig. 2.)

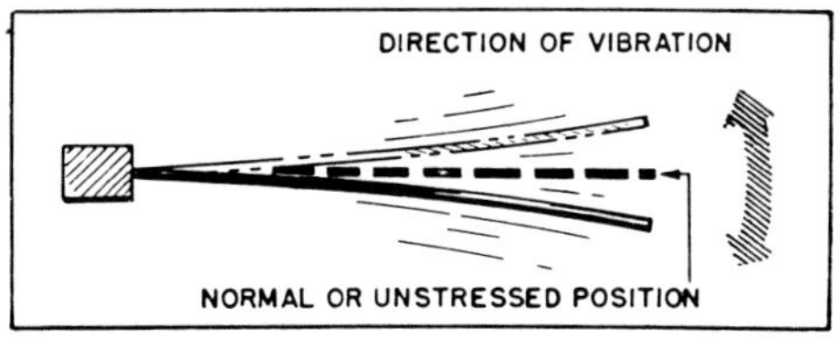

Fig. 1

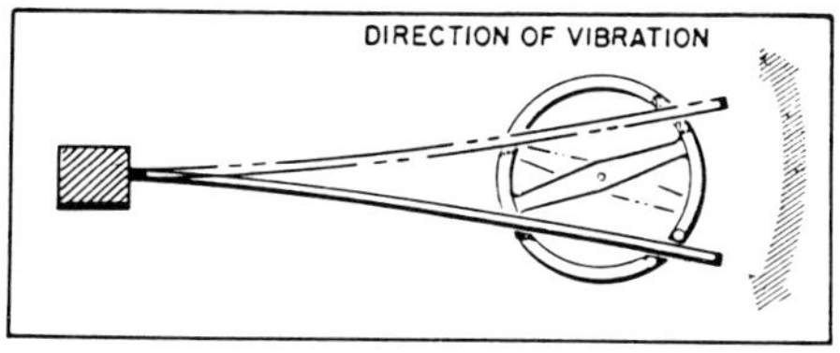

Fig. 2

The pig's bristle was eventually replaced by a steel rod which, because of its delicate nature, was called a hairspring.

Soon it was conceived that if this steel rod were formed into a spiral, and its outer end attached to the pillar plate of a watch and its inner end attached to the balance wheel and then vibrated, it would vibrate a number of times per second, depending upon the length and strength of the spring and the weight of the balance wheel. (Fig. 3.)

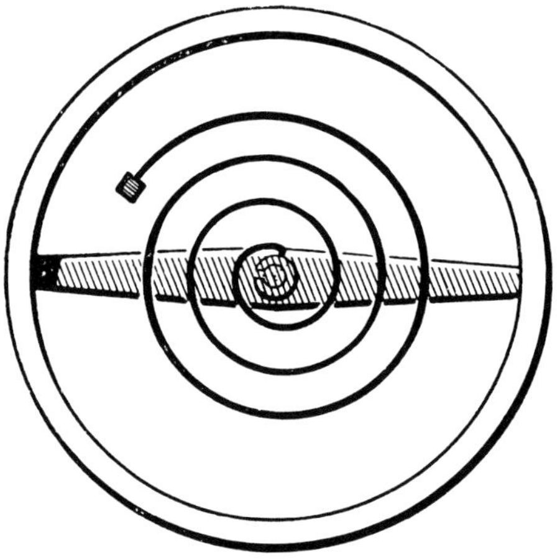

Fig. 3

To understand better the principle involved, let us consider the three simple vibrating rods shown in Fig. 4. Each rod is of the same thickness and width, but differ from each other only in length. Their rates of vibration will differ. Because of a certain law of elasticity, the longer rod will vibrate a fewer number of times in a given time than either of the shorter rods.

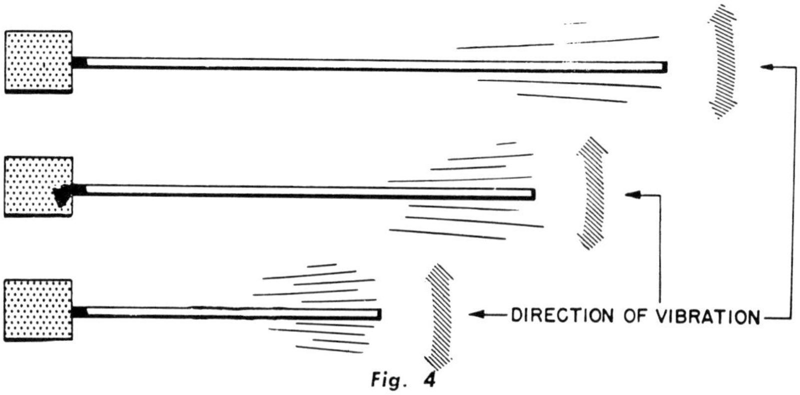

Fig. 4

The rate of vibration of a rod will also depend upon its thickness. If the three vibrating rods in Fig. 5 were of the same length but differed in thickness, we would find that the thinner the rod, the lower its rate of vibration; conversely, the thicker the rod, the faster it would vibrate.

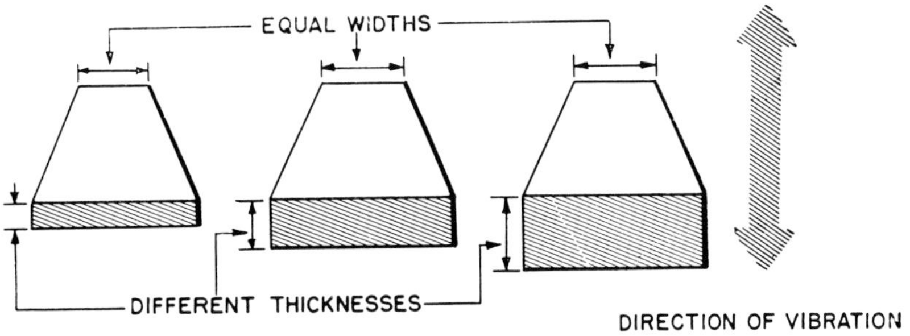

Fig. 5

The elastic principles involved in a simple vibrating rod apply also to a vibrating hairspring. When vibrating a hairspring, it is the length, width and thickness together which influence the selection of the proper spring for a given balance wheel. In the trade, the thickness dimension is referred to as the "strength" of the spring. (Fig. 6.) When springs are received from the manufacturer or materials dealer, the strength is usually indicated by a number. These numbers range from 18 to 96 in the Swiss system. The lower the number, the thicker and stronger the spring.

Because all stock hairsprings of a given number are not of exactly the same strength, the repairman is faced with the task of selecting the proper spring for a given watch. The means for doing this is called "hairspring vibrating." Vibrating is the process of selecting a hairspring with the proper length and strength so as to cause a given balance wheel to vibrate a certain number of times per hour.

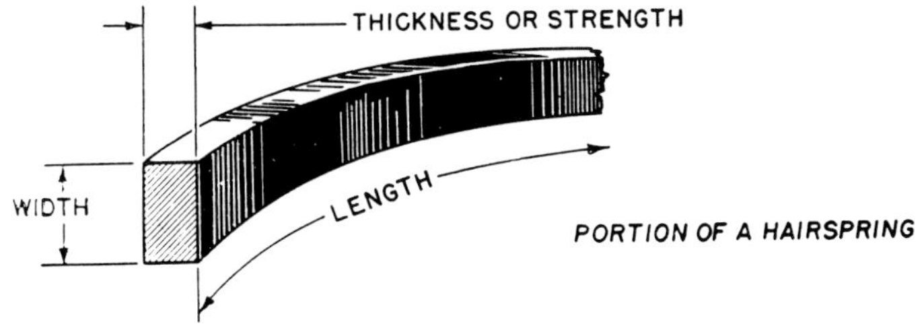

Fig. 6

PRELIMINARY CHECK

Assume that a spring is to be selected for an average watch whose balance vibrates 18,000 times per hour or 5 times per second. Before proceeding with the actual vibrating process, time and labor can be saved by a preliminary check on the spring that will help to determine if it is long and strong enough. The balance wheel assemblies used in this training unit are those that have been developed in No. 6 Training Unit. The preliminary check is carried out by grasping the hairspring with a pair of fine pointed tweezers a short distance from the outer end and then permitting the balance assembly to hang by its own weight. (Fig. 7.) The next step is to count the spaces between the coils downward from the point of suspension. If the count is 8 or 9 and if there are no less than 12 and no greater than 14 whole coils between the tweezer and the pinning-in place at the collet, the spring is suitable for the particular balance wheel. It will be necessary to try several places along the spring before one will be found that will fall within the above limits. When that place is located, it will be found that the diameter of the spring will be approximately equal to the radius of the balance wheel. (Fig. 8.)

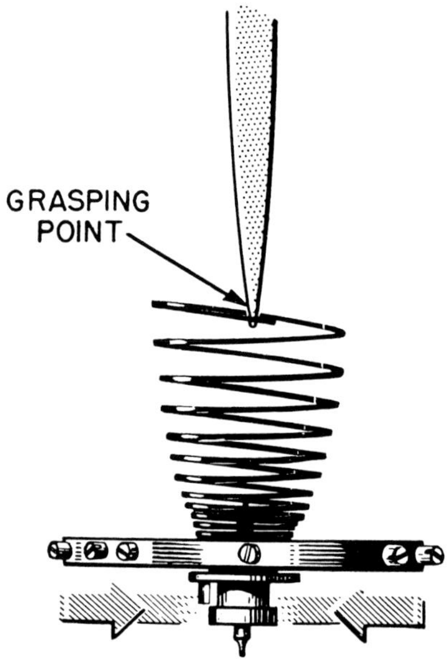

Fig. 7

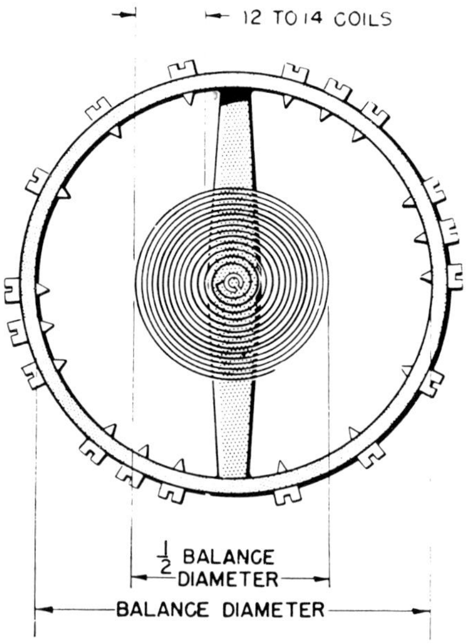

Fig. 8

METHODS OF VIBRATING

1. Oscillation Counting.
2. Master Balance.

Oscillation Counting

It was stated before that the balance wheel of an average watch vibrates 18,000 times per hour. This is 300 times per minute. Because of the physical difficulty encountered when attempting to count 300 vibrations in a given minute, a simpler way is to count every other vibration. When doing it this way, the eye is kept attentive to one of the two ends of the balance arm. A mental note is taken as to where one end of the balance arm is when it is at an extreme end of a swing. Each time that that particular end reaches the extreme end of its swing, the balance has made one complete oscillation or two vibrations. If, for example, it is observed that the count is 150 per minute, then the hairspring has the correct length and strength.

The procedure for finding the vibrating point of a hairspring, which point is finally located between the regulator pins on the balance bridge of the watch, is as follows. Pick up the hairspring with the tweezers at the point found in the preliminary check. Lower the entire assembly until the lower balance pivot just touches the crystal of a watch possessing a second hand. (Fig. 9.) The next step is to start the balance wheel vibrating. This may be done by turning the wheel between the thumb and forefinger against the tension of the hairspring and then suddenly releasing it. With the balance wheel vibrating in the near vicinity of the watch's second hand, it is easy to count the vibrations and keep track of the passing of a given number of seconds. Observe one end of the balance arm and note the point where it reaches an extreme in its angular travel. This extreme will be the reference used in counting every other vibration. When the second hand reaches any convenient starting point, simply start counting every other vibration of the balance wheel, beginning with zero.

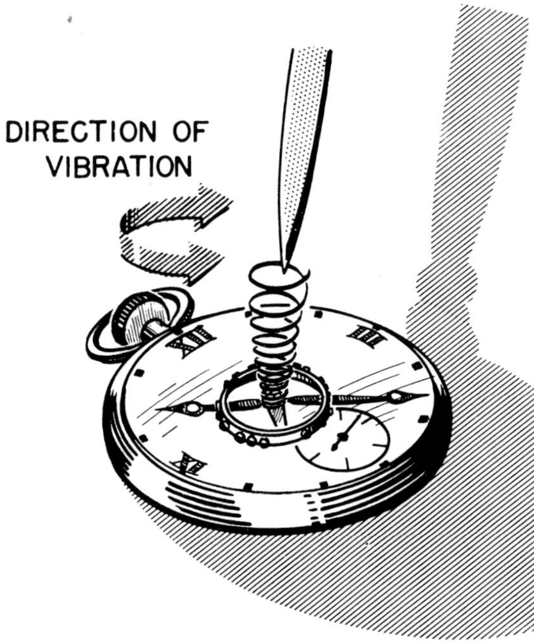

Fig. 9

For a vibrating hairspring whose count is to be 18,000 vibrations or 9,000 oscillations per hour, the coincidence of count with the second hand will be as follows. (Every other vibration is considered to be a count of one.)

2.5 counts	1 second	25 counts	10 seconds
5 counts	2 seconds	75 counts	30 seconds
10 counts	4 seconds	150 counts	60 seconds

Because of the above relationship, it will be possible to determine if a given balance wheel and hairspring are vibrating 18,000 times per hour by simply observing if a count of 25 coincides with a 10 second interval. However, this short method should not serve in finally deciding the exact location of the vibrating point. For more accurate results, several checks over a full minute should be conducted when the probable vibrating point is found by the short 10 seconds count.

If the count is greater than 150 for a minute or greater than 25 for 10 seconds, it is because the length of the hairspring is too short.

If the count is less than 150 for a minute or less than 25 for 10 seconds, it is because the hairspring is too long.

The correct vibrating point and thus the correct length of hairspring is found when the balance wheel oscillates for a count of 150 per minute.

This point on the hairspring which is held between the tweezers is finally located between the regulator pins of the watch. In short, the active length of a hairspring terminates at the regulator pins which is the point of vibration. The actual length of a hairspring ends at the stud. There must be enough spring extending beyond the vibrating point for studding purposes. (Fig. 10.) Therefore the outer coil must be broken off at a point which is equal to the distance from the regulator pins to the stud plus enough to go through the stud. This distance is estimated by actual observation of the balance bridge. Be sure that the regulator is centered. Another and more accurate way of determining where to cut the spring is to take a portion of another spring which has been cut exactly to equal the required length and use it as a gauge.

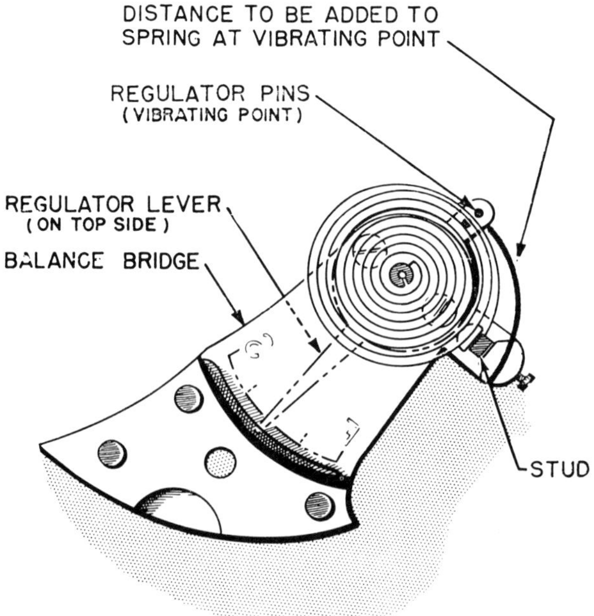

Fig. 10

MASTER BALANCE

In this method, the rate of oscillation of the test balance wheel and hairspring is compared with that of a master balance or vibrator. This vibrator is a tool containing a balance wheel and hairspring which have been very accurately adjusted to vibrate 18,000 times per hour.

The master balance is enclosed in a glass covered case for protection. (Fig. 11.) The vibrator is equipped with tweezers whose jaws are opened by pressing a plunger on its top. When the test hairspring is properly placed between the tweezer jaws, it will be securely held when the pressure on the plunger is released. The tweezers are supported vertically by an arm connected to the main body of the vibrator. The tweezer arm can be raised or lowered.

The process of comparing the rate of vibration of a test balance wheel assembly with that of the master is begun by carefully inserting the test spring in the vibrator tweezers. The next requirement is to lower the tweezer until the lower pivot of the balance wheel is brought into contact with the glass cover above the master. This contact will steady the test balance when it is vibrating and make it easier to observe.

Fig. 11

The vibrator is so constructed that its tweezer arm can be swung and also moved in or out, which feature permits the test balance to be placed directly over the master balance. After the test balance is brought into the correct position, its arms must be brought into alignment with those of the master balance. This is done by turning the vibrator tweezer in its friction support. There is a lever at the bottom part of the vibrator which, when it is depressed and released, gives the starting impulse to both the master and test balance wheels at the same time. If the arms of each balance keep in exact synchronism with each other, it indicates that the test hairspring is held at the exact point of vibration. Always start vibrating at the outside end of the hairspring and work in towards the center.

The chances are that the correct vibrating point will not be found at the first attempt. Until the correct point is found, the observer must be alert to detect which of the two balance wheels steps out of alignment first. The wheel which moves out first is the faster of the two.

If the test balance steps out of synchronism first, then it is apparent that the active length of its spring is too short. The master balance is then stopped by depressing a small button at the side of the glass top case. The holding point of the vibrator tweezer is then reset so as to increase the active length of the test hairspring and the whole procedure explained above is then duplicated. This correction will bring the two balance wheels more nearly in step. Continue to make corrections until the two remain perfectly synchronized.

Suppose the master balance steps out ahead of the test balance. We know then that the test balance is swinging slower than the master. In this case the active length of the spring must be shortened until the point of vibration is found that allows the balance wheel to keep in exact synchronism with the master.

Then, and then only, will it be certain that the correct point of vibration has been found and that the test balance and hairspring will serve accurately as the time keeping element in the watch.

The final step consists in cutting off the spring at the correct distance from the vibrating point. This procedure has been described above under "Oscillation Counting."

The hairspring is now ready to be studded and overcoiled preparatory to installation in the watch.

TRAINING UNIT

NUMBER 8

★

OVERCOILING

OVERCOILING

Overcoiling is a manipulative operation whereby the outer coil of a hairspring is so formed that when it is operating in a watch, the following conditions will be present:

1. The center of gravity of the spring will consistently tend to coincide with the axis of the balance staff.

2. As the spring winds and unwinds, its coils will remain concentric with the axis of the balance staff and its elastic force will increase and decrease in proportion to the angle of rotation of the balance wheel from the line of centers.

3. The spring will not cause the balance pivots at any point in their angular movement to exert a side thrust or pressure against their jeweled bearings.

4. The balance unit will make each swing or vibration in the same amount of time regardless of whether an impulse causes it to travel through a small or a large arc.

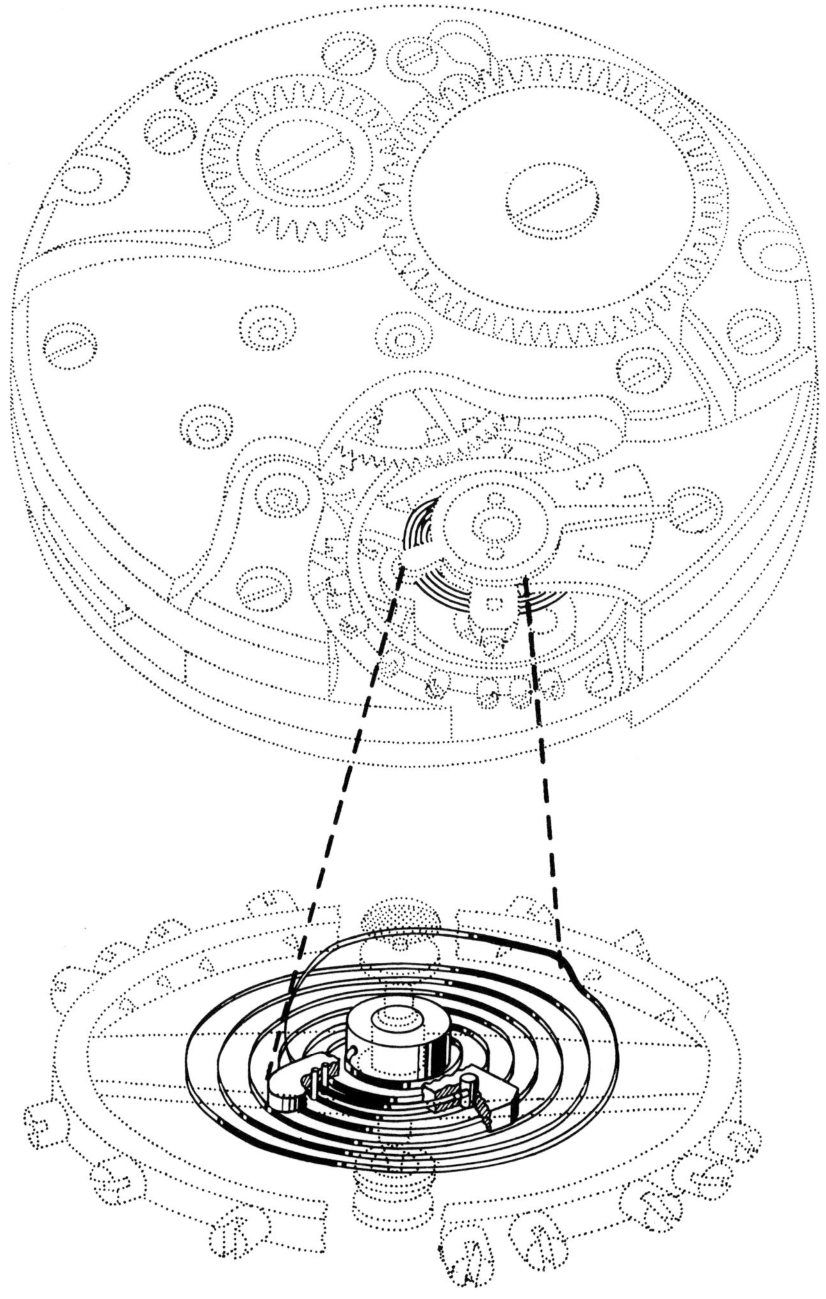

BULOVA SCHOOL *of* WATCHMAKING

SUBJECT:

Practical Development of the Overcoil

OBJECTIVE:

To Develop Skill in Overcoil Work

NEW TOOLS REQUIRED:

1. Regular Overcoiling Tweezers
2. Knee Bend Overcoiling Tweezers

(The overcoiling tweezers are used generally in mass production work, but using two tweezers would be just as functional.

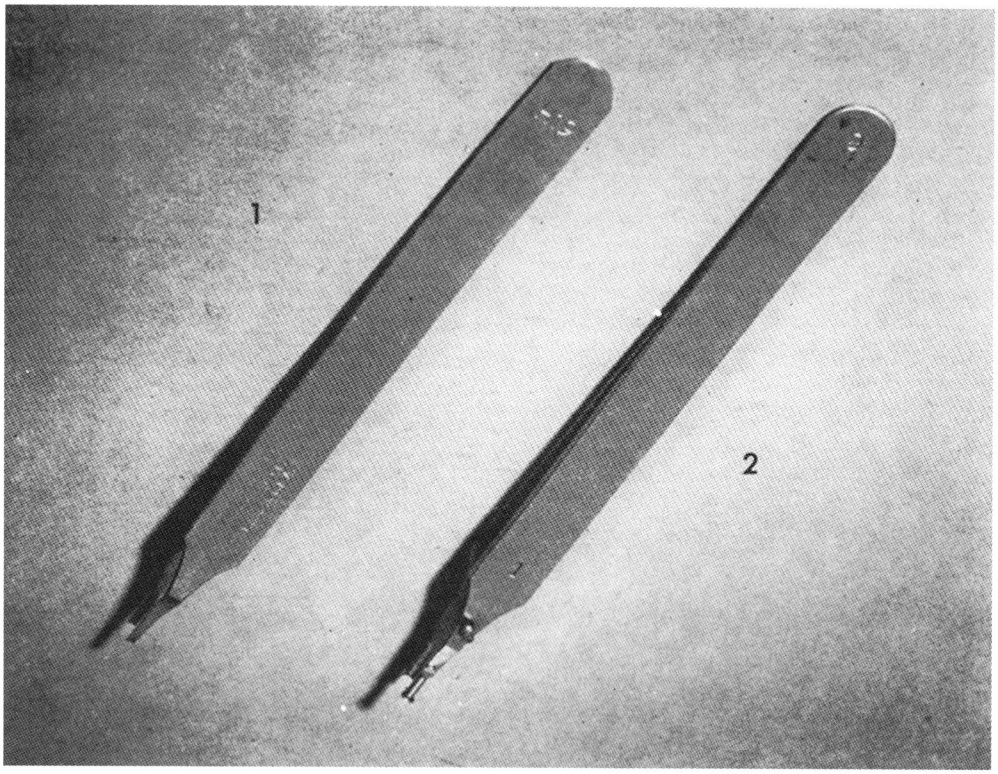

THE ISOCHRONAL HAIRSPRING

If a watch is in good working order and fully wound, the balance wheel may possibly swing in an arc as much as 270 degrees in a single vibration, Fig. 1. As the watch runs down, the swing gets less and less, Fig. 2. But regardless of this change in motion as time passes, it is required of the watch neither to lose nor gain. If this characteristic is present in a watch, then it may be said that it possesses an isochronal hairspring. An isochronal error exists in a hairspring if it causes the balance wheel to swing at a different rate in the high arcs than in the low arcs of motion.

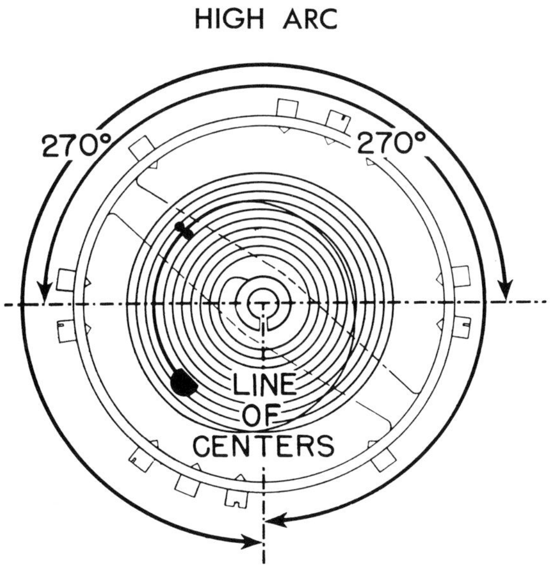

Fig. 1

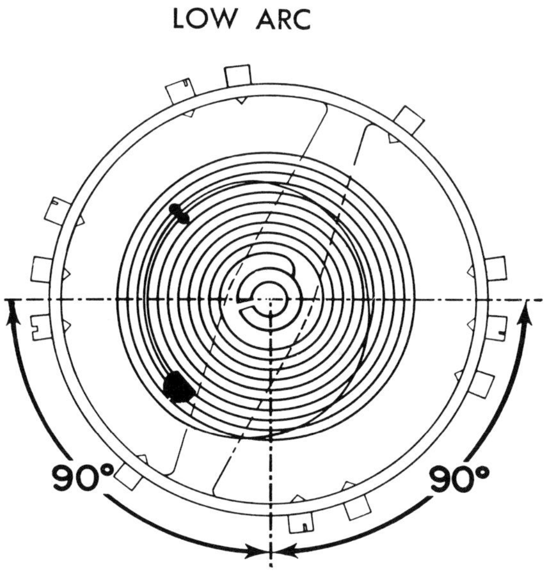

Fig. 2

THE FLAT HAIRSPRING

There is a characteristic present in a flat hairspring as it vibrates, i.e., winds and unwinds, that causes the coils to bunch together first on one side of the balance staff, then on the other. Fig. 3 and Fig. 4. Because of this action, it is apparent that a side thrust or pressure is brought to bear between the balance pivots and their jewel bearings. As the spring is wound up, the pressure is exerted in a direction away from the regulator pins. As the spring unwinds, the pressure is toward the regulator pins. This pressure causes an unequal turning action, or torque on the balance staff. The greater the swing of the balance, the greater will be the side thrust and the resulting friction. This means that the energy received by the jewel pin from the pallet does not all go into winding and unwinding the hairspring. Some of it is dissipated in overcoming friction. Because of this, the force exerted by the hairspring on the balance staff will not be directly proportional to the angle through which the balance staff is turned.

Because of this characteristic, a watch containing such a hairspring would tend to gain in the high arcs and lose in the low. In short, it would not be isochronal.

SPRING WOUND

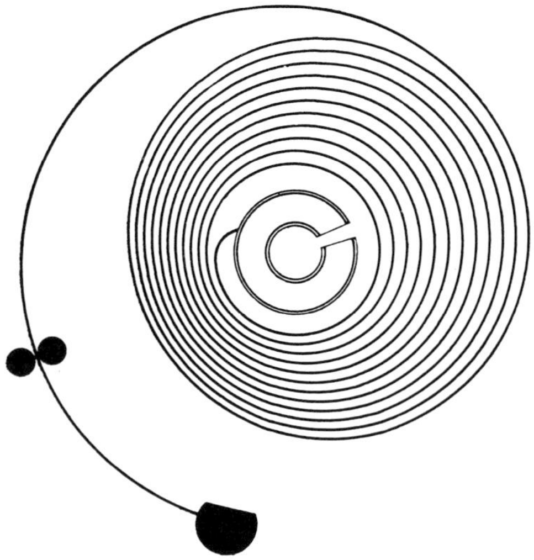

Fig. 3

SPRING UNWOUND

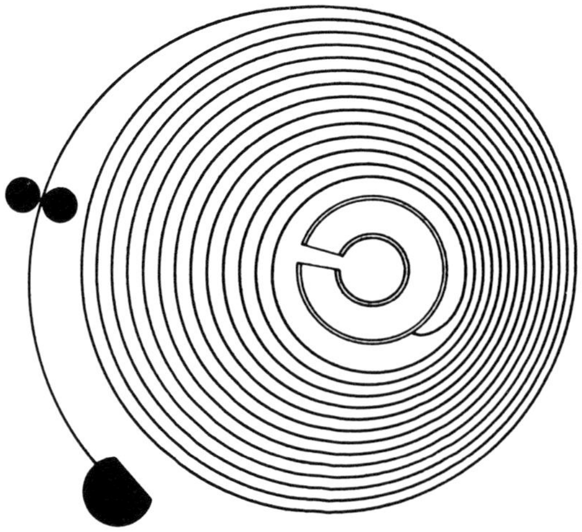

Fig. 4

THE BREGUET OR OVERCOIL HAIRSPRING

The idea for the overcoil came from the horologist Breguet who reasoned that if the outer coil of a flat hairspring was bent up and laid over the top of the main body of the spring, then the spring would wind and unwind more concentrically. As a result, the spring would not exert a side thrust and the force on the balance would be proportional to the angle through which it turned.

However, through experimentation, it was found that an isochronal error opposite to that produced by a flat hairspring was present when the overcoil was laid straight across the spring as shown in Fig. 5. The hairspring in this condition will produce a faster rate in the low arcs than in the high arcs. That is to say, a watch with such a spring will run slow when first wound and will gain as it runs down.

It is obvious that betweeen the two extremes just described, there must be a place to lay the overcoil which will cause the balance wheel to perform its vibrations in the same period of time regardless of whether it swings in a small or large arc of motion.

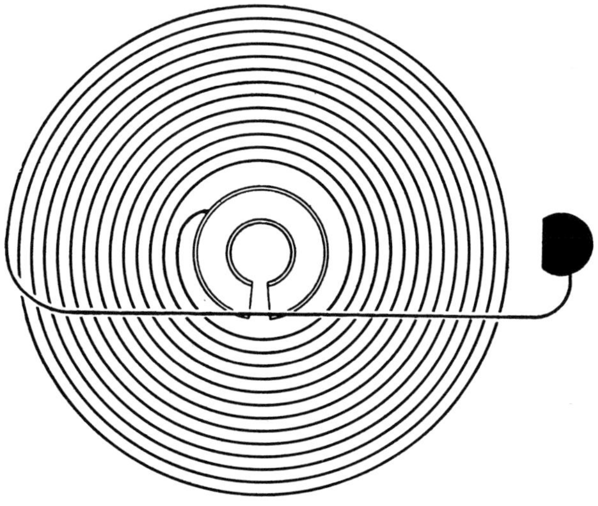

Fig. 5

In Fig. 6 are shown three basic types of overcoils that are in common use today. If any one of these types is selected for a given watch, then due consideration must be had for the location of portion AB of the overcoil relative to the balance staff. If AB should be located too close to the balance staff, the hairspring will cause the watch to have a gaining rate in the low arcs and a losing rate in the high arcs of motion. If AB should be located too far from the balance staff, then the hairspring will cause the watch to have a losing rate in the low arcs and a gaining rate in the high arcs of motion.

In the watch, an efficient overcoil keeps the other coils of the hairspring concentric as they wind and unwind, thus removing side thrust which is a major contributing cause of isochronal errors.

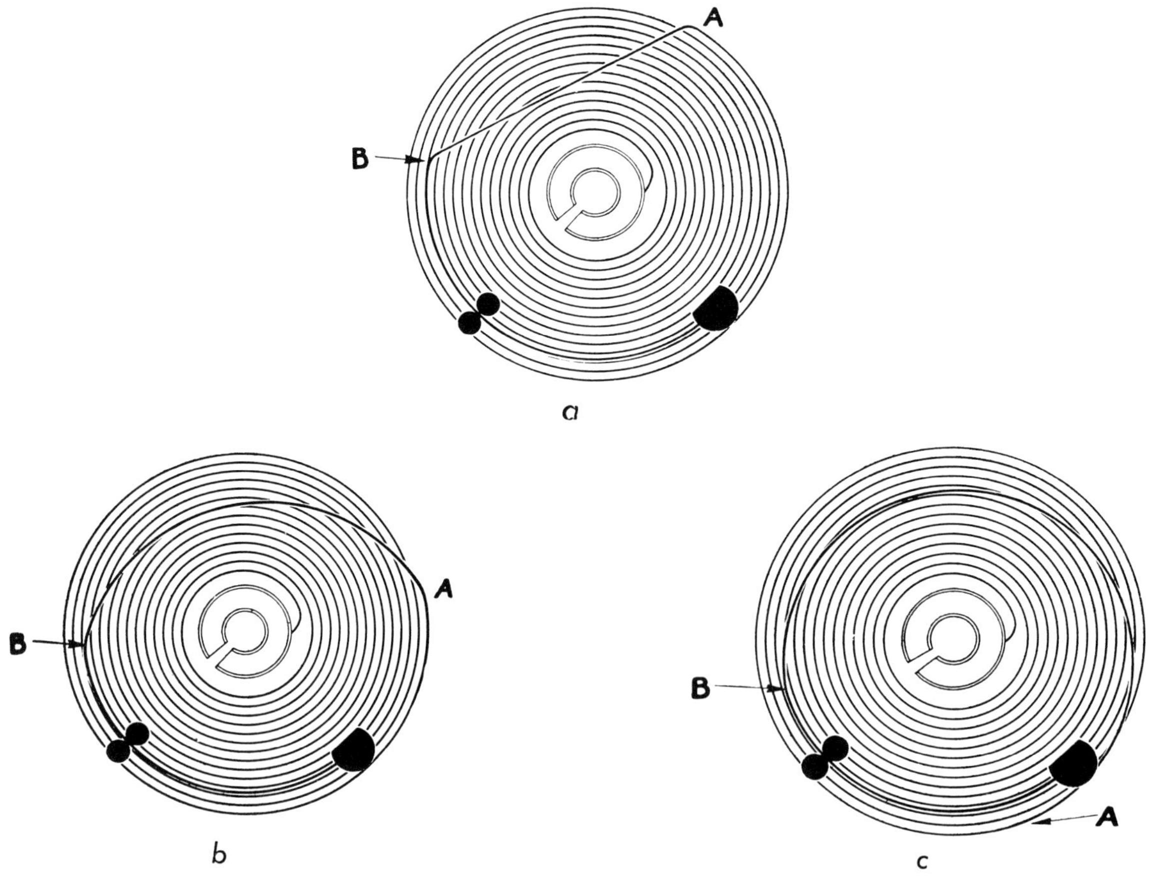

Fig. 6

FORMING THE OVERCOIL

Overcoil Raised with a Knee Bend Overcoil Tweeezers

The Overcoil will be developed according to the type shown in Fig. 6b. To accomplish a knee bend, a special tool shown in Fig. 7 is used. This is a knee bend overcoil tweezers. These tweezers will permit a variety of adjustments for use with many sizes of springs. Two adjustments must be made on the tweezers in order to prepare it for use. The width of the slot in the tweezers is adjusted by the slide and small set screw located at A, Fig. 7. The other adjustment is made with set screw B, Fig. 7. This screw regulates the distance apart the tweezer points will be when closed. The setting made in this latter adjustment determines the angle that the overcoil will rise from the main body of the spring.

Bear in mind that too sharp a bend will weaken or possibly fracture the spring. By experimenting with a discarded spring similar to the one to be overcoiled, the proper adjustments can be made accurately beforehand. An adjustment at B, Fig. 7, which will produce approximately a 25 degree rise from the main body of the spring, will be suitable.

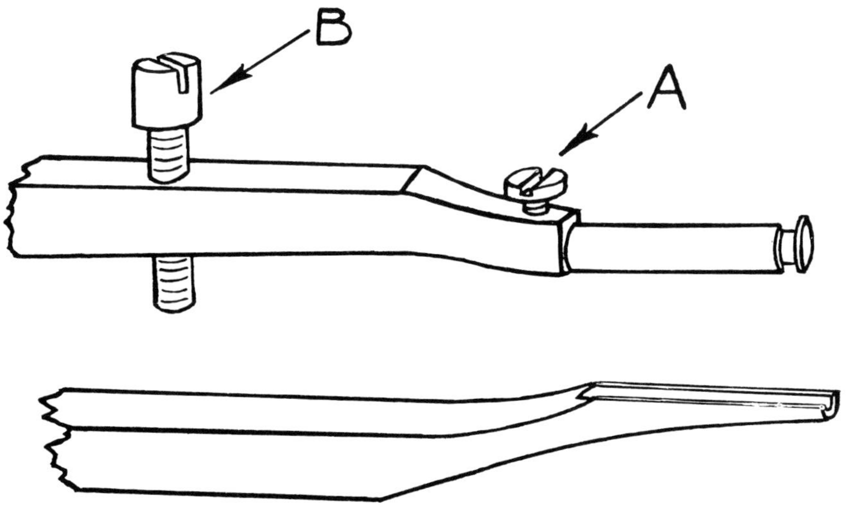

KNEE BEND OVERCOILING TWEEZERS

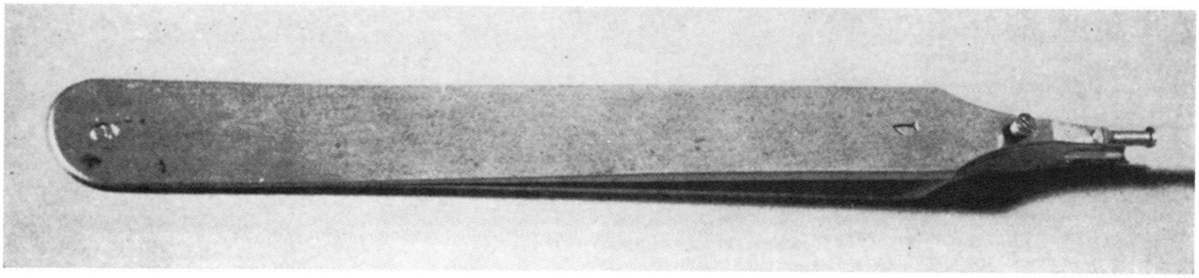

Fig. 7

The first bend to be made in forming the overcoil is at A, Fig. 8, located about 270 degrees from the stud. (This angular measurement will vary with various sizes and types of balance units). The spring is suspended from a pair of ordinary tweezers held at T approximately 45 degrees to the right of A, Fig. 8. While in this condition, the hairspring is inserted into the slotted end of the overcoiling tweezers which are applied to the spring, with the slotted part above the concave part of the tweezers. The tweezers should be pressed together firmly until the set screw prevents further bending. When the spring is entirely released from the tweezers, it should have the appearance to that shown in Fig. 8C.

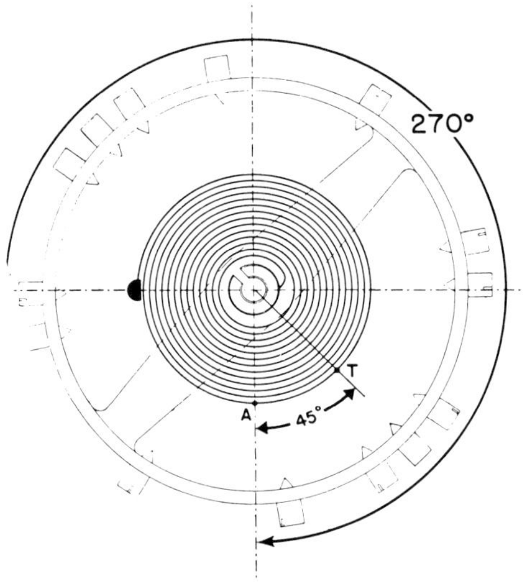

Fig. 8A

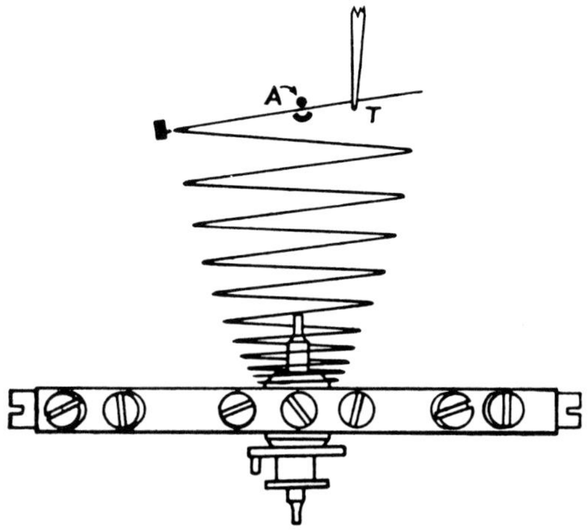

Fig. 8B

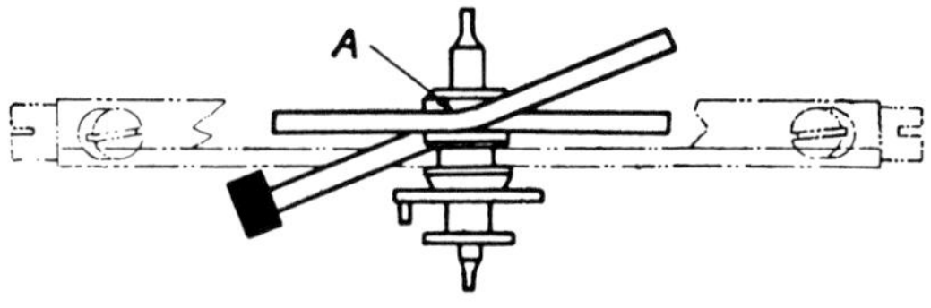

Fig. 8C

The second bend is at B, located to the right of A, Fig. 9. The purpose of this bend is to bring the overcoil into a plane parallel with the main body of the spring. The location of point B is determined by the desired height of the overcoil above the main body of the hairspring. For present purposes the height of the overcoil is to be about 2½ times the width of the hairspring itself. This bend is accomplished by suspending the balance and spring from a pair of ordinary tweezers held at point T, Fig. 9A approximately 45 degrees from point B. The knee bend overcoiling tweezers is applied in reverse from the way it was used to form bend A, i.e., with the concave part above the slotted part of the tweezers. As before, the spring is inserted into the slotted end of the tweezers and the tweezers are then pressed firmly together as far as they will go.

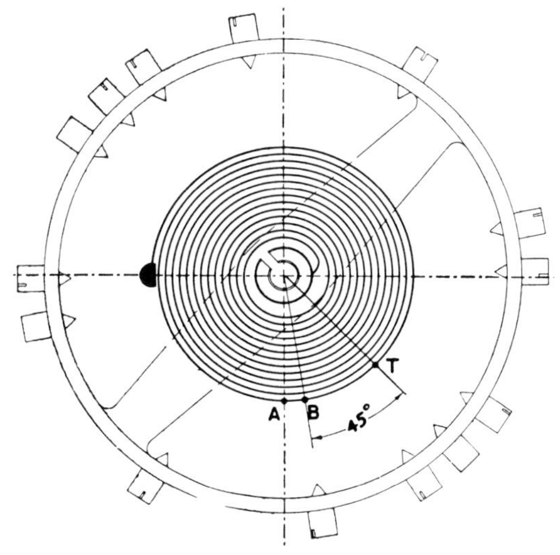

Fig. 9A

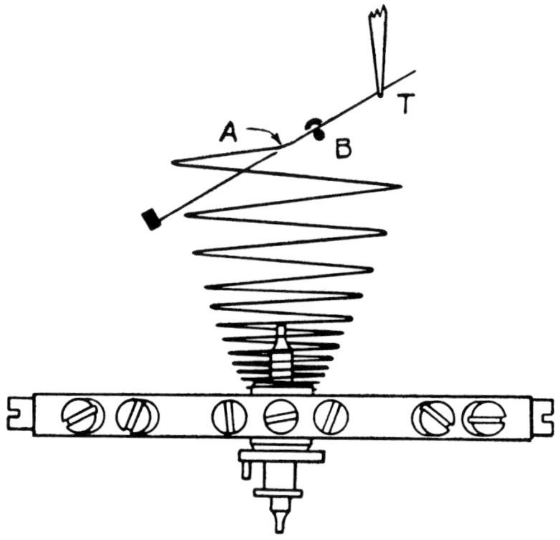

Fig. 9C

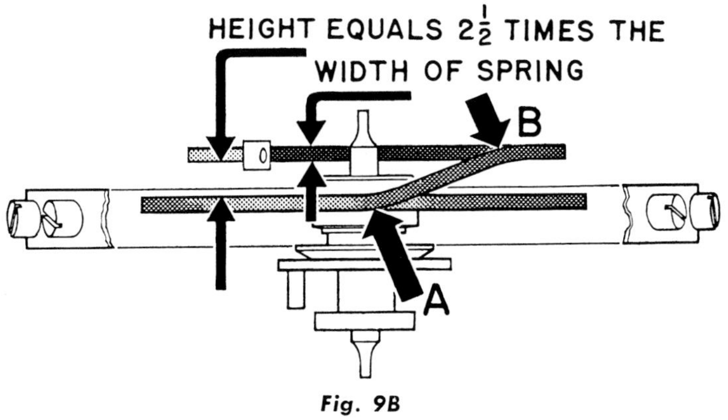

Fig. 9B

It will be found that, after bending the spring at B, the overcoil will not be exactly parallel with the main body of the spring. The point 180 degrees from B will be lower than it should be. A simple bend using two ordinary tweezers will provide the remedy. Grasp the spring a little to the left of bend B with a pair of tweezers held upright in the left hand. Then with the second tweezers held upright in the right hand, grasp the overcoil a trifle to the right of B. By simply swaying the right hand tweezer toward yourself, the overcoil can be brought to assume a perfect parallel position above the main body of the spring.

KNEE BEND METHOD #2

PRESSING HAIRSPRING INTO A SMALL PIECE OF WOOD. 1″ x 2″ x ¼″ THICK WOULD BE SUGGESTED.

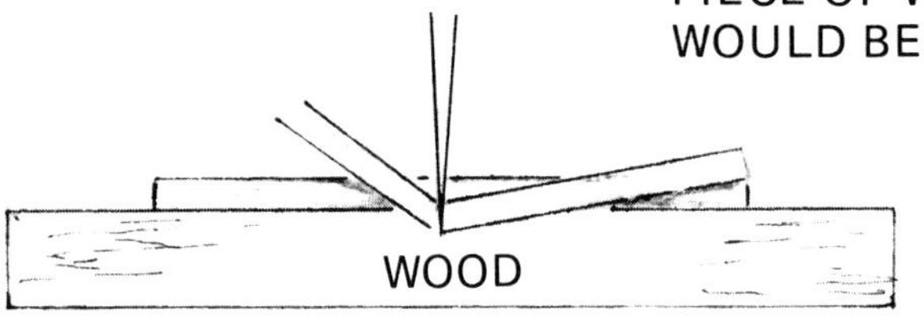

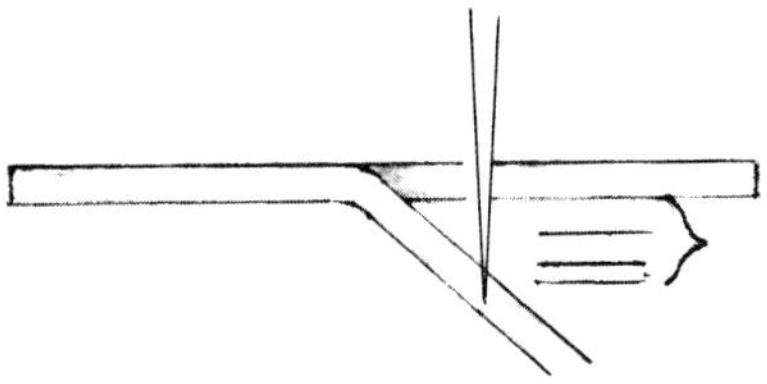

2½ COILS HIGH

TWEEZER PLACEMENT FOR SECOND BEND INTO THE WOOD.

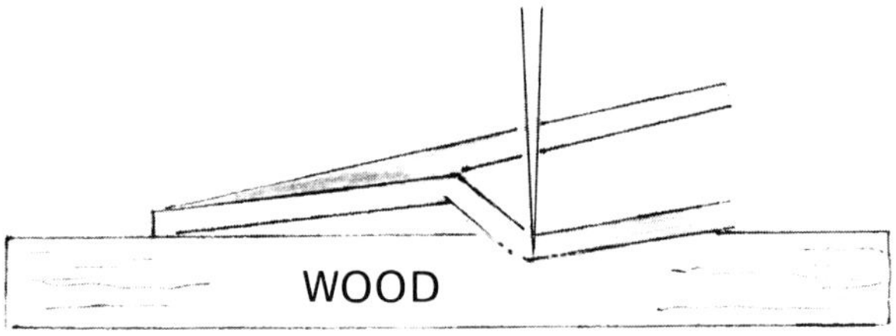

SECOND STEP FORMED

The next bend occurs at C, Fig. 11, using a pair of regular overcoiling tweezers, This style of tweezers has its ends made in various widths for different sizes of springs. The inside of one of the legs is convex, while the opposite leg is concave and when they are closed, they fit closely together.

Bend C is located about 45 degrees from bend A, Fig. 11 and is executed by grasping the spring at C with the convex leg of the tweezers nearest the balance staff and squeezing upon the tweezers until the overcoil at D moves to nearly halfway betweeen the outer coil and the collet. This position of the overcoil has been selected at random, the main purpose at present being to get practice in forming overcoils and not in developing an isochronal overcoil.

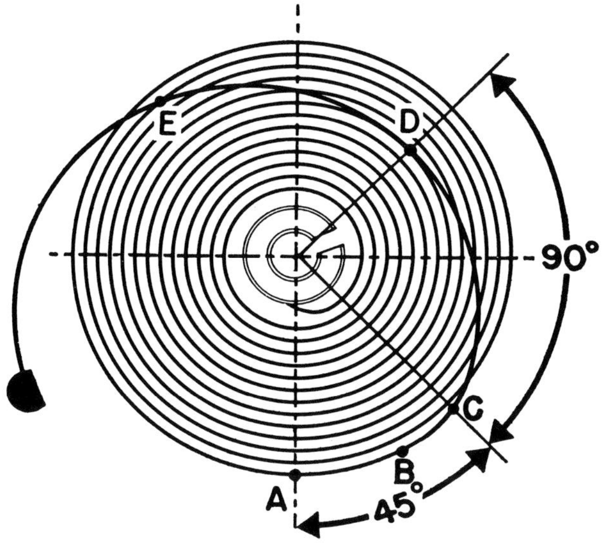

Fig. 11

The next bend is placed at E, Fig. 11 and is also performed with the regular overcoiling tweezers. Point E is arbitrarily selected where the overcoil crosses the 2nd coil counting from the outer coil. Since no attempt is being made to fit the hairspring to a given watch, this procedure will do. However in passing, note that for a given watch, the location of point E is determined by the distance from the hole jewel in the balance bridge to the regulator pins.

Forming the regulator circle is the final step in developing an overcoil. Since the regulator pins are mechanically held to a fixed distance from the hole jewel, they move in an arc of a circle. Because of this fact, it is necessary that the part of the overcoil that lies in the path of the regulator pins should be the perfect arc of a circle and should be free from any small kinks. If this should be otherwise, then proper regulation of the watch would be impossible.

The actual forming of the arc of a circle from point E to the stud may be performed in either of two ways. One way is to use the regular overcoiling tweezers. Grasp the overcoil at E, Fig. 12, and by carefully controlling the amount of pressure which is applied at successive points between E and the stud, the arc of a circle will result as shown in Fig. 12. Care should be taken to keep the tweezers perpendicular to the spring while it is being formed.

Another method of forming the regulator circle consists in holding the overcoil with a pair of ordinary tweezers at point E and pushing against the spring from the outside of the spring toward its center with another pair of tweezers until that part of the spring between E and where the push is applied conforms to an arc of a circle. Care should be taken not to select the point to be pushed too near where the other tweezers is grasping the overcoil because a kink may result. Part of the regulator circle is now formed. Next, observe where deviation from the circular arc occurs. Grasp the spring with the tweezers at this point, and with the other pair of tweezers, again push inward until another section of the overcoil conforms to the circular arc being formed. This procedure is continued all the way along the spring to the stud and if correctly done, a true circular arc will be produced.

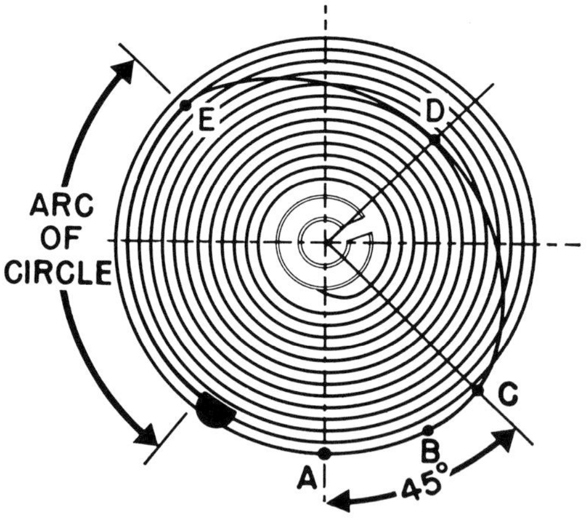

Fig. 12

OVERCOIL RAISED WITH THE GRADUAL BEND METHOD

The first bend upward is made at point A, which is located about 350 degrees from the stud. Fig. 13. As in the knee bend procedure, this measurement will also vary, depending on the requirements of the individual type of balance unit. This bend is made by using two pairs of ordinary tweezers. The spring is not staked to the balance staff which makes it easier to manipulate the gradual bend. The coil is firmly grasped at point A with one of the tweezers held vertically against the bench surface in the left hand. With the second tweezers held vertically in the right hand, grasp the coil about a millimeter to the right of the first tweezers. Hold the left hand tweezers steady and tilt the top of the right hand tweezer toward yourself. This will bend the spring in such a way that point B, 180 degrees away, will be the highest point. A sufficient bend is put in the spring at A until point B rises 2½ times the width of the spring.

For convenience, turn the entire spring clockwise until point B, Fig. 14, is nearest yourself. Then grasp the spring at B with a pair of tweezers held in the right hand and against the bench surface. About one millimeter to the left, grasp the spring with the other pair of tweezers held in the left hand. Be sure that the inner surfaces of the tweezers are in the same plane with that portion of the hairspring to be bent. Hold the left hand steady and tilt the right hand tweezer so that the top of it comes toward yourself. This manipulation bends that part of the overcoil between B and the stud upward, and enough of a bend should be put in the overcoil at B to bring it into a plane parallel to the main body of the spring.

The first bend inward occurs at point C which is located 90 degrees from point A. It is made with the regular overcoiling tweezers. Put sufficient pressure on the tweezers so that point B moves to almost halfway between the collet and the outer coil, Fig. 14. The next bend is placed at D where the overcoil crosses over the second coil. The regular overcoiling tweezer is used here as before.

The procedure of forming the regular circle from point D to the stud may be carried out using either of the two methods described when making a knee bend overcoil. When finished, the spring with overcoil will look like that shown in Fig. 15.

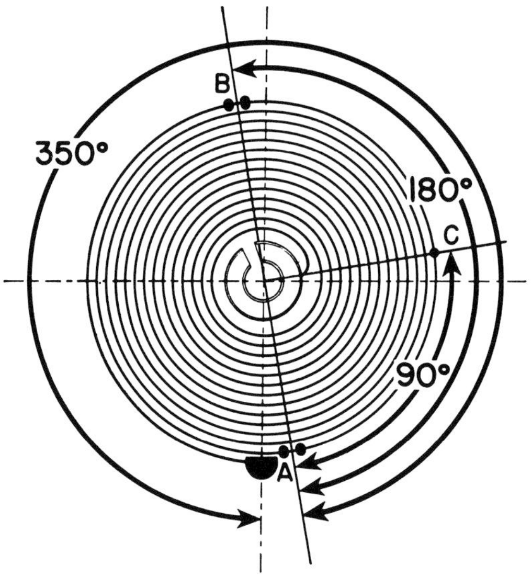

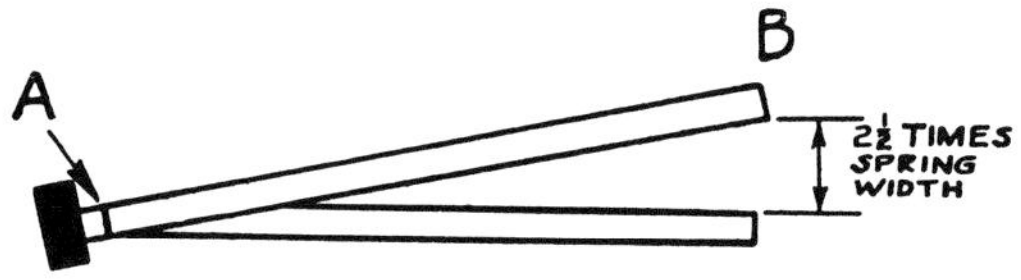

Fig. 13

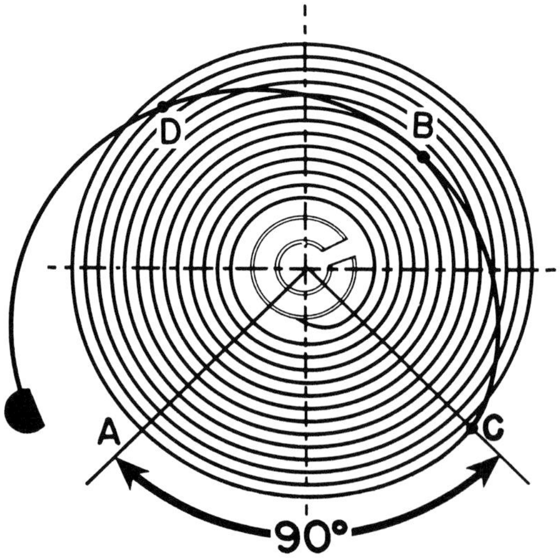

Fig. 14

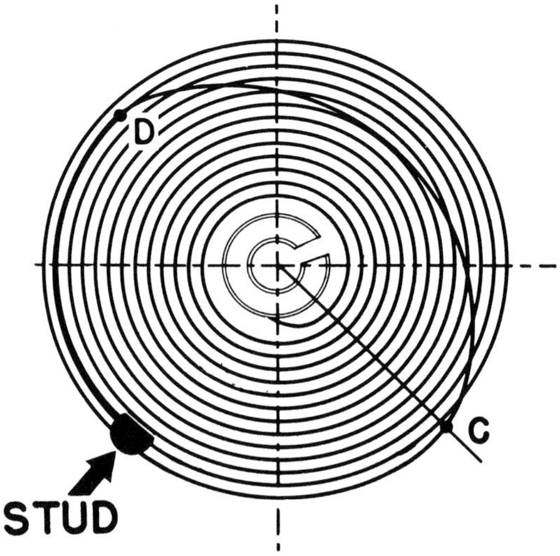

Fig. 15

TRAINING UNIT NUMBER 9

★

WATCH ASSEMBLY

WATCH ASSEMBLY

There are approximately 125 separate parts in the watch of today. Each one has its important function in the complete mechanism which measures intervals of time. When each wheel, each lever, each bridge is fitted and assembled in its proper position, an efficient accurate timepiece has been completed.

Two important requirements of every expert watchmaker are a thorough knowledge of each part and a full understanding of its necessary function in the working watch machine. In our modern civilization, the millions of watches being used daily in every type of activity require occasional servicing by the watchmaker. Servicing may mean a periodic cleaning involving a complete disassembling and reassembling of all the parts of the watch, or it may require a properly fitted replacement part.

Regardless of the type of job necessary, the high grade watch repairer must be able to fit and reassemble all the parts of the watch into an accurate instrument which will satisfy every requirement of its original maker.

BULOVA
WATCH Cº
U.S.A.
17 JEWELS
10AK

BULOVA SCHOOL *of* WATCHMAKING

SUBJECT: Watch Assembly

Watch assembly covers the problem of the timepiece as a whole. For convenience, the technique of this broad subject will be separated into five parts:

1. Mainspring barrel assembly (power unit).
2. Friction jeweling.
3. Wheel train assembly.
4. Winding and setting assembly.
5. Cap jewel assembly and fitting balance pivots.

The pages which follow will first describe in simple brief statements "How the watch functions."

HOW THE WATCH FUNCTIONS

MAJOR UNITS of the WATCH

The watch can be broken down into nine major units, each of which serves its separate function in this machine for time measurement. These nine units are as follows:

1. PLATE AND BRIDGES
2. POWER UNIT
3. TRAIN UNIT
4. JEWELS
5. ESCAPEMENT
6. BALANCE AND HAIRSPRING UNIT
7. WINDING AND SETTING MECHANISM
8. DIAL AND HANDS
9. WATCH CASE

The functions of these units are briefly described and the parts illustrated under each group.

1. The **PLATES AND BRIDGES** form the foundation of the watch and hold all the parts in proper relation to each other.

Fig. 1

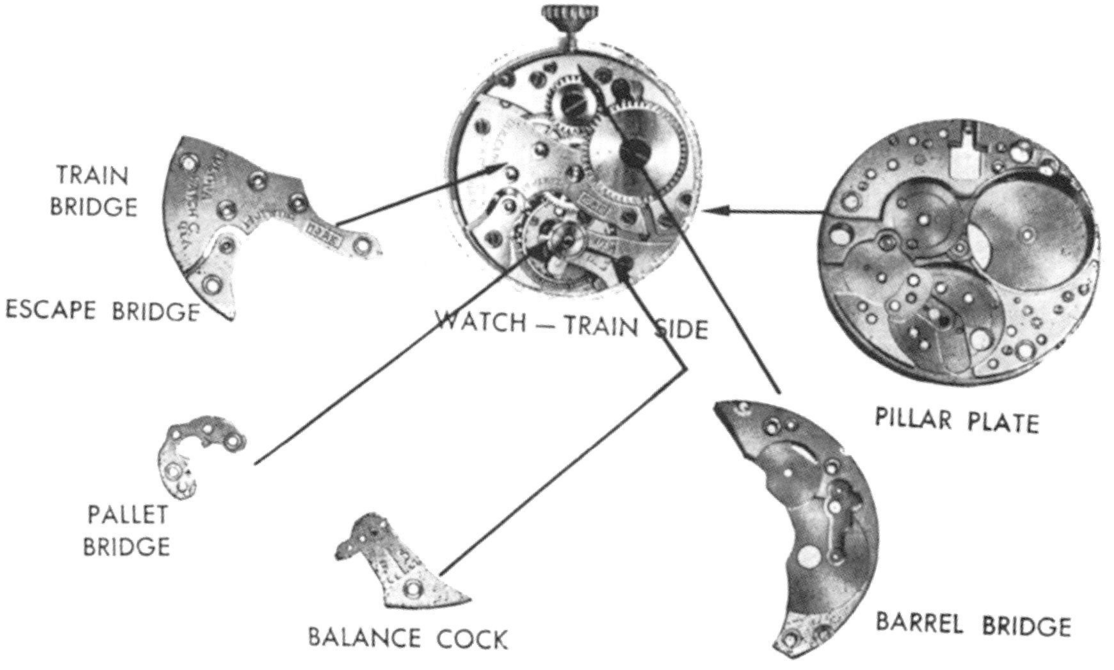

2. The **POWER UNIT** consists of the mainspring coiled around an arbor within a barrel. When the watch is wound by turning the crown, sufficient power is stored up in the wound spring to drive the gears.

Fig. 2

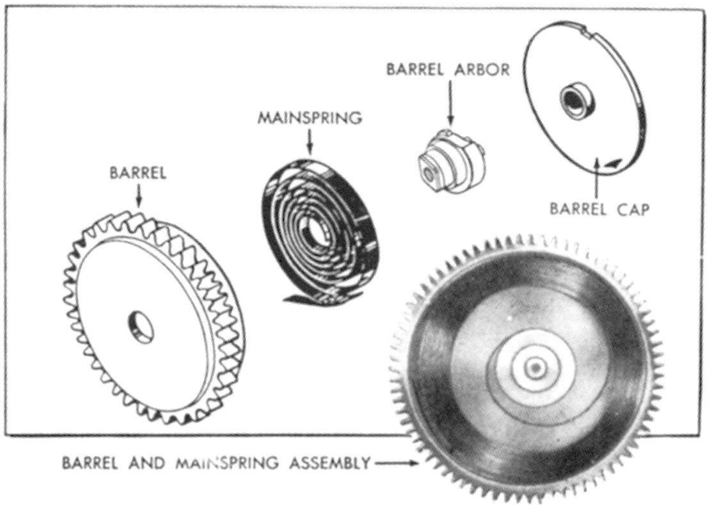

3. The **TRAIN UNIT** which includes the center wheel and pinion, 3rd wheel and pinion, 4th wheel and pinion, escape wheel and pinion, transmits the power from mainspring barrel to pallet and also supports the hands of the watch.

Fig. 3

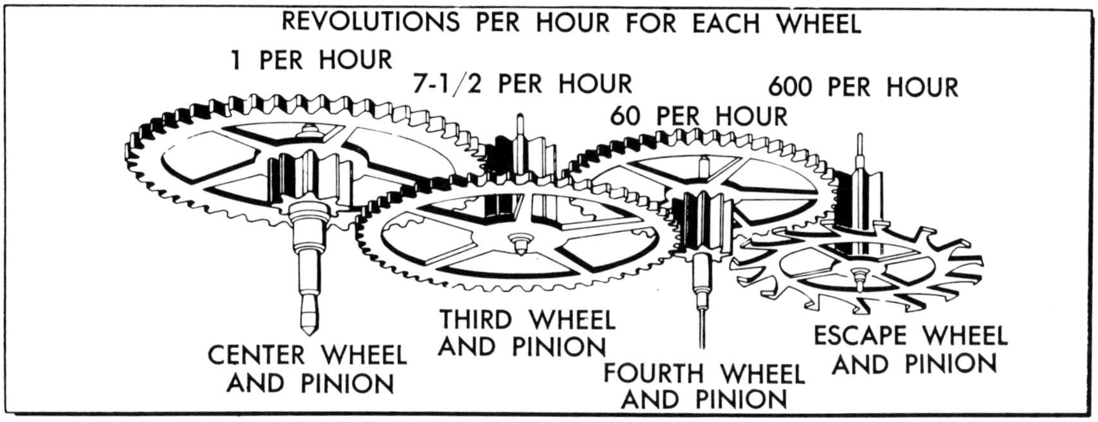

4. The **JEWELS** are located in the plates and bridges, and form the bearings for the train, escapement, and balance pivots. The bearing surfaces of the jewels and pivots are highly polished and thereby reduce friction. The jewels are shaped with an oil reservoir which holds an oil supply sufficient for approximately 8 to 18 months lubrication. In this way, the life of the watch is prolonged.

Fig. 4

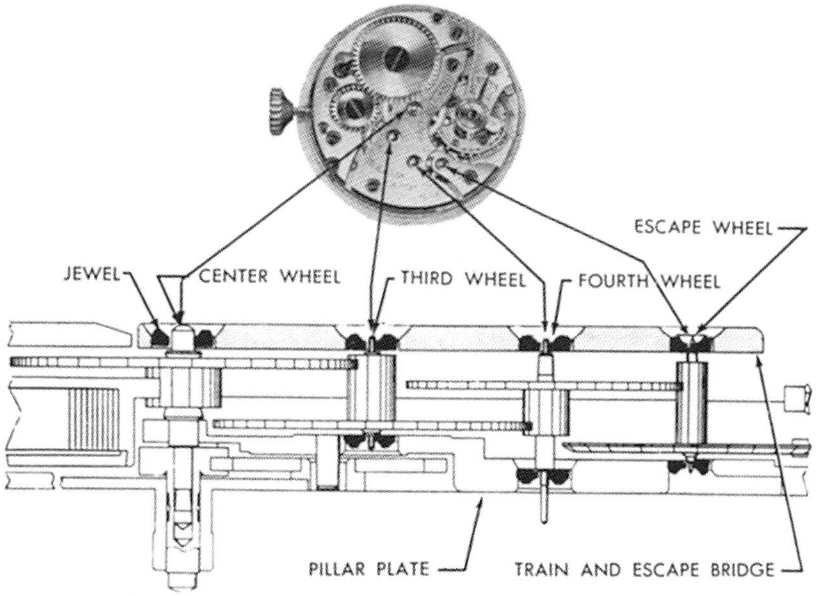

This profile sketch of the Bulova 10 AK Train illustrates the relative positions of the jewels and settings.

5. The ESCAPEMENT changes the rotary motion of the escape wheel to a back and forth motion of the pallet. At the same time, the pallet imparts an impluse to the balance wheel, causing an oscillating or vibratory motion.

Fig. 5

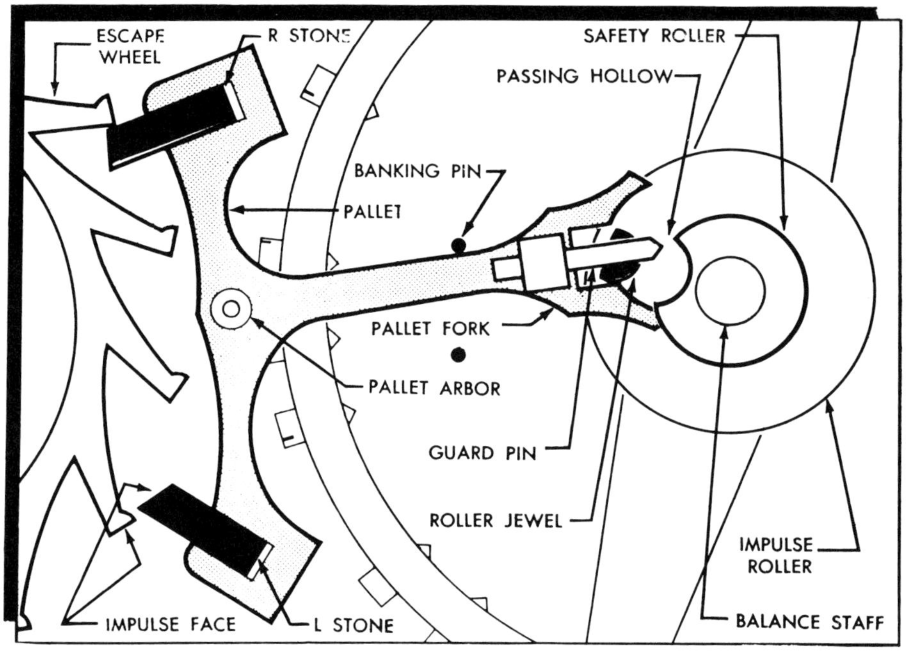

6. Because of the control features of the BALANCE and HAIRSPRING UNIT, it is often termed the "Brains of the Watch." Actual timekeeping is performed by the balance and hairspring whose period of vibration is established by a correct relationship between the weight of balance and strength of hairspring.

Fig. 6

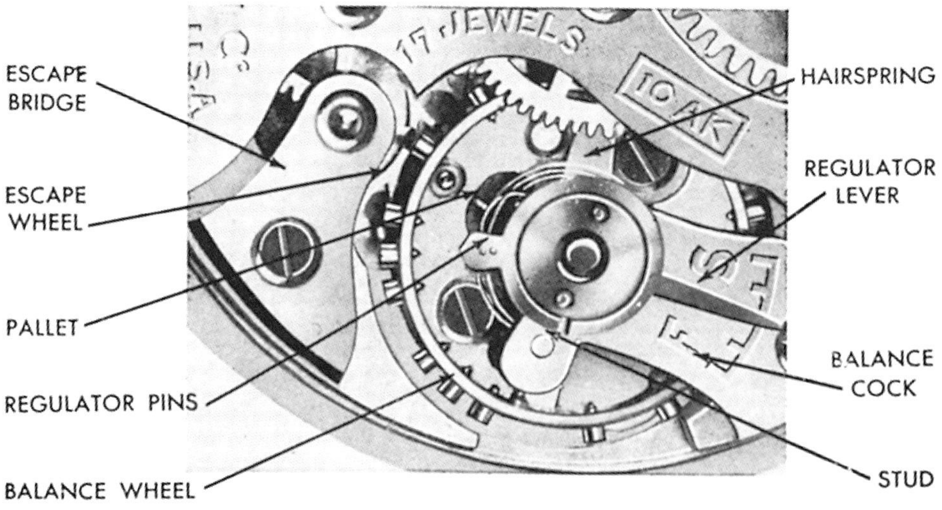

7. WINDING AND SETTING MECHANISM provides the means of storing up energy in the mainspring when the crown and stem are turned clockwise. Also by means of a simple system involving the shifting of a setting lever, clutch lever and clutch wheel, as the stem and crown are pulled out, this unit permits the setting of the watch hands.

Fig. 7

OPERATION OF WINDING AND SETTING MECHANISM

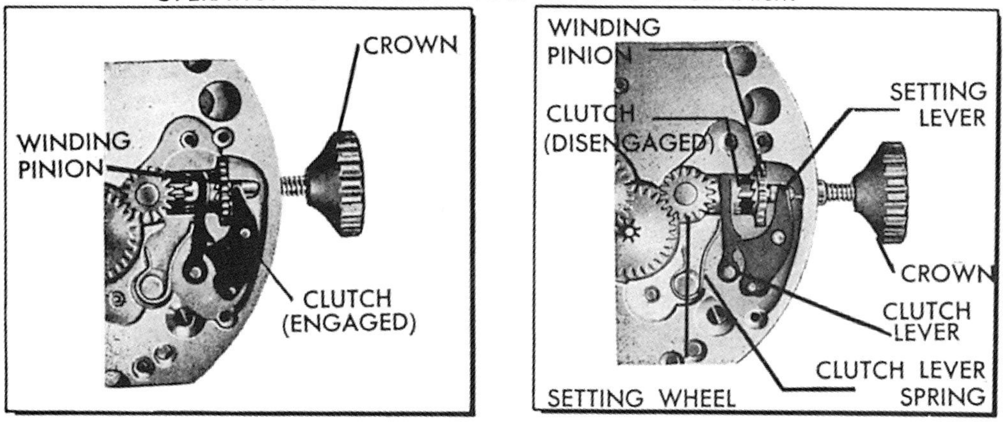

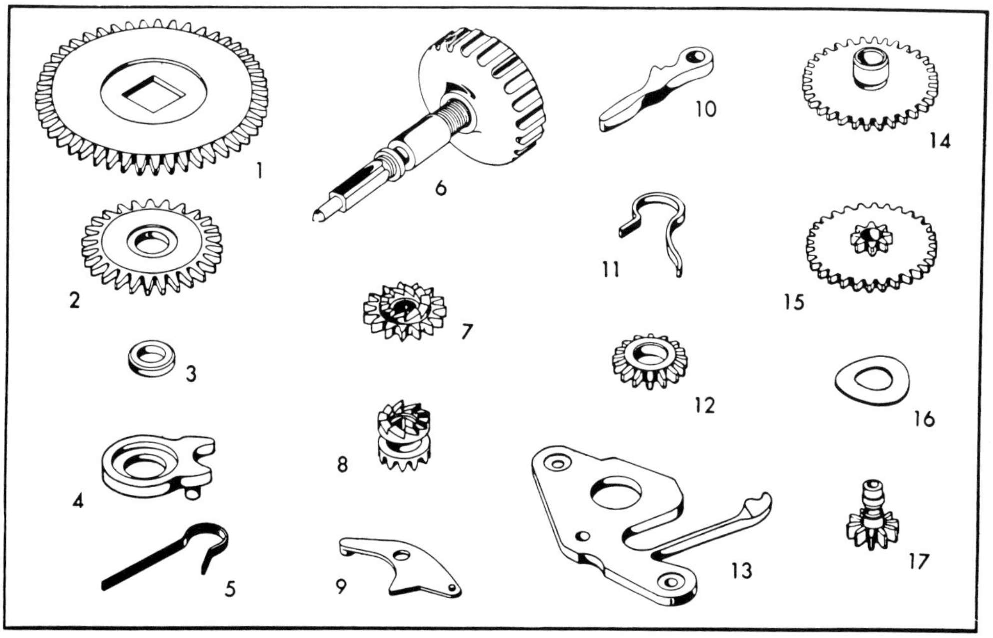

WINDING & SETTING PARTS

1. **Ratchet Wheel**
2. **Crown Wheel**
3. **Crown Wheel Center**
4. **Click**
5. **Click Spring**
6. **Stem and Crown**
7. **Winding Pinion**
8. **Clutch Wheel**
9. **Setting Lever**
10. **Clutch Lever**
11. **Clutch Lever Spring**
12. **Setting Wheel**
13. **Yoke**
14. **Hour Wheel**
15. **Minute Wheel**
16. **Dial Washer**
17. **Cannon Pinion**

8. **DIAL and HANDS** indicate the elapsed time. The dial is usually graduated into hours, minutes, and seconds over which movable indicators called hands are rotated. **Actually, the hands denote the time periods of the balance and hairspring unit.**

Fig. 8

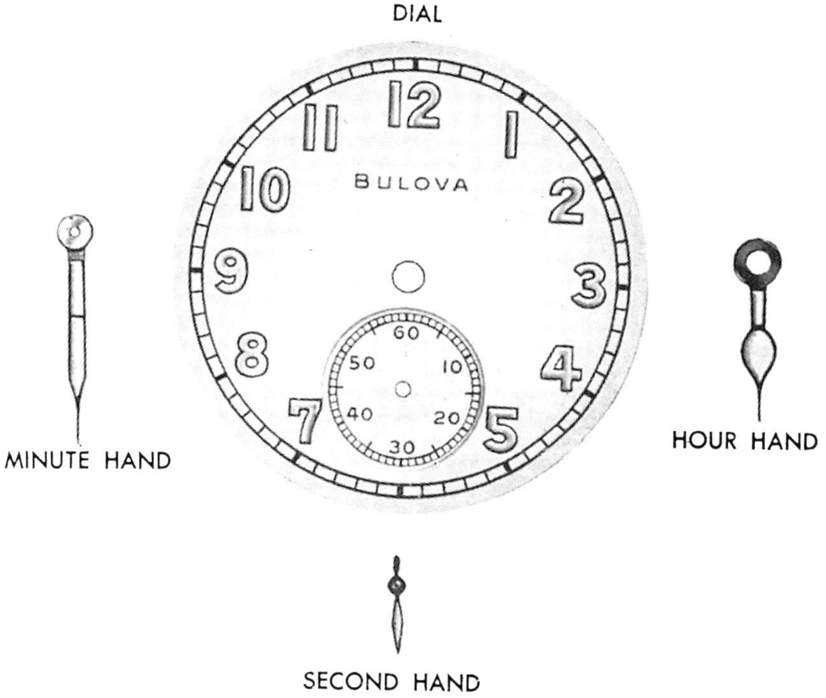

9. The **WATCH CASE** provides the protective covering for the completely assembled movement. Watch cases are made in a large variety of styles and designs and from various materials such as steel, nickel, brass, silver, gold, and platinum. Gold has survived the test of Time, and at present is considered to be the best and most desirable material for watch cases. Recent demands have developed the water and dustproof cases which are generally made of stainless steel.

Fig. 9

In Fig. 10, can be seen two different views of all the operating parts assembled into a complete compact instrument. Although each unit performs its own particular job in the complete assembly, the parts are closely linked together to operate as a coordinated machine.

The bridges are secured to the pillar plate by bridge screws and are accurately positioned by semi-friction tight steady pins. Because the back of the movement has been divided into several bridges, each unit may be disassembled without disturbing the other parts. This arrangement facilitates repair and assembly work.

Fig. 10

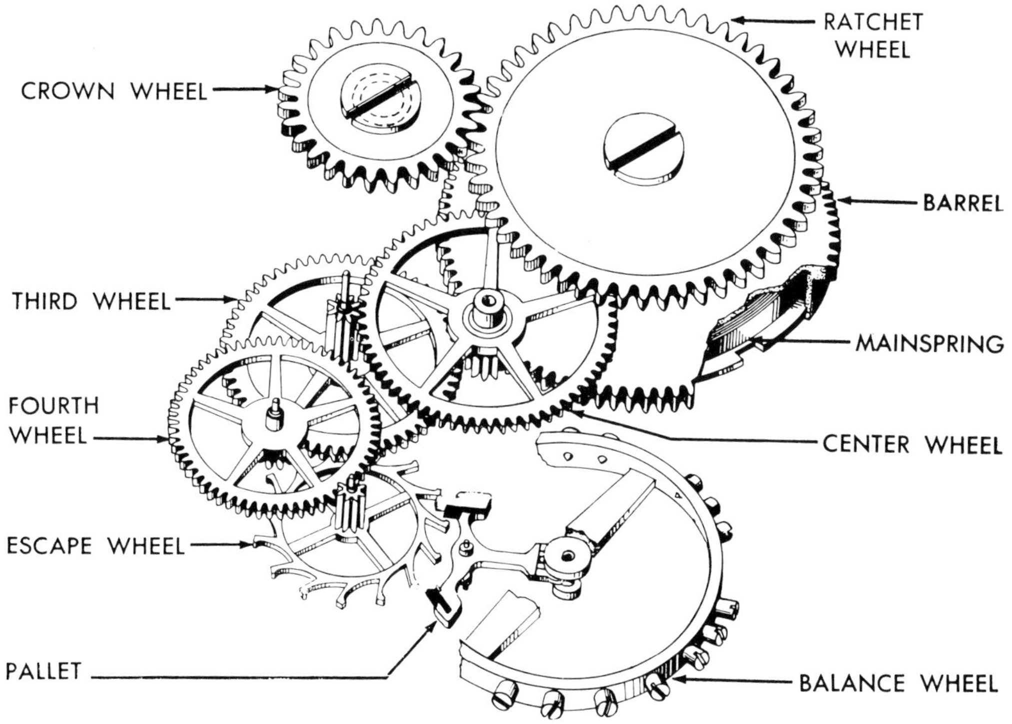

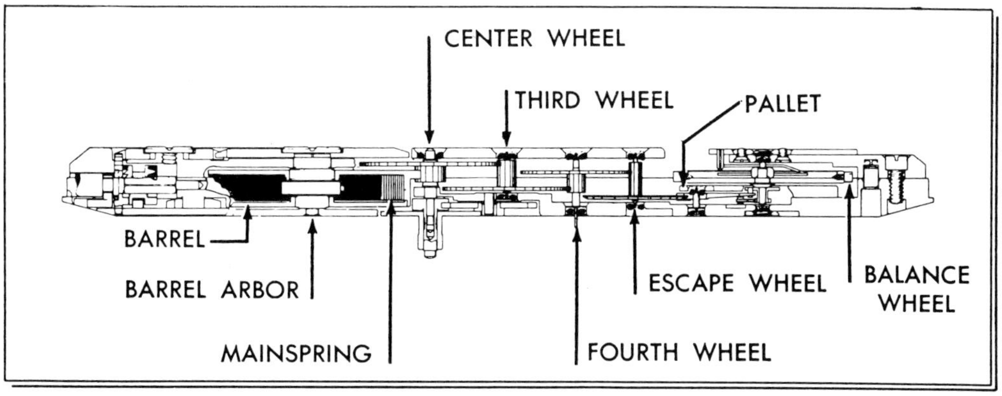

BULOVA SCHOOL *of* WATCHMAKING

TRAINING UNIT
NUMBER 9A

★

MAINSPRING BARREL ASSEMBLY
(Power Unit)

MAINSPRING BARREL ASSEMBLY

The watch like every other engine or mechanical contrivance must depend on a steady source of power in order that it may function. Through a series of evolutionary improvements, the present day mainspring and barrel assembly was developed.

Because the watch is such a minute machine, the slightest interference with the even flow of its mainspring power can readily affect its timekeeping accuracy. Experience has found considerable dissatisfaction with repaired watches because of errors in the power unit.

For this reason, mainspring barrel construction will be carefully explained with special emphasis on correct repair procedure for making replacements and adjustments.

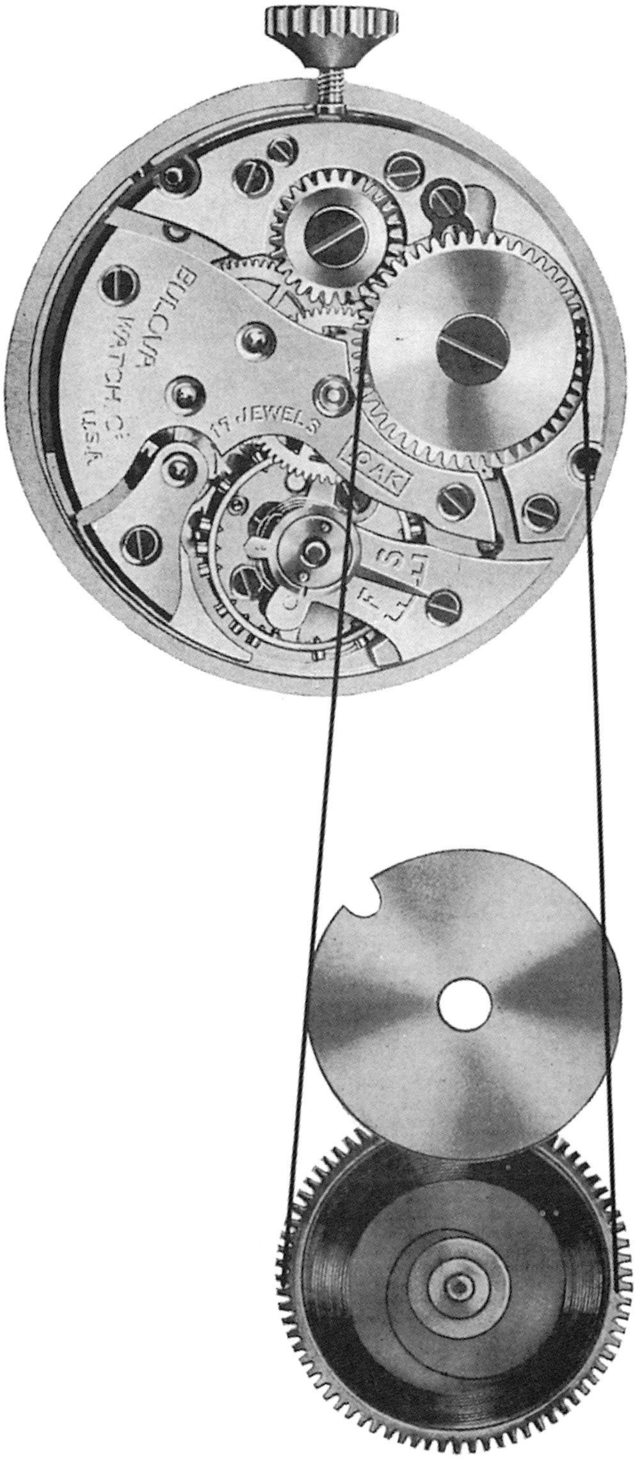
BULOVA
WATCH CO
U.S.A.
17 JEWELS
10AK

BULOVA SCHOOL *of* WATCHMAKING

Subject:

Mainspring Barrel Assembly (Power Unit)

Objectives:

1. To learn construction of modern watch barrel assemblies.
2. To determine and insert the correct size of mainspring in any barrel.
3. To recognize and correct common errors found in mainspring barrel assembly.

New Tools Required:

1. Mainspring winders.
2. Dennison gauge.
3. Set of broaches.
4. Plier for shaping mainspring inner coils.
5. Small files.
6. Plastic blocks.
7. Boxwood punch.

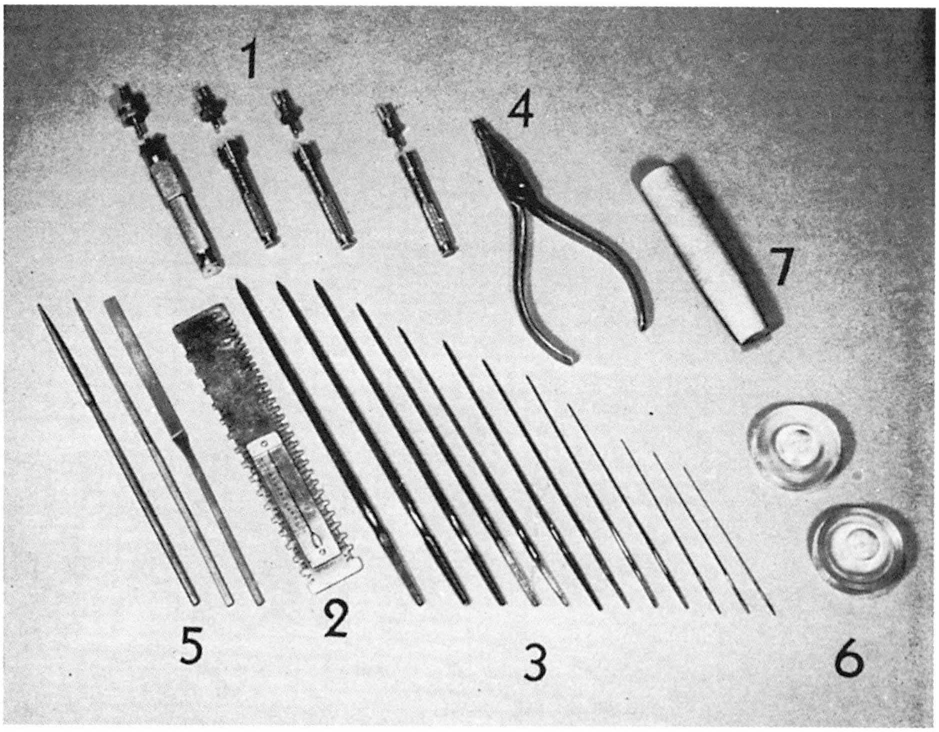

MAINSPRING BARREL ASSEMBLY

Power Unit

There are three types of barrel units used in recent watch construction.

1. THE GOING BARREL.
2. THE MOTOR BARREL.
3. THE SUSPENDED BARREL.

All three will be briefly explained, with special emphasis on the most widely used going barrel.

THE GOING BARREL

Fig. 1 shows a cross section of the going barrel assembiy. The mainspring (A) is wound about an arbor (B) in a barrel (C) to which is hooked the outer end of the mainspring. The arbor remains in a stationary position between the barrel bridge (D) and pillar plate (E), except during the winding process when the ratchet wheel (F) turns the arbor within the barrel. The ratchet wheel is prevented from springing back by a click which is held against the ratchet wheel by a click spring.

The barrel which rotates about the arbor is equipped with epicycloidal gear teeth which are meshed with the center wheel pinion (G). As the spring unwinds, causing the barrel to turn, this driving action furnishes the motive power for the watch.

Fig. 1

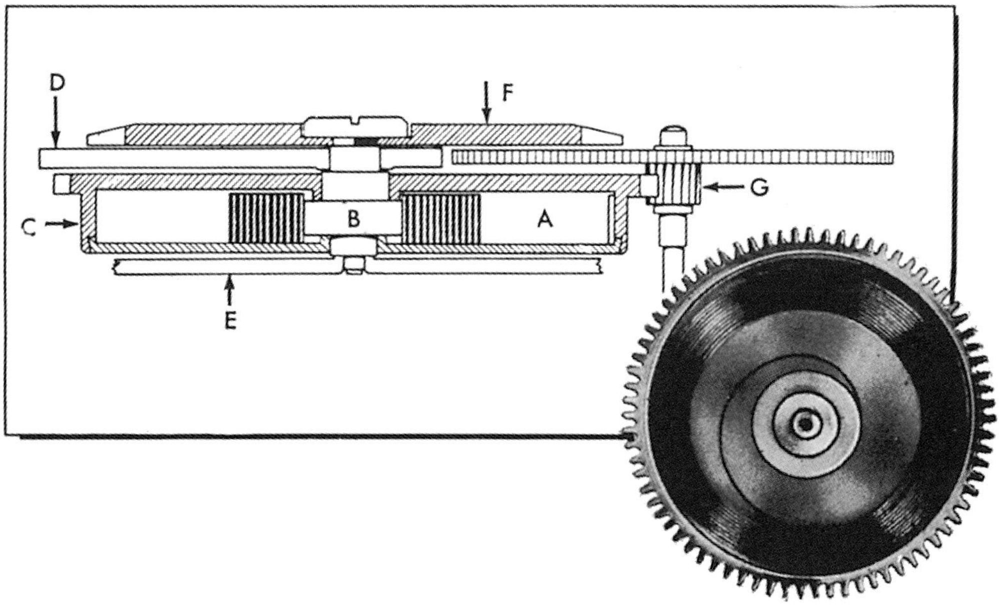

CONSTRUCTION OF THE MAINSPRING BARREL

The barrel which contains the mainspring is capped with a barrel cover which keeps the arbor and mainspring in their respective positions. This cover snaps on the barrel and remains firmly closed due to a slight bezel effect or back taper in the wall of the barrel which matches the taper on the cover's edge. See A and B in Fig. 2.

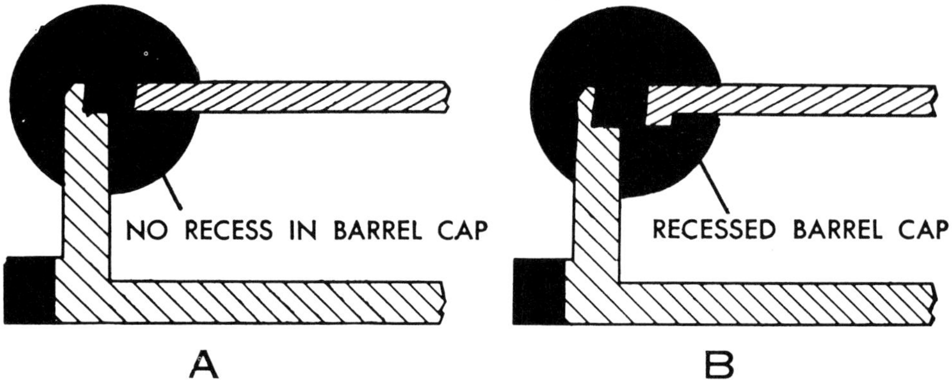

Fig. 2

Many barrel covers such as B in Fig. 2 are also recessed to accommodate part of the mainspring's width. Care must therefore be exercised to prevent the barrel cover from being bent or distorted in any way which might interfere with its remaining closed or with the free movement of the mainspring coils within the barrel.

Located at the holes of the barrel and barrel cover are the shoulders which serve as bushings for the arbor and also control the end and sideshake of the arbor in the barrel. See Fig. 3.

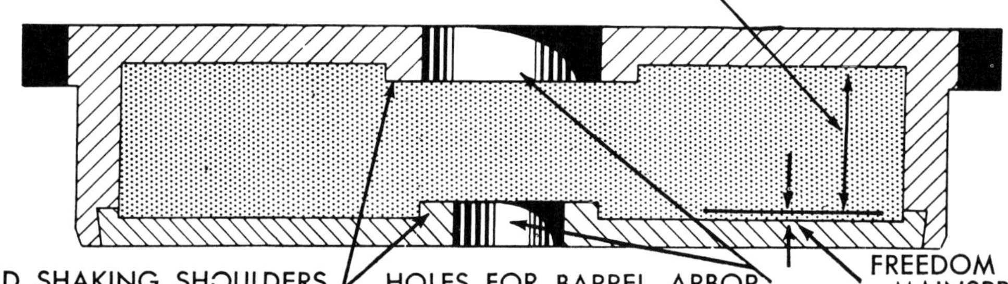

Fig. 3

Since the barrel must be free to revolve about its arbor, the clearance between these two parts must be checked before the mainspring can be inserted. The perpendicular clearance of the arbor between the cover and the bottom of the barrel is known as endshake and should be between .015 and .02 m.m. Sideshake, which is the difference between the diameters of the arbor and barrel hole, should be about .015 m.m.

HOW TO ADJUST ENDSHAKE AND SIDESHAKE OF BARREL ARBOR IN A GOING BARREL

1. To increase the endshake, place barrel on a bench block over a hole smaller in diameter than that of the barrel but considerably larger than the diameter of the arbor. Place a punch on arbor and tap lightly to bend the barrel cover slightly outward, thereby increasing the endshake. See Fig. 4.

Fig. 4

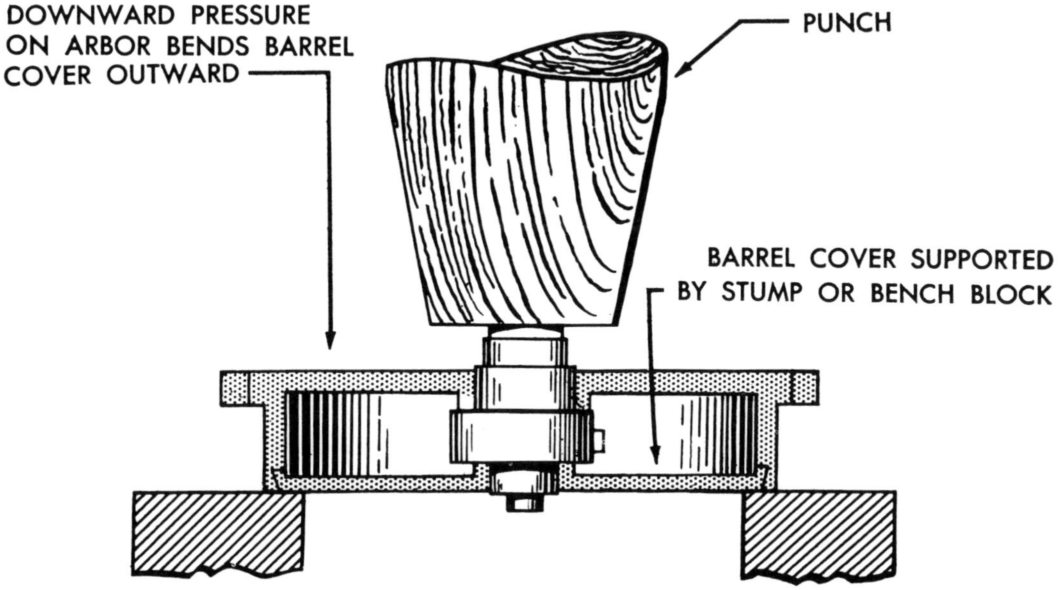

2. To decrease the endshake, determine first whether the cap or the barrel is bent too much by testing each with a straight edge (Fig. 5A). Remove outward bend by tapping a rounded piece of wood against the bent metal as in Fig. 5 (B). This will decrease the endshake.

Fig. 5

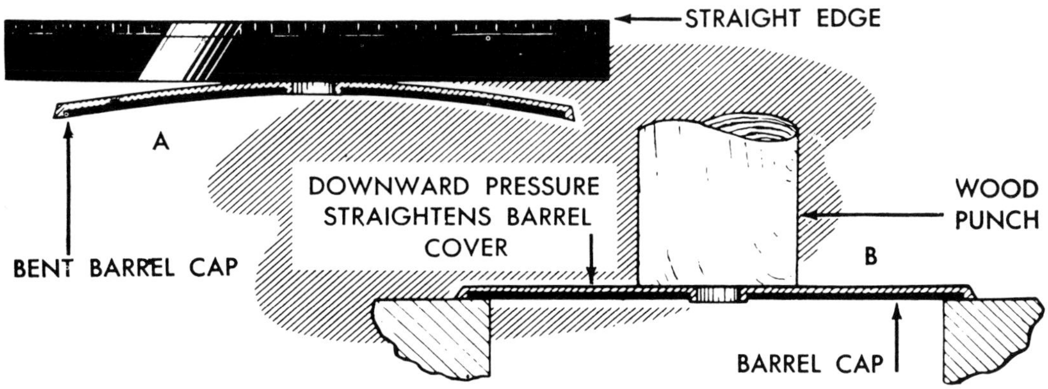

3. A. To eliminate excessive sideshake, the hole may be slightly closed by using a rounded punch against the barrel hub which has been placed bottom up on the staking stump. See Fig. 6.

Fig. 6

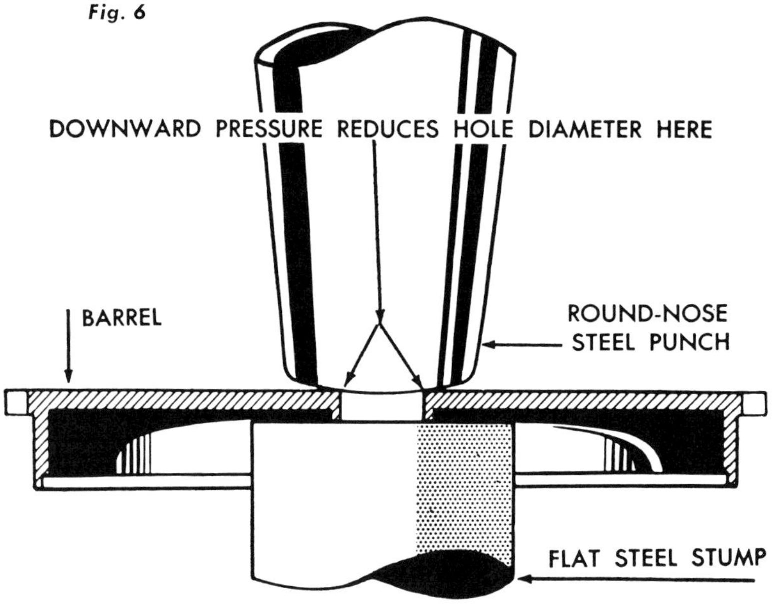

3. B. If the hub of the barrel is unusually high, the hole should be closed somewhat with a taper mouth closing punch on staking stump. The punch should tend to close the walls of the hub toward its center without touching the bottom of the barrel as in Fig. 7.

Fig. 7

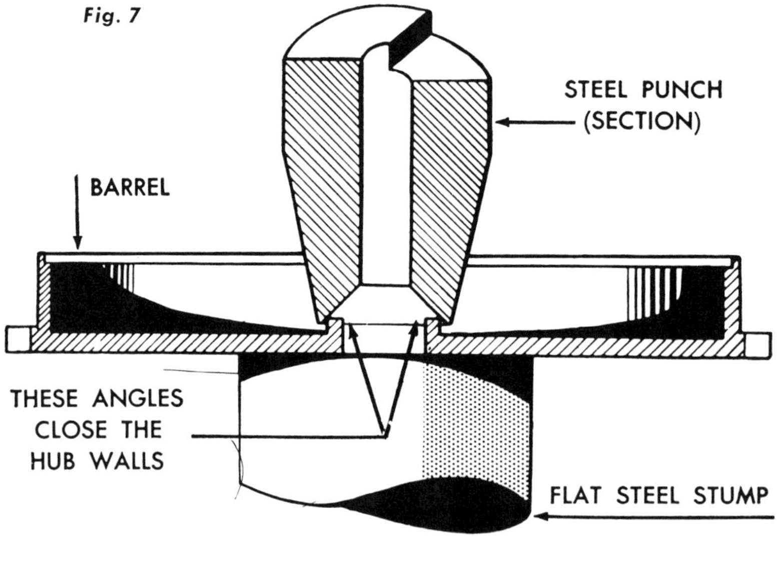

4 If the sideshake is insufficient, the hole may be slightly enlarged by using a reamer as in Fig. 8. The barrel should be reamed from the outside to keep the thrust at proper point on the pivot and to form a reservoir for the oil (insert).

Fig. 8

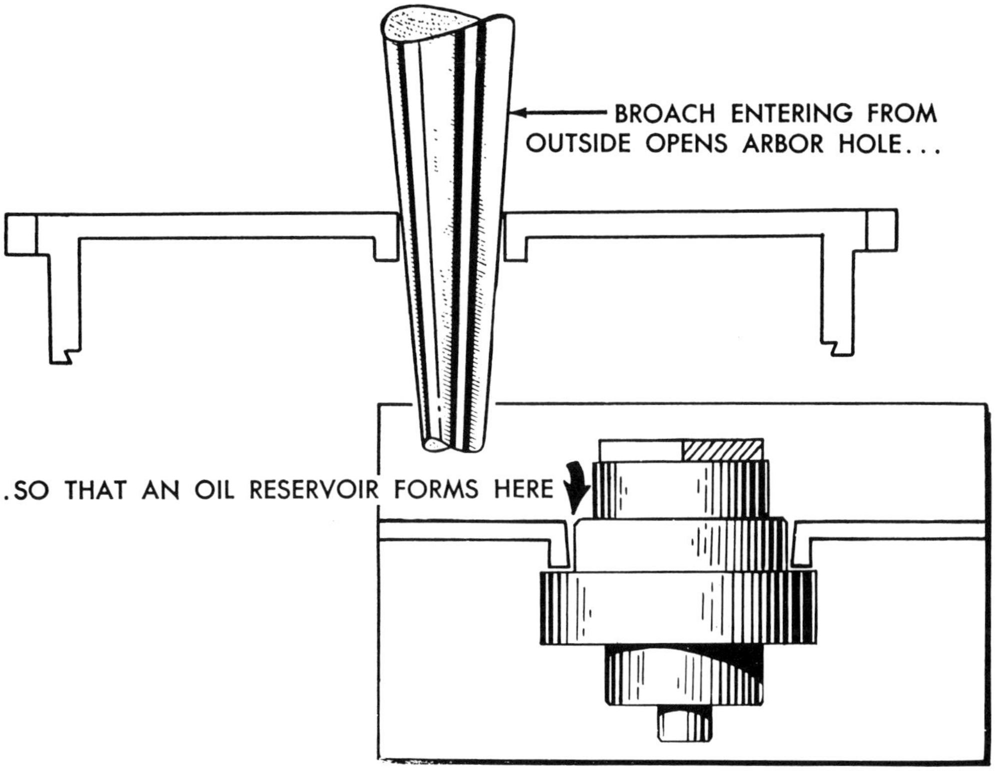

In addition to establishing correct shakes, these following points must also be checked before inserting the mainspring in the barrel:

1. The barrel must run true on its arbor.
2. The barrel and cover must be flat. (Check with a straight edge.)
3. The barrel must be absolutely clean.

MOTOR BARRELS

In the Hamilton type of motor barrel assembly shown below, there are really two arbors: the barrel arbor (Fig. 9, K) which is permanently staked to the barrel (G) and the ratchet wheel arbor (E) which fits freely over the barrel arbor. The inner terminal of the mainspring is hooked on to the ratchet wheel arbor which turns only during the winding process as the ratchet wheel (C) is being turned with it. The barrel arbor, of course, rotates with the geared barrel during the unwinding of the mainspring and serves only as a shaft for the barrel within the ratchet wheel arbor. As in the going barrel assembly, the driving power of the mainspring (J) is transmitted from the barrel through the wheel train to run the watch.

Fig. 9

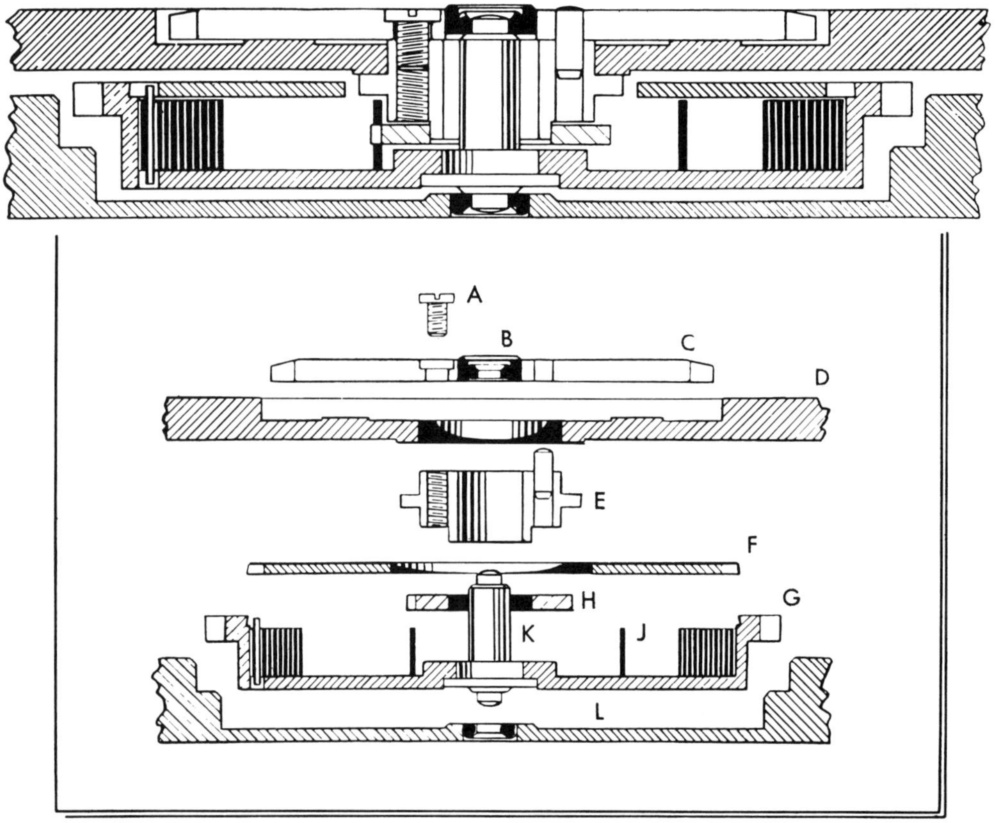

A—Ratchet Wheel Screw
B—Barrel Upper Jewel and Setting
C—Ratchet Wheel
D—Barrel Bridge
E—Ratchet Wheel Arbor
F—Barrel Cap
G—Barrel
H—Mainspring Hook
J—Mainspring
K—Barrel Arbor
L—Pillar Plate

2. In the Waltham non-jeweled type of motor barrel (Fig. 10), the geared section of the barrel known as the main wheel (A) is separated from the lower portion or shell (B). The main wheel is staked to the arbor hook (C) which fastens to the inner terminal of the mainspring (D). The arbor (E) is squared on top and bottom so as to remain fixed to the ratchet wheel and shell. When the mainspring is wound, the ratchet wheel (F) turns the arbor and shell. However, as the mainspring energy is released, these parts now remain stationary, but the main wheel, which is geared to the wheel train, delivers the driving power for the watch mechanism.

Fig. 10

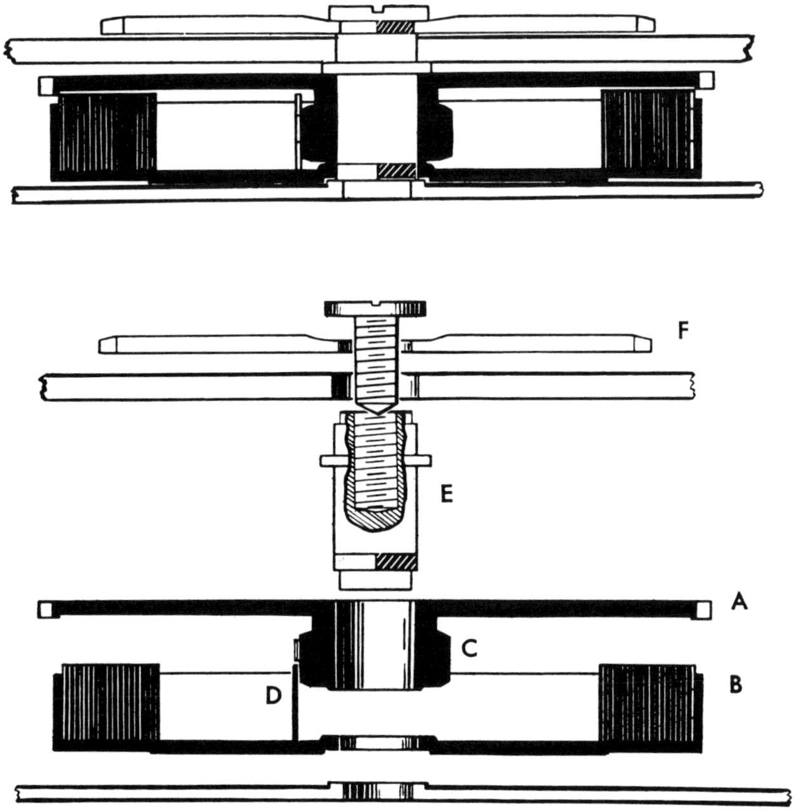

3. In the Waltham jeweled motor barrel (Fig. 11), the main wheel which turns on the barrel arbor and drives the wheel train is fitted with jewels. The arbor consists of two parts, an outside member (A) and an inside member (B) which when screwed together keep the steel barrel fixed to the arbor. As in the non-jeweled Waltham motor barrel assembly, the ratchet wheel which is mounted on the top square of the arbor turns the arbor when the mainspring is wound.

Fig. 11

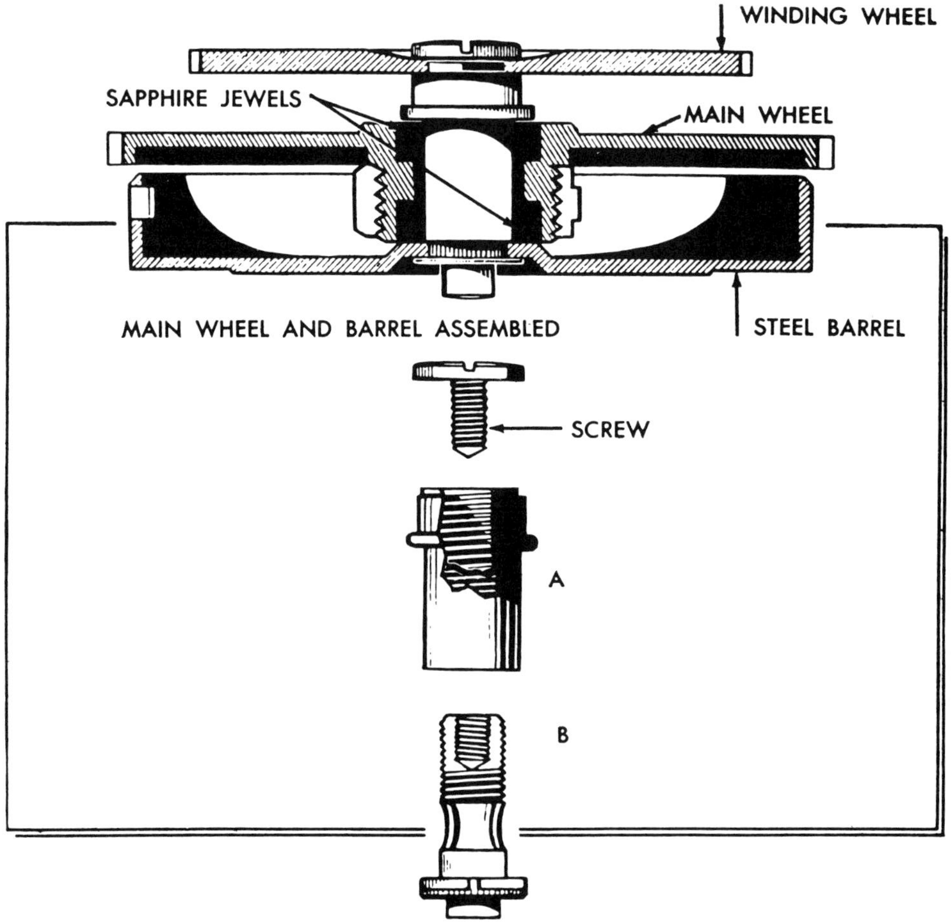

When disassembling, the barrel assembly is reached by removing the ratchet wheel screw, ratchet wheel and barrel bridge. To disassemble the barrel, place the assembly in the left hand, and using the ratchet wheel as a key on the square of the arbor, turn backwards until the parts separate.

SUSPENDED BARREL

The suspended or hanging barrel although more popular in many of the older watch models is still being utilized today in very thin Swiss watches. As illustrated in Fig. 12, the assembly of barrel, arbor, and mainspring of a suspended barrel is similar in construction to that of the much used going barrel. However, in the suspended barrel assembly, the arbor is suspended from the barrel bridge without any support at the lower end of the arbor. A solid ratchet wheel held firmly between plates on the barrel bridge keeps the suspended barrel arbor in a true upright position.

The stored up energy in the wound mainspring is delivered to the wheel train by means of the barrel rotating about its arbor, exactly as in the going barrel arrangement.

Fig. 12

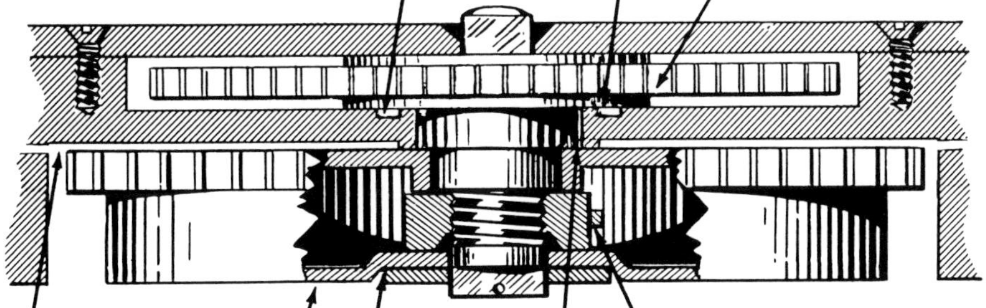

INSERTING THE MAINSPRING

When it is certain that the barrel and arbor fit together properly, the necessary size and type of mainspring must be selected for insertion into the barrel. Listed below are five of the most common types of mainspring ends for hooking on to the barrel:

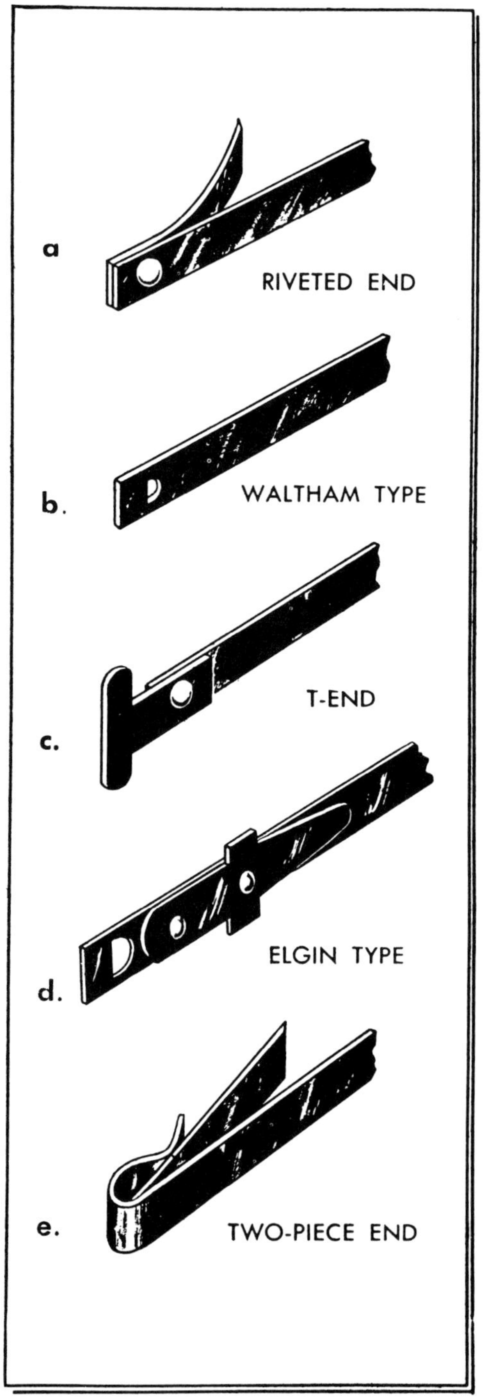

a. The turned up tongue which is a piece of spring riveted to the end of the mainspring. This sharpened tongue is arrested by a recess in the wall of the barrel.

b. The Waltham type of mainspring wherein a small opening in the end of the spring fastens on to a hook which has been punched from the wall of the barrel.

c. The T end wherein both ends of the T fit into corresponding holes in the barrel and barrel cover.

d. The Elgin type consisting of an opening and a broad T brace at the end of the mainspring. This hooks on to the wall of the barrel and also fits into corresponding slots in the barrel and cover.

e. The mainspring with an extra piece of spring inserted under its turned over end. It is this extra tongue which hooks on to the wall of the barrel.

The thickness and width of mainsprings are measured in millimeters with a small micrometer. See Figs. 18 and 19.

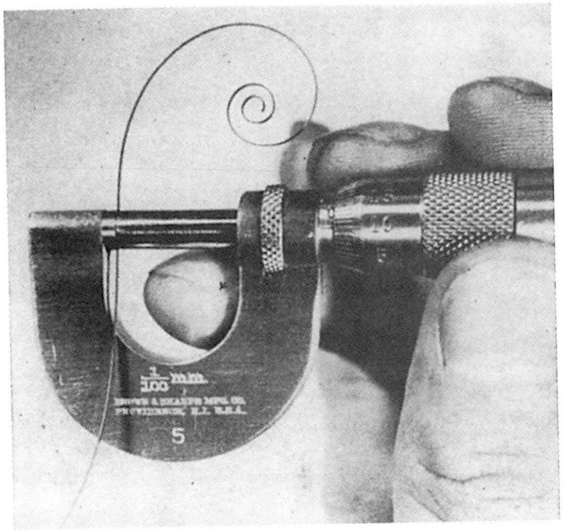

Fig. 18

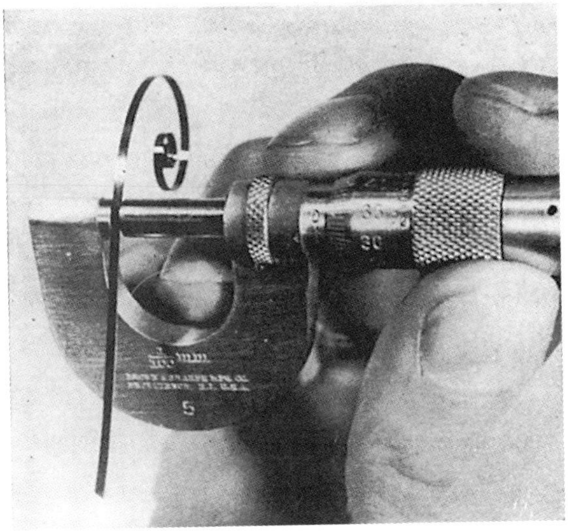

Fig. 19

If a micrometer is not available, a Dennison mainspring gauge (Fig. 20) may be used. This tool measures the width and thickness of the mainspring in terms of a numbering system which corresponds to that used on packaged mainspring replacements. For true accuracy, this method cannot compare to a micrometer measurement.

Fig. 20

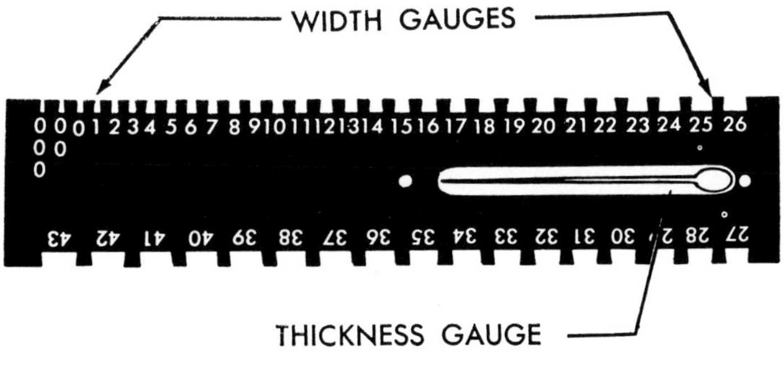

When the watch is a standard make, the most dependable method is to use a genuine factory-made spring as specified by the manufacturer for the specific model. However, there will be many situations where the maker will be unknown, and a reliable method to ascertain the proper thickness and length of the mainspring for the barrel and watch must be understood.

Before computing the size of mainspring, the effects of a spring too long or too short and too strong or too weak should be explained.

1. A spring that is too long will not have enough space in the barrel in which to unwind with the result that the watch will not run its maximum number of hours. In Fig. 21 is shown a barrel with a spring that is too long. Compare this with a proper sized spring as depicted in Fig. 1.

2. A spring that is too short will have insufficient coils to unwind, with the result that it will be released too quickly, and again this watch will not run its maximum number of hours. If Fig. 22 will be compared with Fig. 1, this condition will be readily apparent.

3. A spring that is too thick and therefore too strong (since the thickness of the spring determines its strength) will cause the motion of the watch to increase beyond safe limits, resulting in the banking of the jewel pin against the pallet fork. Accurate regulation of a watch under such conditions is impossible.

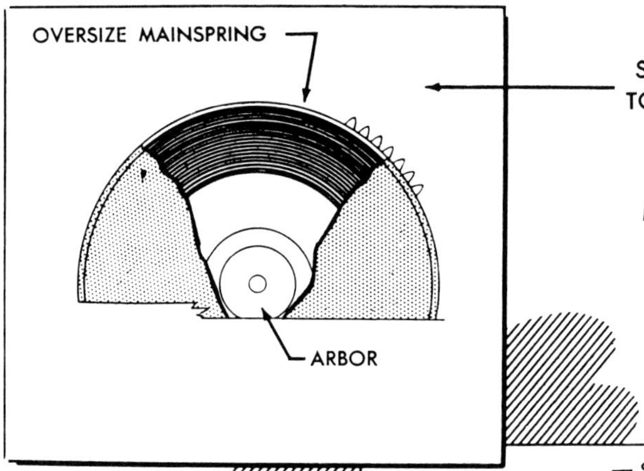

Fig. 21

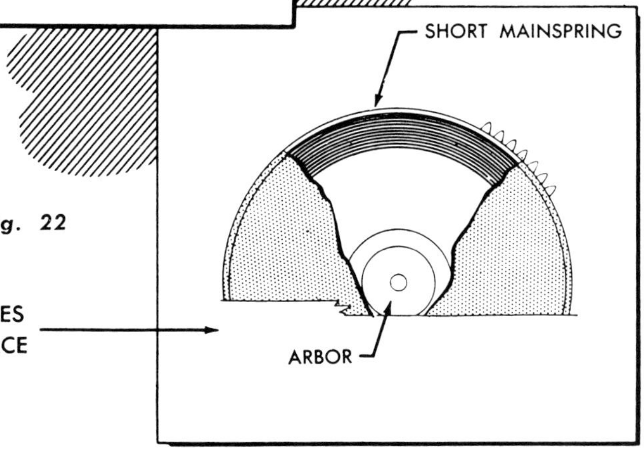

Fig. 22

4. If the spring is too thin and therefore too weak, it will not have sufficient power to run the watch. When the spring is fully wound, the balance motion may at first be satisfactory, but after the weak spring has run a few hours, there will not be enough tension to maintain a good running motion of the balance wheel.

After determining the mainspring thickness, the correct length of spring to give the maximum number of barrel turns can be quickly calculated by W. Dodgion's equation as follows:

$$\frac{(B + A) \times (B - A)}{64.6 \times T} = \text{Length of mainspring in } \textit{inches}$$

wherein A — arbor diameter in millimeters

B — inside diameter of barrel in millimeters

T — thickness of mainspring in millimeters

For example:

Barrel diameter = 16.8 m.m.
Arbor diameter = 5 m.m.
Spring thickness = .18 m.m.

$$\frac{(16.8 + 5) \times (16.8 - 5)}{64.6 \times .18} = \frac{(21.8) \times (11.8)}{64.6 \times .18} = 22.1 \text{ inches}$$

HOW TO CHECK WITH "UP AND DOWN CIRCLE"

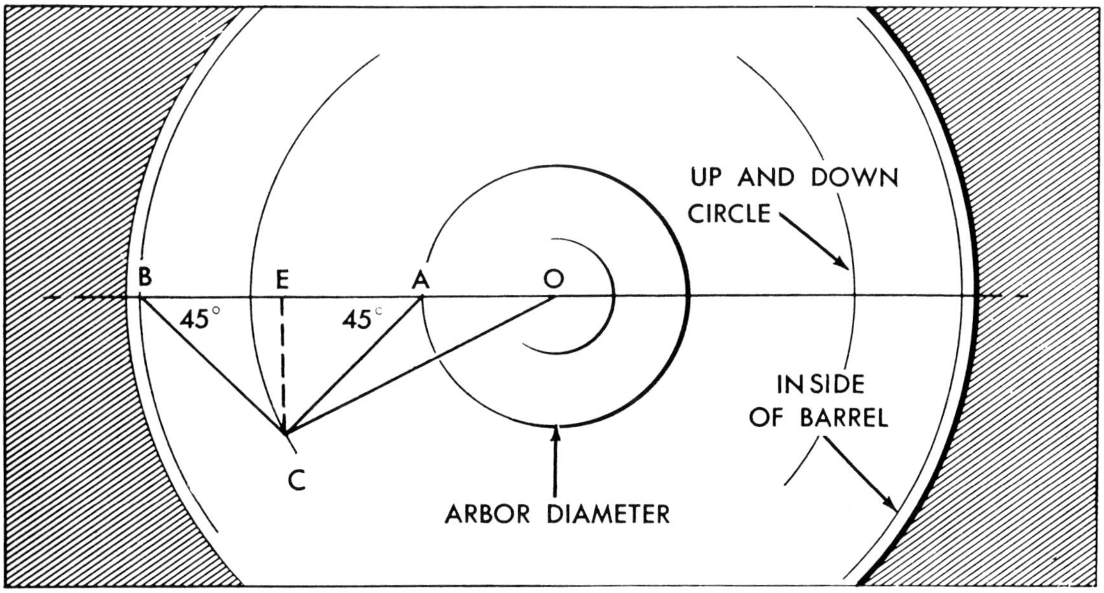

Fig. 23

1. Draw barrel and arbor diameters 10 times full size.
2. Draw horizontal diameter through point 0.
3. From point A lay out AC with 45 degree triangle.
4. From point B lay out BC with 45 degree triangle.
5. Lines AC and BC will meet at point C.
6. Point C gives position of "up and down circle."
7. Draw line CO from point C to center point O.
8. Measure CO and divide by 10.
9. Set depthing tool to this size and scribe "up and down circle" in barrel.
10. Wind spring in barrel.
11. Last coil should just meet "up and down circle."

In determining the proper thickness of the mainspring for a given barrel, the following rules should apply:

1. With a given arbor diameter, the inside barrel diameter should be approximately three times as great. One half the net difference in area between the barrel and the arbor should be the area covered by the spring.

2. The number of barrel revolutions possible with a given spring is equal to the difference in the number of coils between the wound up and the run down positions of the spring.

3. For a barrel arbor of a given size, the mainspring thickness should vary from 1/26 to 1/34 of the arbor diameter, depending upon the quality of the movement. In the table that follows it can readily be seen that the better watch movement requires the weaker spring.

GUIDE TO MAINSPRING THICKNESS PROPORTIONS

Pocket Watches

7 - 15 Jewels - mainspring thickness - 1/26 arbor diameter

15 - 17 jewels - mainspring thickness - 1/28 arbor diameter

17 - 19 jewels - mainspring thickness - 1/30 arbor diameter

19 - 21 jewels - mainspring thickness - 1/32 arbor diameter

21 - 23 jewels - mainspring thickness - 1/34 arbor diameter

Wrist Watches

7 - 15 jewels - mainspring thickness - 1/28 arbor diameter

15 - 17 jewels - mainspring thickness - 1/30 arbor diameter

17 - 19 jewels - mainspring thickness - 1/34 arbor diameter

These figures can in no way be considered as binding but will serve as a guide where data from the manufacturer may not be available. Of course, any irregularity in the depthing, poor pivots, incorrect banking, etc., will void all the proportions.

4. With a spring which occupies one-half the net difference in area between the barrel and arbor, the last coil when wound up and the first coil when down will both lie on a common circle called the "up and down circle."

5. The diameter of the "up and down circle" is approximately equal to ¾ of the inside barrel diameter if, and only if, the barrel is three times the arbor diameter.

INSERTING THE MAINSPRING IN ITS BARREL

With the proper thickness and length of mainspring selected and the fitting of the arbor in its barrel already checked, the mainspring is now ready for insertion into the barrel. For this operation it is necessary to use a mainspring winder. Any attempt to feed the spring into the barrel with the fingers will deform or distort the spring, rendering it unfit for reliable service.

The winders are made in a range of sizes to accommodate the various sizes of mainspring barrels. Extra flexibility is provided by the movable jaws on the winder which increase or decrease the diameter of the winder's barrel to conform to the barrel of the watch.

Fig. 24

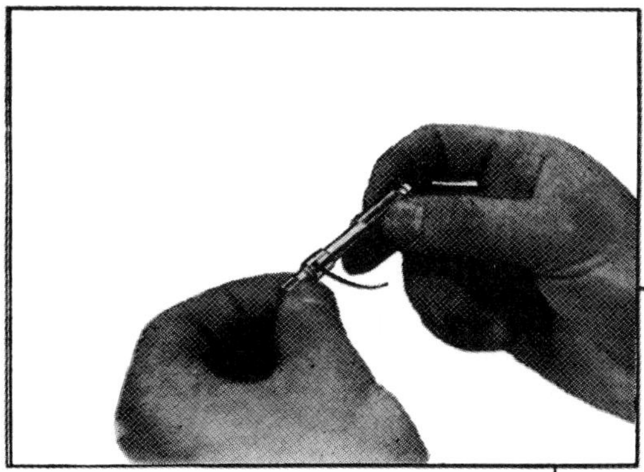

WINDING MAINSPRING IN MAINSPRING WINDER

Fig. 25

EJECTOR FORCES MAINSPRING OUT

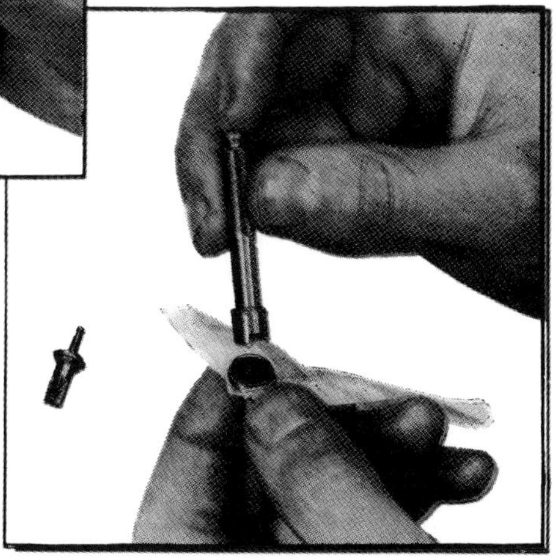

TRANSFERRING MAINSPRING TO BARREL

1. Select proper size mainspring winder.

2. Adjust movable jaws of the winder to the size of the barrel.

3. Check direction to wind the spring. (Note position of hook on arbor.) For insertion, the mainspring must be wound in a reverse direction to its position in the barrel.

4. Attach inner coil of spring to arbor of winder.

5. Wind up spring within barrel of winder.

6. Place winder over barrel so that end of spring is at hooking end of barrel. (If it is the tongue type, the tongue must be bent away from spring.)

7. Insert mainspring into barrel.

8. Lubricate mainspring and arbor with heavy oil. (A light oil would be forced from the spring by the pressure between the coils of a fully wound spring. This would cause a binding which would result in uneven power and inaccurate regulation.)

9. Snap on barrel cover with fingers after carefully lining up cover with scratch mark on side of the barrel. Note the tissue protecting the barrel from finger prints. See Fig. 26.

10. If the cover fits rather tightly, use a brush handle to apply sufficient force to snap cover into place. See Fig. 26.

Fig. 26

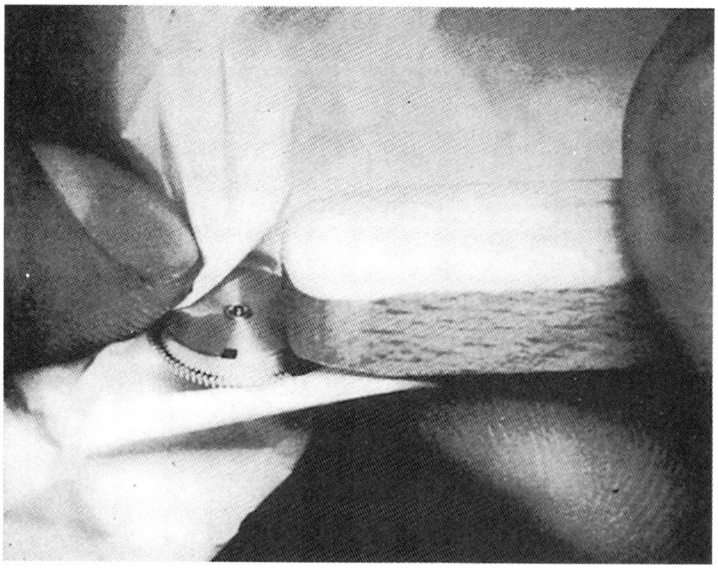

11. Check barrel carefully to remove dirt and fingerprints.

When a T brace is inserted and the ends extend out of the barrel and cover, the projections should be made flush with the barrel and head by the tapping method shown in Fig. 27 instead of by filing. As the flat punch is tapped against the projection brace, supported by a steel block, both projections will be flattened down even with the barrel and cover.

Fig. 27

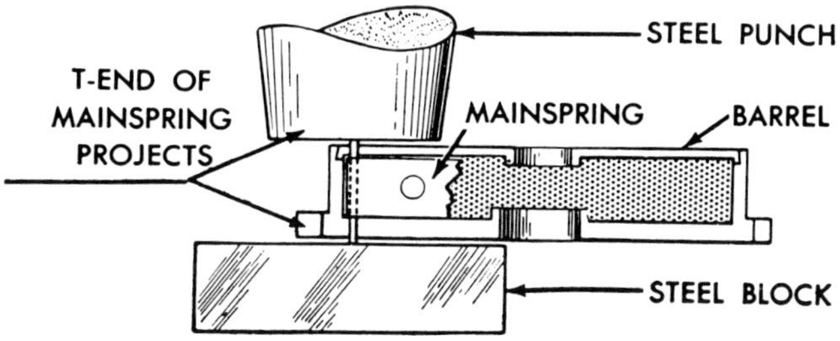

When the spring slips off the hook or slot in the barrel, it may be necessary to file the hook or brace so that the resulting angle will pull the spring tight to the barrel wall. See Fig. 28.

Fig. 28

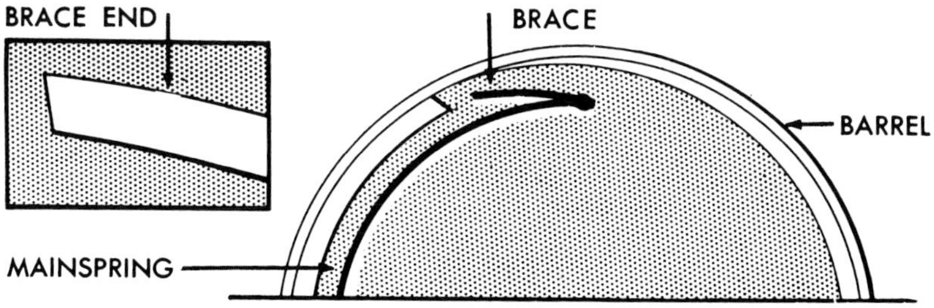

If the mainspring unhooks at the arbor, the inner coil may be improperly formed or too wide. A special shaping plier shown in Fig. 29 is helpful in bending the inner coil so that it will be circular and slightly smaller in diameter than the arbor. This will cause the spring to maintain a constant pressure on the arbor.

Fig. 29

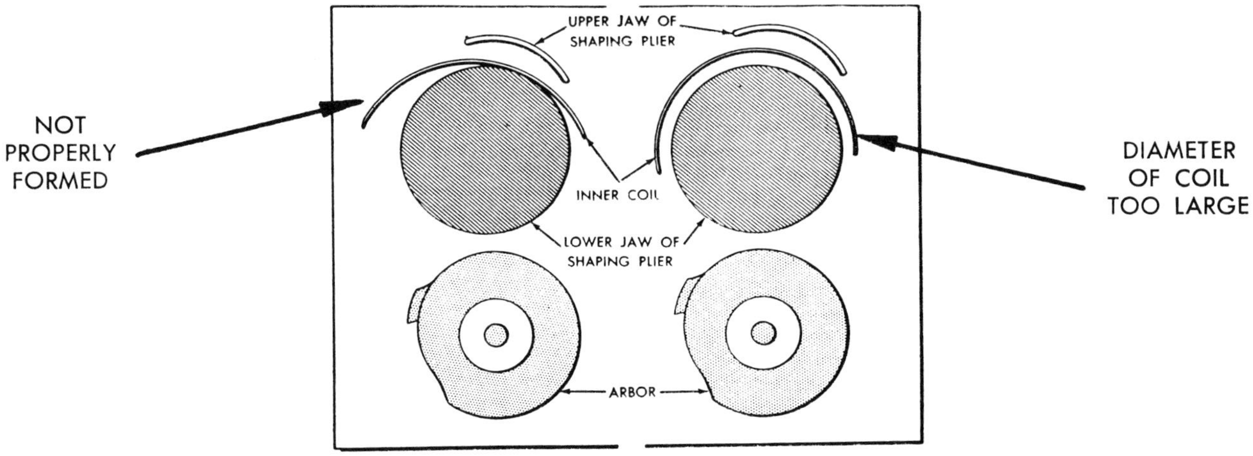

Sometimes, a poorly finished end of the mainspring may rub against the barrel or cover, causing poor balance motion and erratic rate. A similar condition may result if the hole in the inside end of the mainspring is not in the center of the spring. When the hole is placed over the hook on the barrel arbor, the end of the spring may rub against the barrel or cover. (Fig. 30.)

Fig. 30

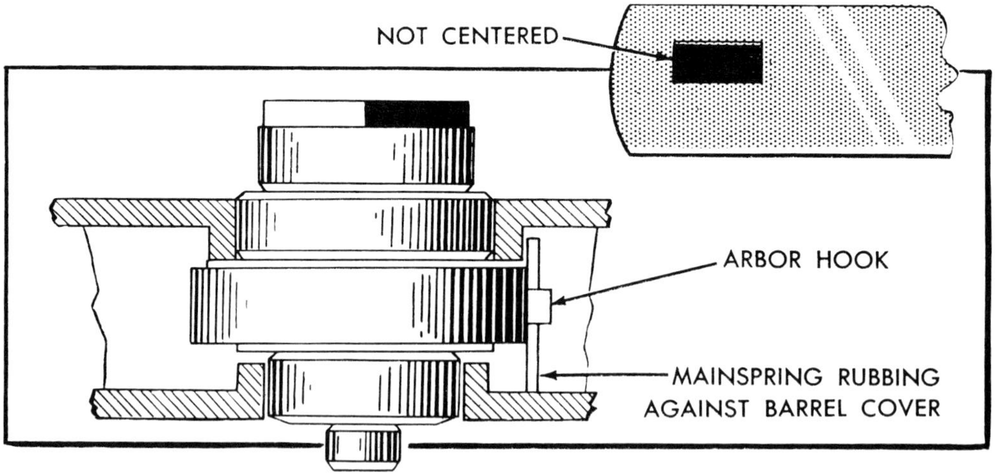

The hole in mainspring end should be in the center and slightly larger than the hook on the arbor.

If it becomes necessary to check the depth of the barrel for proper clearance of the mainspring, this may be accomplished with the sliding Vernier gauge or ordinary micrometer.

In the Vernier method (Fig. 31), the tongue of the gauge is used in getting the depth of the barrel (A) and the depth of the cover recess (B). The two measurements, (A) and (B), are added to determine the total space for the mainspring (C). The mainspring selected should be .1 m.m. less than the mainspring space (C).

If the barrel cover is not recessed for the mainspring, then the barrel and cover depth (A) only, is gauged.

Fig. 31

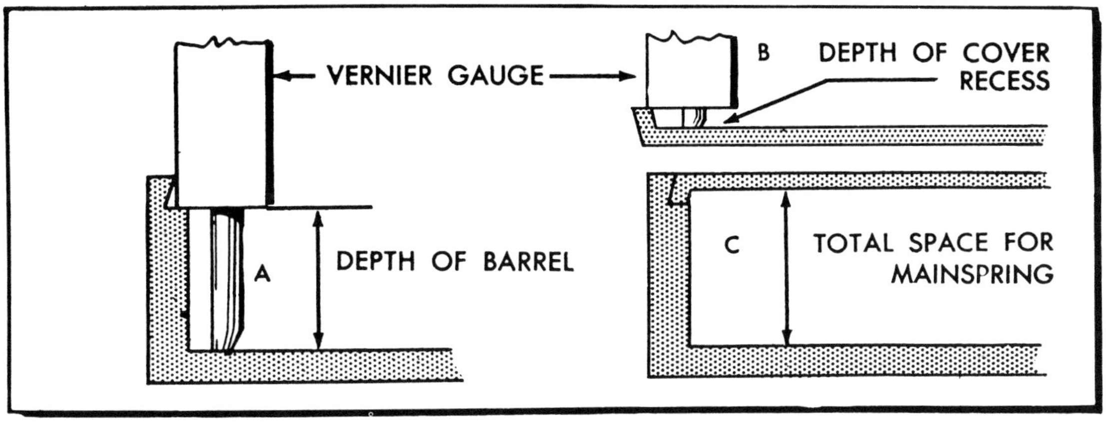

In the micrometer method (Fig. 32), the thickness of the capped barrel (A) is first gauged, and from this is deducted the thickness of the barrel (C) and cover (B). The net difference less at least .1 m.m. for clearance should be the proper width of the mainspring inserted in barrel (D).

Fig. 32

$$A - (B + C) = D$$

in which

A — total thickness of barrel

B — thickness of barrel cover

C — thickness of barrel bottom

D — space allowance for mainspring and .1 m.m. clearance

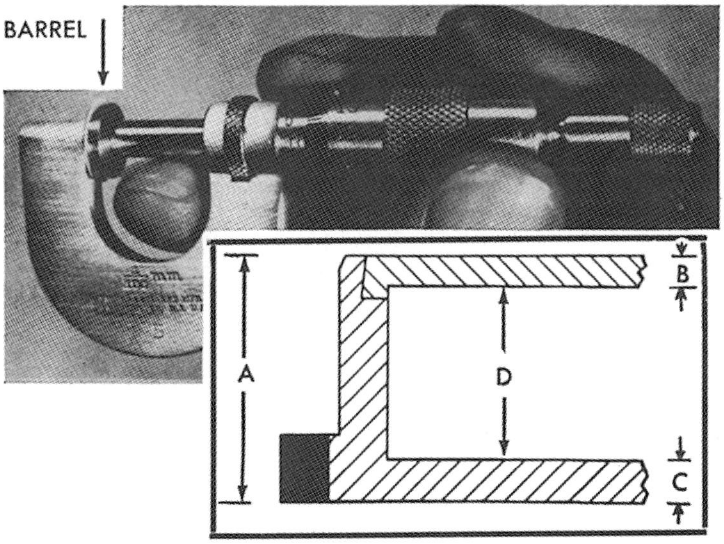

The number of revolutions of the barrel is important in so far as it becomes necessary to store sufficient energy in a spring to keep the train and balance wheel in motion for a 24 hour period and a sufficient excess to take care of irregular winding. 36 to 40 hours is the average running time of most wrist watches.

Fig. 33

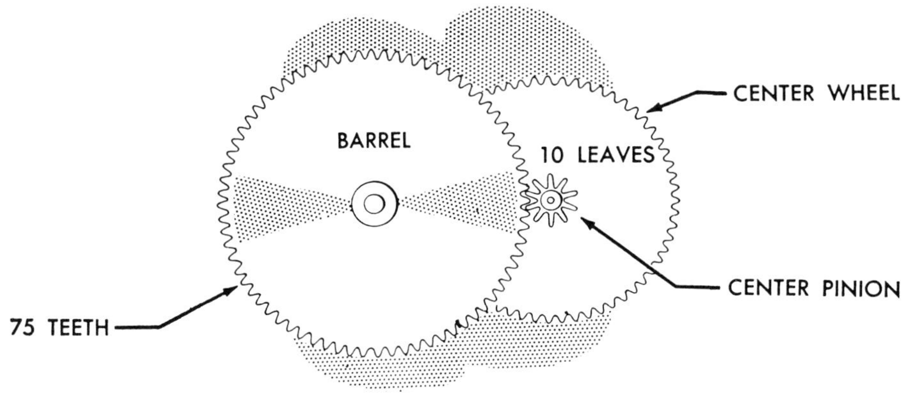

In Fig. 33 it can be seen that there are 75 teeth in the barrel and 10 in the center wheel pinion, or a gear ratio of 7½ to 1 between barrel and center wheel. Since the center wheel which supports the minute hand rotates 24 times in a 24 hour period, the mainspring barrel will move 10/75 as much or 3.2 times in a full day. This would mean that a 10 AK Bulova model barrel would make about 6.4 turns for a 48 hour run.

As an over-all check on the mainspring assembly, the following simple procedure should be followed:

1. Tighten a pin vise to the arbor of the assembled barrel, mainspring and arbor.
2. Count the turns that the spring allows the arbor to make within the barrel.
3. If this is 6.4 turns in a 10 AK Bulova model, or at the very least, 4 turns for any other manufacturer's watch (using a different gear ratio), the spring is certain to develop its maximum number of turns in the barrel.

ASSEMBLY OF BARREL BETWEEN PLATES

The mainspring assembly is now ready for mounting between the pillar plate and barrel bridge. The clearance of the arbor between the plates should be limited to an endshake of .015 m.m. and a sideshake of .015 m.m. The holes in the plates must be perfectly aligned so that the barrel will revolve in a plane parallel with the pillar plate and bridge. If the barrel is not true on the arbor, it will wobble or rise and fall. If the arbor is out of upright in the plates, the barrel won't wobble but will turn in a plane which is not parallel to the plates. Either condition may result in a binding of the barrel against the bridge or plate as illustrated in Fig. 34.

Fig. 34

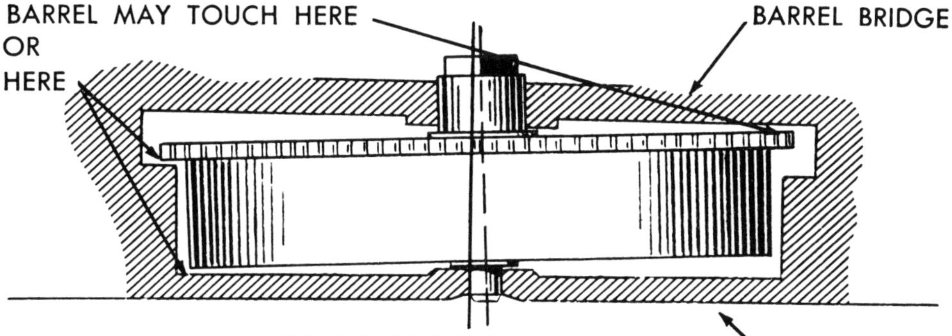

Before putting mainspring bridge into position, it will be necessary at this point to insert the setting lever screw into its proper position. The adjoining parts such as center, 3rd and 4th wheels must be carefully checked to insure that they do not rub against the barrel. A bent wheel could easily interfere with an even flow of power as shown in Fig. 35.

Fig. 35

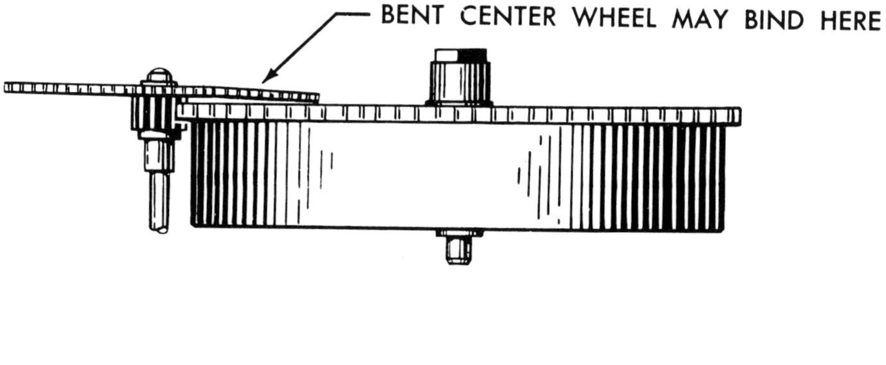

STOP WORKS

Occasionally, the watch repairman will encounter a barrel equipped with a stop works mechanism. For this reason, the subject will be briefly discussed to complete the full story on mainspring assembly.

Stop works which have been used on the going barrels of better grade Swiss watches for many years have three definite advantages:

1. Utilize the most uniform part of the mainspring by eliminating excess power of mainspring when almost fully wound up and the weakness of the spring when almost run down.

2. The mainspring is prevented from being wound too tight with the result that increased pressure on the barrel teeth is avoided.

3. The coils of the mainspring cannot be squeezed too tightly so that the oil between the coils could be forced out.

The most popular type of stop works known as the Maltese Cross is illustrated in Fig. 36. Here, a finger fitted on the square of the barrel arbor moves a star wheel one notch each time the barrel makes a complete revolution about its arbor. When four turns have been made, the finger jams against the convex curve on the fifth tooth of the star wheel, thus preventing the barrel from any further rotation. This stop action takes place in the winding up process immediately after the arbor has made four turns in the barrel, and again when the power is being released on completion of four revolutions of the barrel. In this way, the most constant power of the mainspring is utilized.

Fig. 36

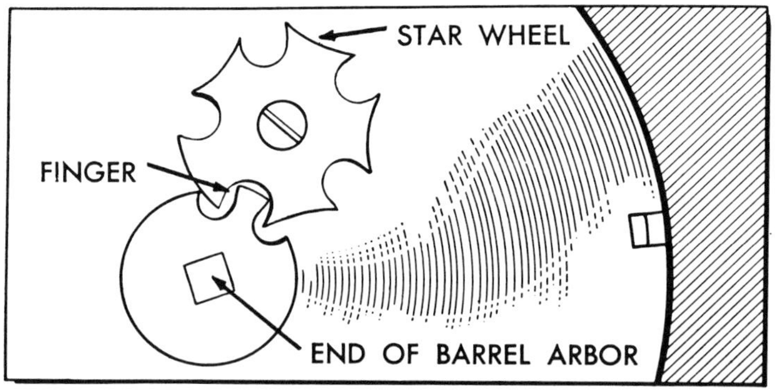

HOW TO SET STOP WORKS

1. Determine how many turns the arbor will make in the barrel by winding with a pin vise.
2. If the star wheel has five teeth (usual construction), subtract four from the number of windings.
3. Divide the surplus by three. One-third of the surplus is let down from the fully wound mainspring in the barrel, and the remaining two-thirds will be subtracted from the run down winding of the spring. For example, if the barrel has 5½ turns, the power would be released ½ turn before setting stop works. The last weak full turn will then be eliminated.
4. With the spring fully wound up in the pin vise, release ⅓ of the surplus and set finger against the stop of the maltese cross.

TRAINING UNIT
NUMBER 9B

FRICTION JEWELING

Jeweled bearings are a very essential part of modern watches. The quality of a watch is usually judged by the number of jewels employed in its manufacture. Some of the advantages for the wide use of jewels in watches and instruments are:

1. Reduction of friction.
2. Less wear on the pivots and bearings.
3. More efficient utilization of available power.
4. Longer life to the watch.

Synthetic ruby and sapphire jewels which are used in the better grade of watches are made up in a variety of shapes, diameters and hole sizes.

BULOVA
WATCH CO
U.S.A.
JEWELS
10AK

Training Unit Number 9B

BULOVA SCHOOL *of* WATCHMAKING

Subject:

Friction Jeweling.

Objectives:

1. To familiarize the student with types of bearing jewels and jeweling tools.
2. To teach modern friction jeweling methods.

New Tools Required:

1. Jeweling Tool
2. Set of jeweling reamers
3. Set of jeweling pushers
4. Assortment of stumps
5. Countersinking Tool

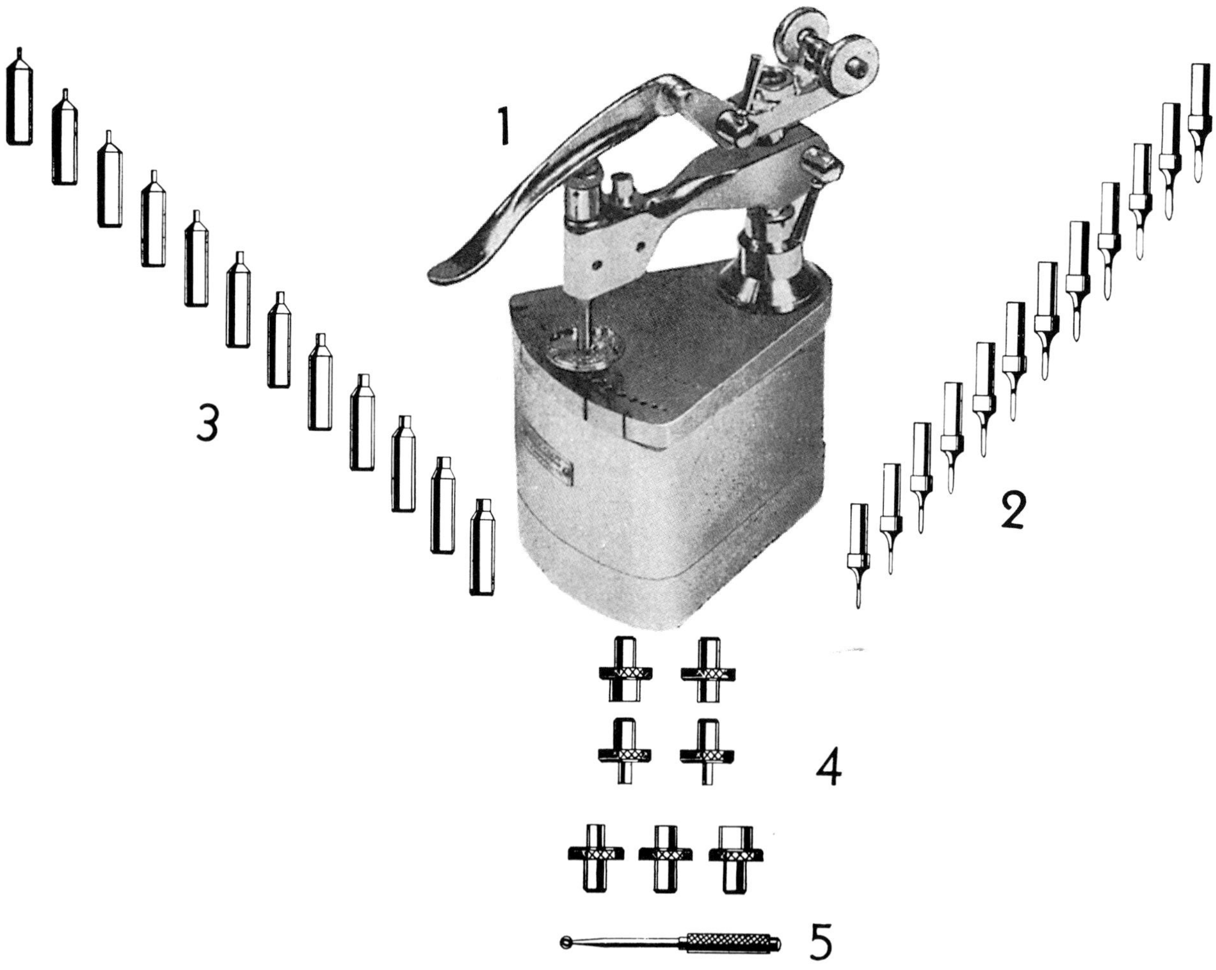

TYPES OF BEARING JEWELS

All bearing jewels are made either with straight edges for friction setting or with beveled edges for the burnished-in type of setting. The straight edged jewel is known as the friction jewel (Fig. 1) and the beveled jewel is termed the bezel or burnished-in type jewel. (Fig. 2).

Fig. 1 *Fig. 2*

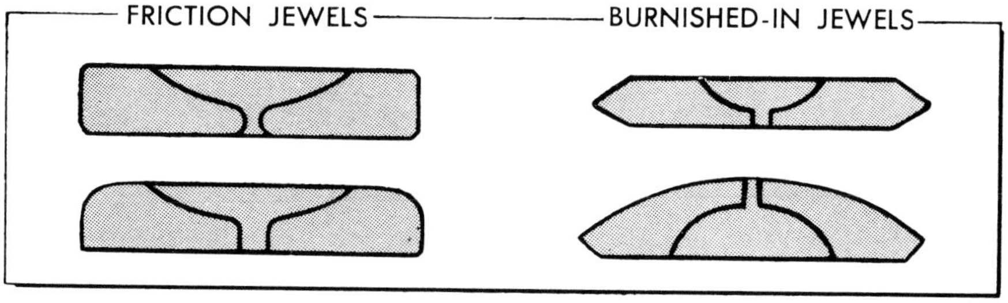

Jewels may have flat or curved surfaces as shown in Figs. 3 and 4. Curved jewels are more favorable to balance, escape wheel and pallet arbor pivots because of the oiling advantages.

To explain this further, oil or any liquid has a tendency to flow into a narrow space. For example, if two plates were held together so that their edges touched on one side and were slightly separated at the other side, a drop of oil between the plates would tend to run towards the touching sides. This is known as capillary action, and has a direct bearing on the use of curved jewels.

When a curved jewel is placed against a flat jewel, the oil will have a greater cohesion at the pivot than with a flat jewel. Also more oil may be stored, which in larger quantity may offer better resistance to chemical change and evaporation.

Train wheels which are generally uncapped use flat surfaced jewels, but balance, escape wheel and pallet arbor jewels may be flat or curved.

Fig. 3 *Fig. 4*

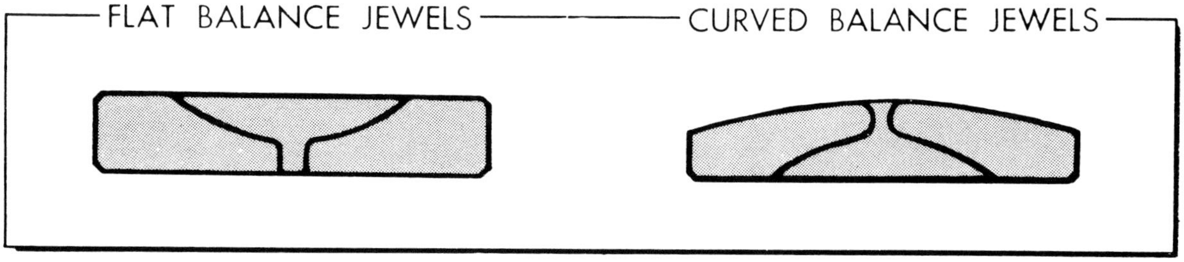

There are two types of holes in jewel bearings:

1. The regular straight hole which makes a sharp corner with the top and bottom surface of the jewel. See Fig. 5.

2. The olive hole which is made by grinding off the sharp corners at the top and bottom of the hole. See Fig. 6.

This process has the effect of reducing the area of contact between the jewel hole sides and the pivot which turns in the hole.

Straight holes are usually used for train pivots, and olive holes are preferred for escape wheel, pallet and balance pivots.

Fig. 5 *Fig. 6*

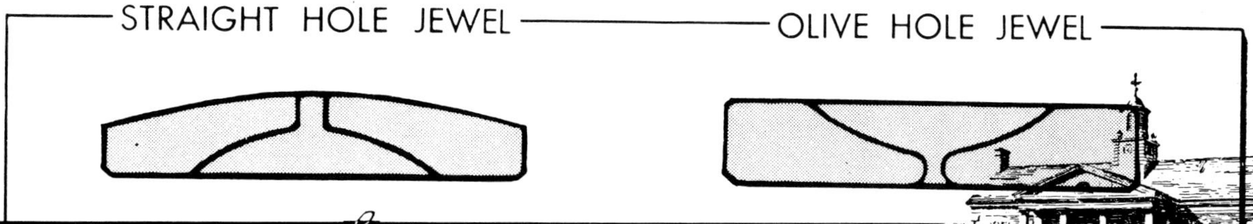

Balance hole jewels are always backed with cap jewels (also called endstones) such as those shown in Fig. 7, and in the higher jeweled watches, the escape wheel and pallet arbor jewels are also made this way. Fig. 8 shows how the olive hole jewel and endstone assembly appears in the watch plate.

These endstones make it possible for the pivot of the balance or arbor to rotate on its end when in a horizontal position. A steady lubrication for the pivot is dependant on the staying power or capillary action of the oil at the pivot. This, of course, will be considerably influenced by the type of endstone used. The flat endstone shown at A, Fig. 7 provides a flat surface on which the pivot can rotate, but the oil may run off into the seam between the metal setting and jewel. As mentioned above, the curved endstone shown at B in Fig. 7 surpasses the flat endstone, but the proper theoretical curve must be maintained. Practical experience has proven such a consistency in the curve is often found to be lacking, with the result that the pivot may not be rotating at the dead center of the curved surface. In other words, the pivot end will tend to wear on one side because it is not centered over the flattest or most central part of the curved endstone.

However, best results have been obtained with the latest development in endstones shown at C, Fig. 7. With this type of endstone, a roomy, flat surface is always present, and if the oil should tend to flow towards the metal setting, then the capillary action of the oil will keep it from going beyond the start of the beveled edge of the endstone.

Fig. 7

TYPES OF ENDSTONE

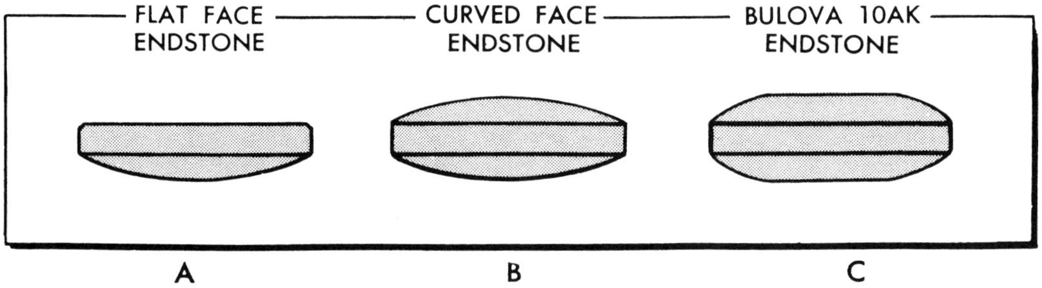

Fig. 8

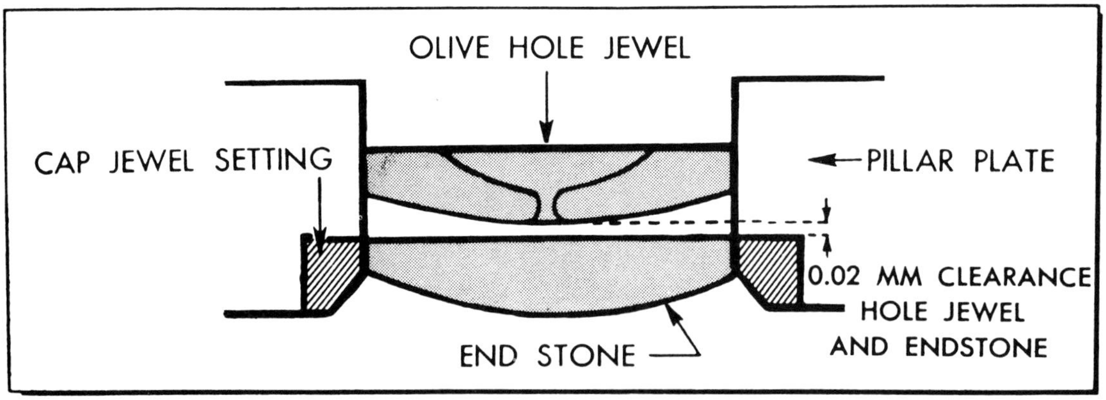

In the last few years, the leading watch manufacturers have adopted the improved system of friction jeweling in place of the older and less efficient method of setting the jewels in individual brass or gold settings. Not only is the friction jewel process a time saver for the manufacturer and watch repairer alike, but the friction jewel is more rugged than its predecessor, the bezel jewel. Being much thicker throughout its entire diameter, the friction jewel will be less apt to chip or crack than the older type (see Fig. 2) with its thin edge. Another valuable advantage of the friction jewel is the ease with which it may be shifted in the plate hole to obtain the desired endshake.

The jeweling tool consists of a lever device, a set of reamers, a set of pushers, and an assortment of stumps. Before inserting the jewel, the hole in the plate must be reamed out perfectly straight by a precision-gauged self-centering reamer. These reamers are numbered to correspond to the diameters of the jewels which may be purchased for this purpose. Actually, the reamer is .01 mm smaller in diameter than the jewel. The jewel can then be pressed to any required depth by means of the jeweling lever which applies a pressure of 30 to 35 lbs. to force the jewel in and 6 to 15 lbs. to push out the jewel. A micrometer stop adjustment or depth gauge controls the depth to which the jewel is to be inserted.

Various types of jeweling tools are available for the purpose of removing and replacing broken jewels. All of them are good and serve the purpose well. Fig. 9 shows a jeweling tool and its accessories.

Fig. 9

JEWELING TOOL

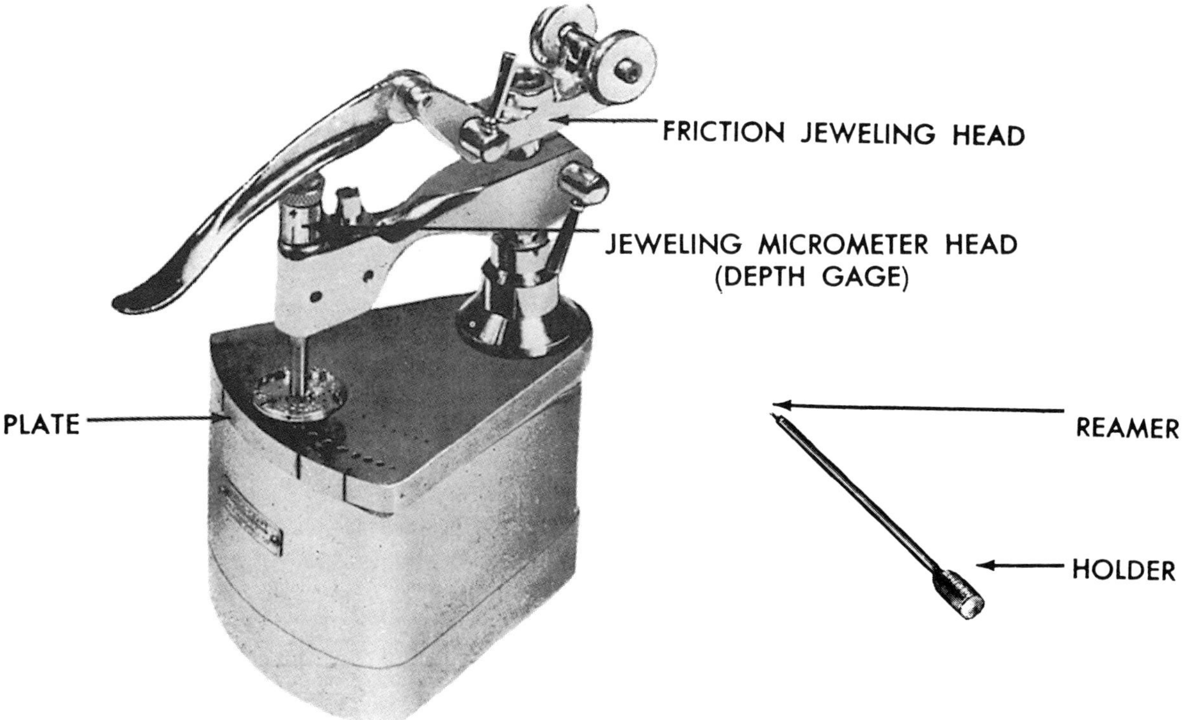

HOW TO USE THE FRICTION JEWELING TOOL

A. *When the plate is being fitted with a friction jewel for the first time,*

Select a self-centering reamer which corresponds to the diameter of the available jewel. Its number will be the same as that on the jewel bottle. (If no system is being used the reamer must be .01 mm smaller than the jewel.)

Insert reamer in tool and ream plate hole to size.

If the jewel is more than .10 mm larger in diameter than the plate opening, step up the reaming job .10 mm at a time.

Fig. 10

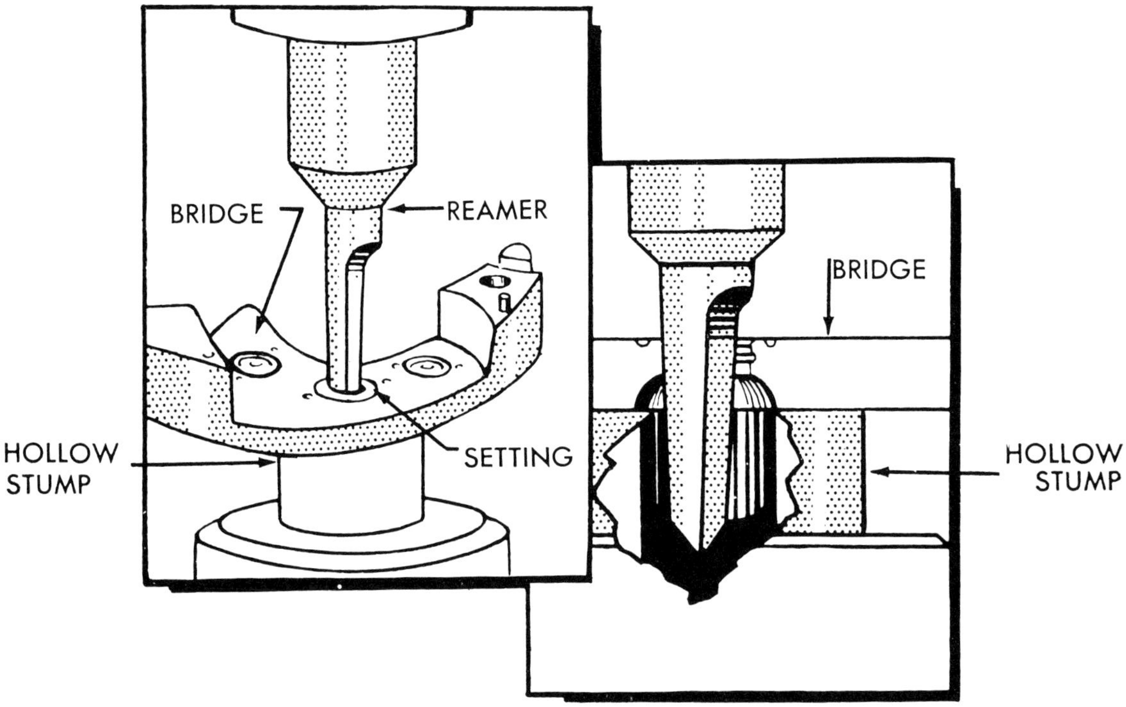

Remove burr at edge of the hole with a countersinking tool. (See Fig. 11).

Fig. 11

Wheel Countersink for finishing a reamed hole

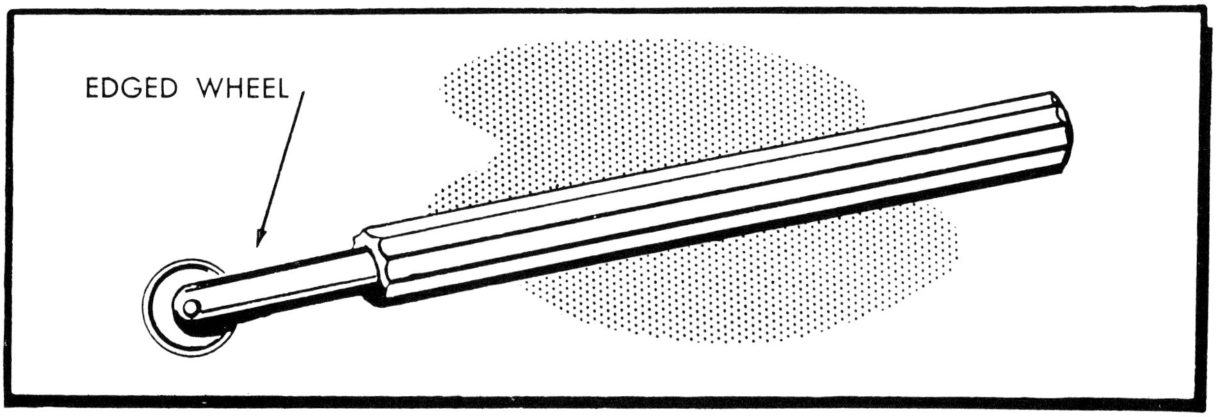

Select proper jewel to correspond with reamer number. Pivot hole must be checked with diameter of wheel pivot for proper side shake.

When the face of the jewel is to be *flush* with the plate or bridge insert jewel by using a pusher larger in diameter than the plate hole (A-B Fig. 12).

When the face of the jewel is below flush with the plate or bridge, use a pusher smaller in diameter than the jewel (C, Fig. 12).

Fig. 12

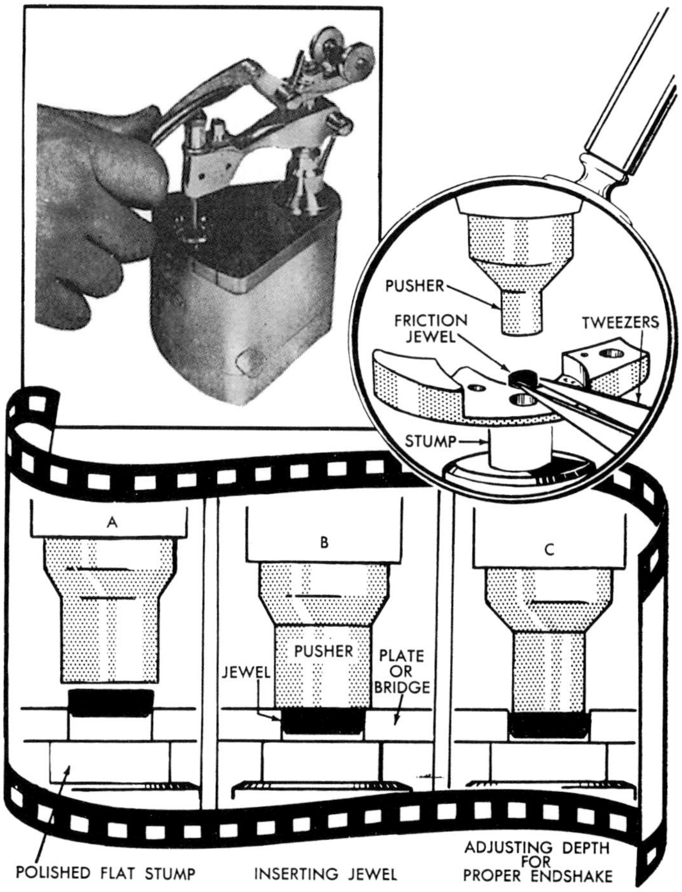

Check endshake of wheel mounted between the plates.

Endshake can be adjusted by raising or lowering the jewel in the *pillar* plate. (Keep bridge jewels the same depth).

Examine jewel for chips or cracks. Also check to see that the jewel is parallel with the plate.

B. *When a broken jewel is being replaced,*

1. Insert the pusher so that it rests on the old jewel, and then adjust the micrometer stop. (See Fig. 13.)

Fig. 13

MEASURING THE DEPTH OF BROKEN JEWEL

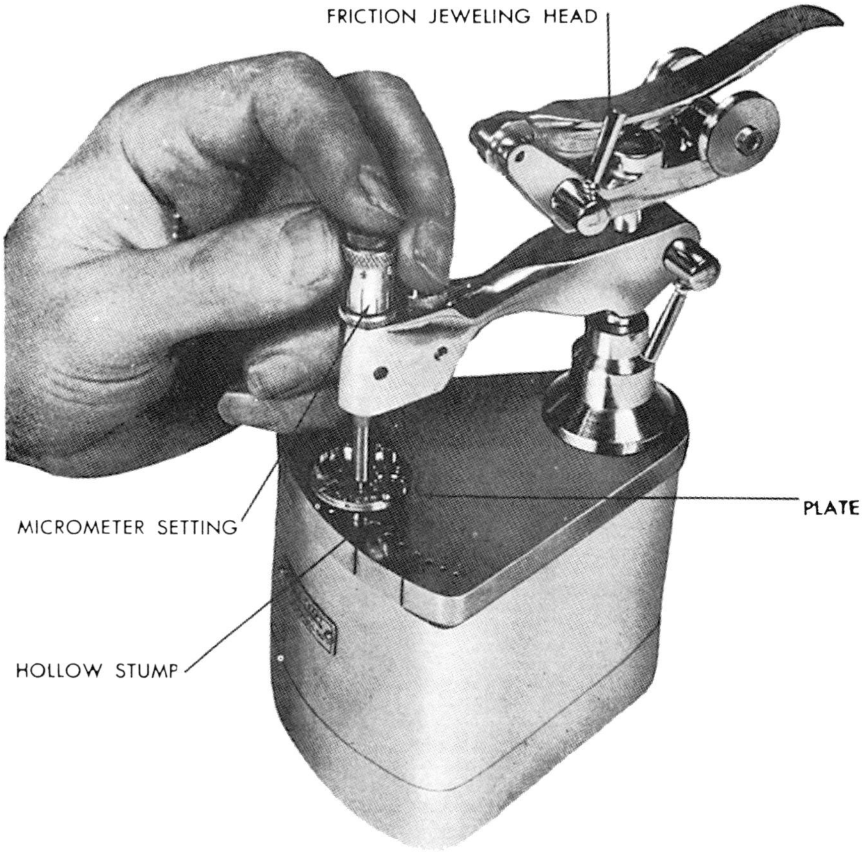

2. Remove pusher and cracked jewel.
3. Insert reamer which is just slightly la.ger than hole.
4. Ream hole to size. (See Fig. 10).
5. Remove burr with countersink.
6. Select proper fitting jewel to correspond with reamer number.
7. Insert jewel to depth determined by micrometer stop.

The remaining steps are exactly the same as those wherein the jewel is being inserted in a plate for the first time.

In many models of watches, instead of inserting the jewel directly into the plate and bridges, the jewel is first mounted in a setting and then the jewel and setting assembly is fitted friction tight into the plate. In such cases, the procedure of inserting the jewel in the setting or adjustment of endshake by raising or lowering the jewel setting in the pillar plate is accomplished in exactly the same way as the unmounted friction jewel. See Fig. 14.

Fig. 14

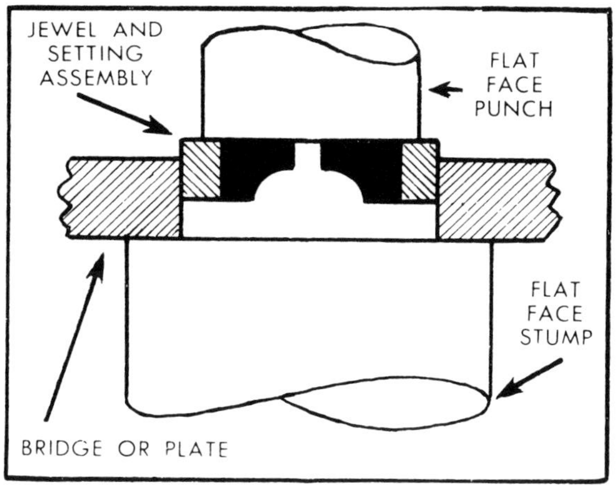

However, when a balance hole jewel is being replaced in its old setting, a clearance of .01 to .02 mm must be allowed from the bottom of the hole to the outer edge of the setting. This will insure the proper capillary action of the oil when the endstone is in place against the pivot.

In Fig. 15, it is demonstrated how careful measurements must be made so that the jewel is .02 mm below flush with the balance endstone cap.

Fig. 15

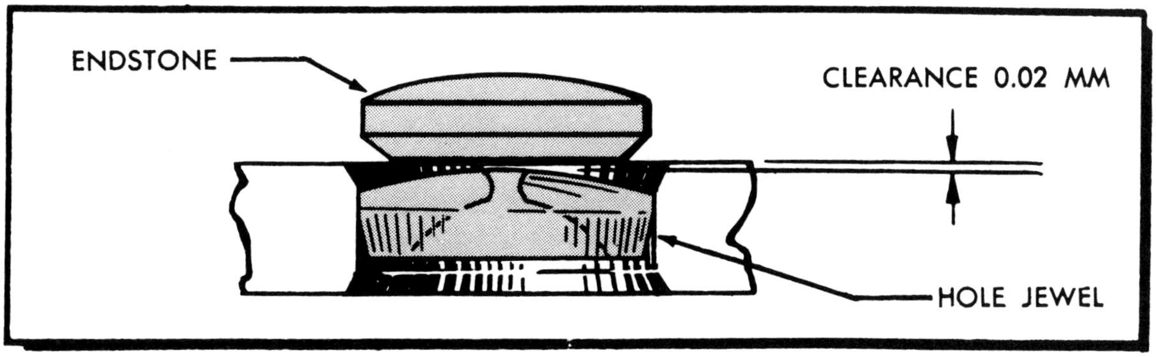

Occasionally it will be found that even a genuine factory made jewel setting will appear to be loose fitting. This can be due to careless handling or frequent removal and replacement of the setting. In such cases, the setting may be "spread" by tapping it on a staking stump with an inside taper punch. As shown in Fig. 16, the diameter of the punch must be wider than that of the jewel setting. The edge of the setting is thus rolled over, increasing its diameter sufficiently to make the necessary friction.

Fig. 16

"Spreading" a Jewel Setting

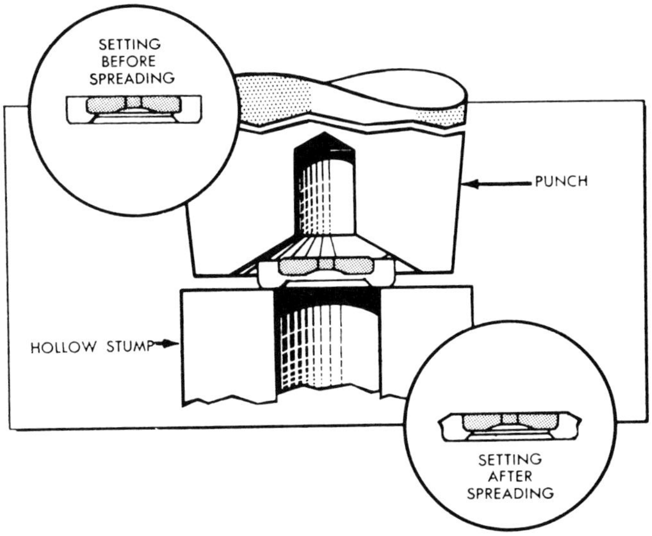

Fig. 17 Definitions

The terms "outside" and "inside" in the following description refer to the locations shown in figure 17. Inside designates a position between the plates and bridge.

Fig. 17

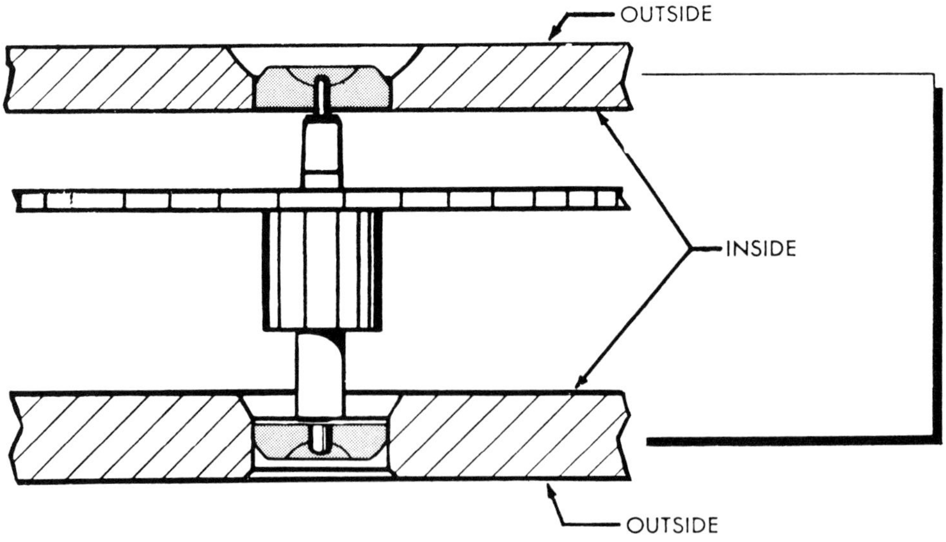

When replacing plate or bar jewels, it is good practice to remove the old jewels by driving from the outside as illustrated in Fig. 18.

Fig. 18

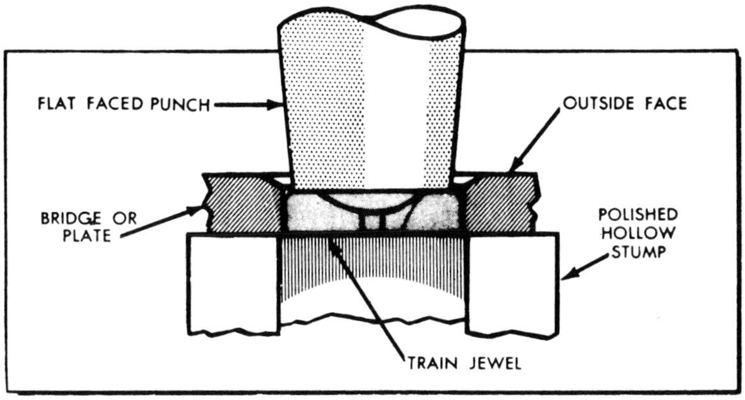

The new bar jewel should be inserted from the inside as shown in Fig. 19.

Fig. 19

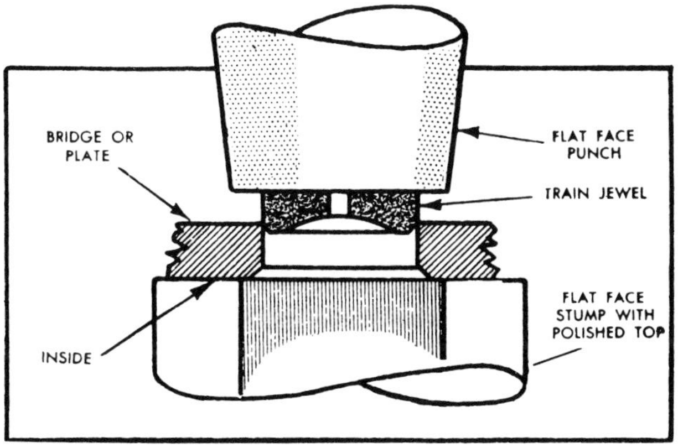

Balance olive hole jewels should be driven out from the inside. (See Fig. 20).

Fig. 20

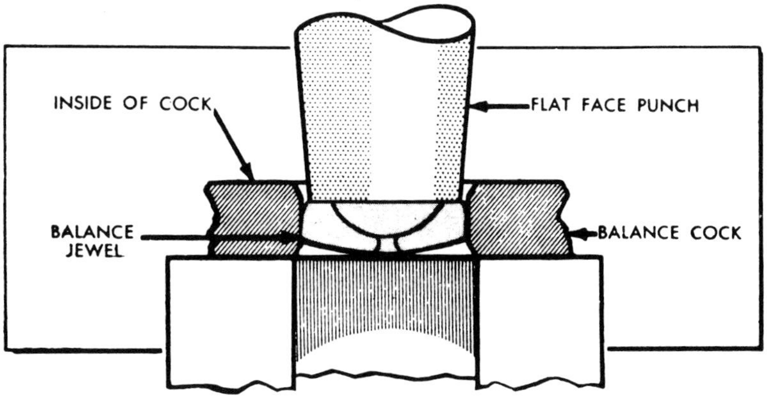

When balance olive hole jewels are being replaced, they should be pushed from the outside with a flat punch with a hole in the center slightly larger than the size of the hole in the jewel. (See Fig. 21.)

Fig. 21

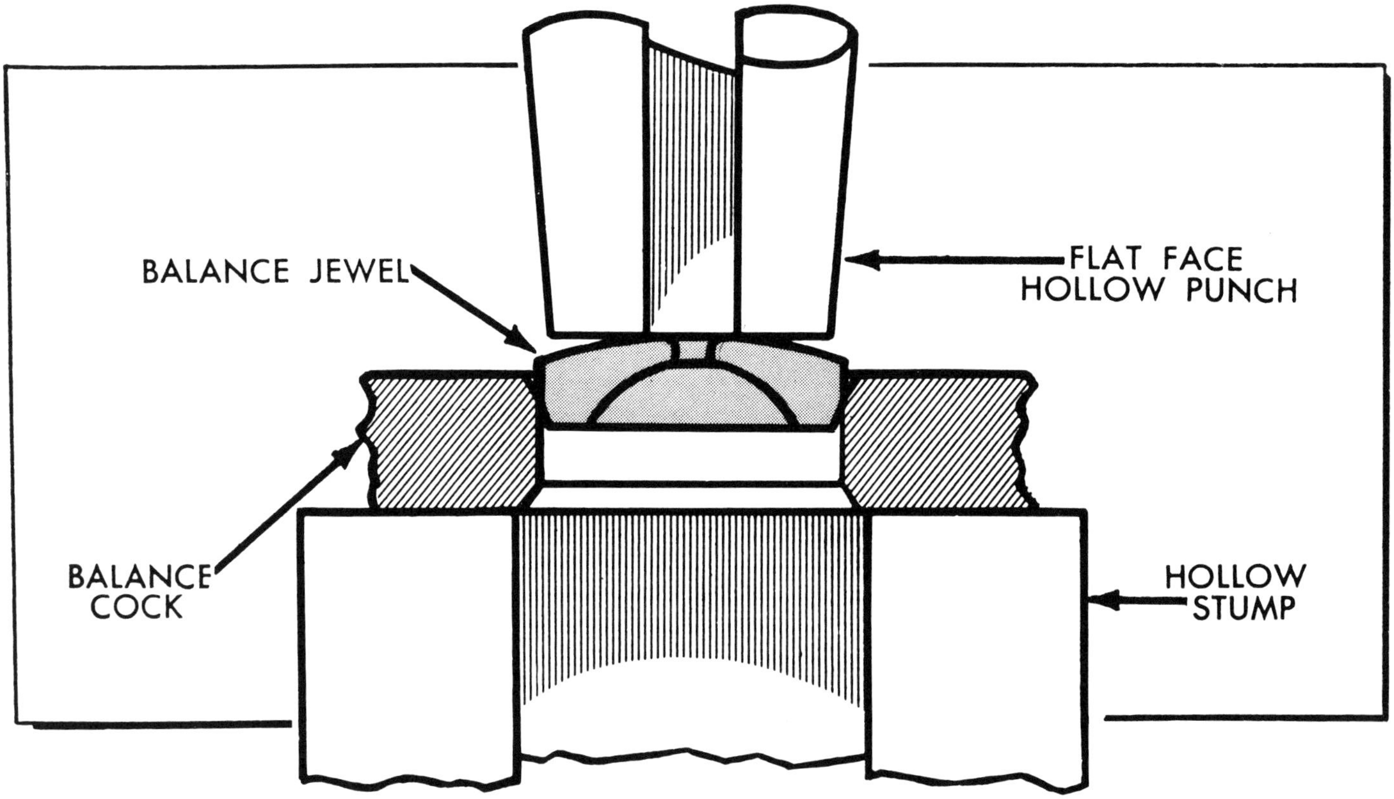

With careful handling on a standard staking or friction jeweling tool, it will be seldom necessary to ream or modify the plate and bridge jewel holes. Wherever a bridge comes in contact with the stump as in Figs. 18 and 19, the finish will be preserved if a stump with a highly polished flat surface is used.

TRAINING UNIT
NUMBER 9C

★

WHEEL TRAIN ASSEMBLY

TRAIN WHEELS
TRANSMISSION

The purpose of the wheel train is to carry or transmit the power of the mainspring to the escapement. The train consists of four wheels which are the center, third, fourth, and escape wheels. These wheels are mounted on pinions or staffs which in turn are supported by pivots running in jewels or bushings. The plates and bridges hold the train wheels in their respective positions. By a system of gear ratios and a control unit known as the balance and hairspring unit, these train wheels are driven by the mainspring at varying rates of speed which determine the time of a watch. Actually, the movable markers known as the watch hands are connected to auxiliary parts of the train and so indicate time elapsed down to the second.

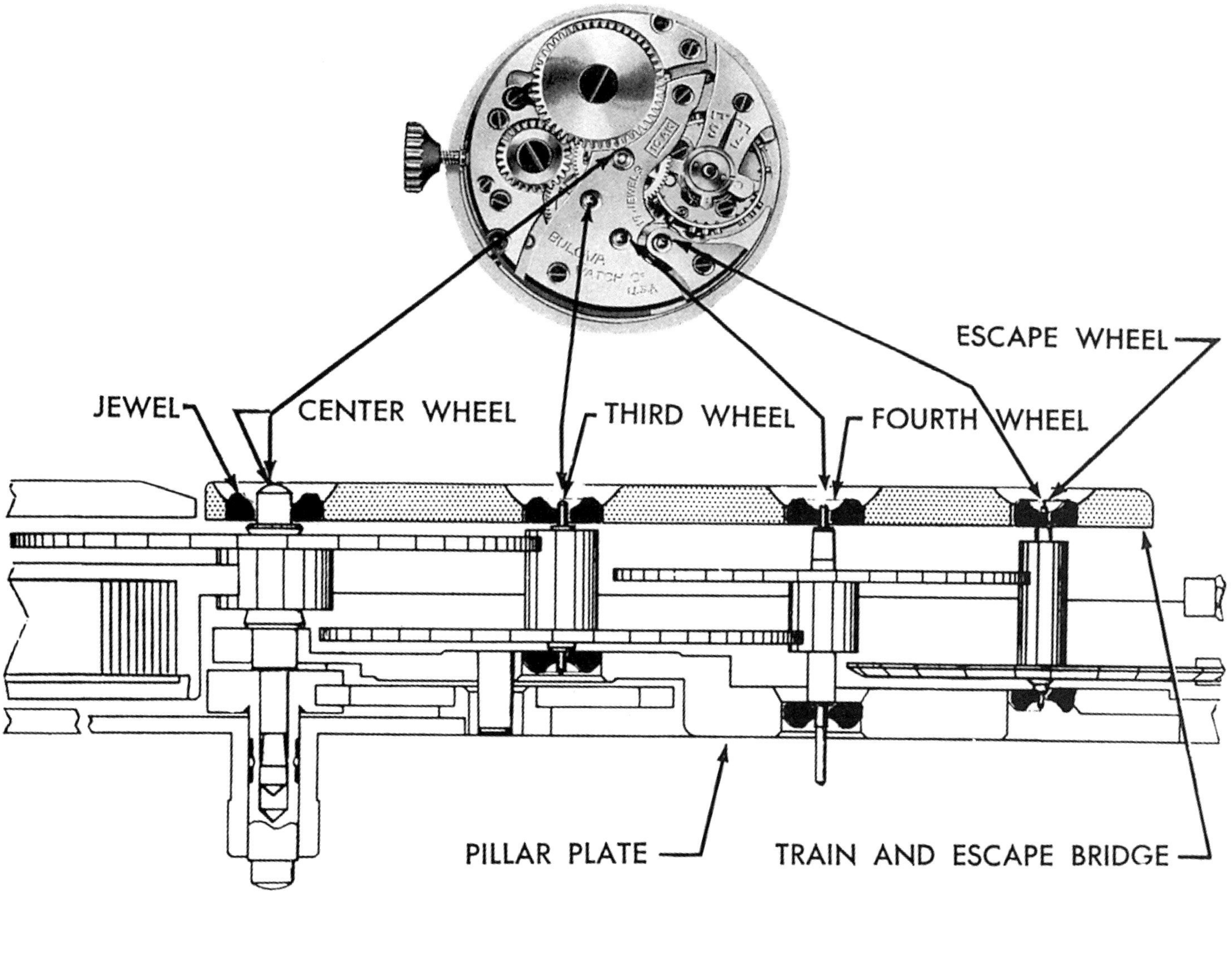
BULOVA
WATCH CO
U.S.A.
17 JEWELS
ESCAPE WHEEL
JEWEL
CENTER WHEEL
THIRD WHEEL
FOURTH WHEEL
PILLAR PLATE
TRAIN AND ESCAPE BRIDGE

BULOVA SCHOOL *of* WATCHMAKING

SUBJECT: The Wheel Train Assembly

OBJECTIVES:

1. To learn the principles of gearing and the gear ratios of the watch.
2. To become familiar with the many repair problems of the wheel train and their solutions.

In discussing gear ratios, velocities, and assembly of the wheel train, the Bulova 10AK movement will be used to illustrate these principles and procedures.

BASIC PRINCIPLES OF GEARING

Before discussing the wheel train of the watch, a few explanatory remarks on gears and gear ratios should prove helpful. For example, if it were practical to engage two perfectly smooth rollers of one inch circumference each, which we will label A and B respectively as in Fig. 1, it will readily be seen that roller A will drive roller B exactly the same distance. In other words, a 1 to 1 ratio between A and B will be set up.

Fig. 1

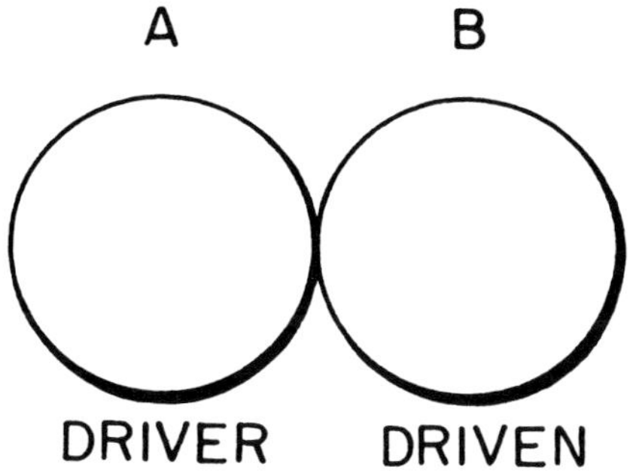

Now if a roller of two inch circumference were substituted for A, as in Fig. 2, B would travel twice as much or make 2 revolutions to each revolution of A. Here a 2 to 1 ratio between the speeds of A and B will be established.

Fig. 2

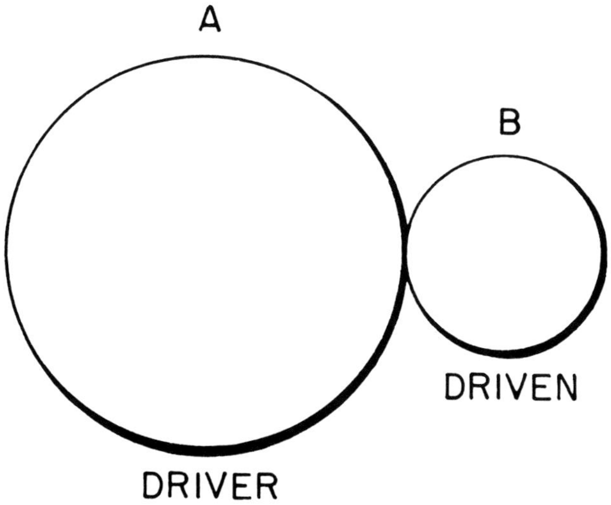

Likewise, if a roller of 2 inch circumference is used for B and A's circumference is still one inch, as in Fig. 3, then B will rotate at ½ the speed of A. In this case, the ratio between A and B would now be 1 to 2.

Fig. 3

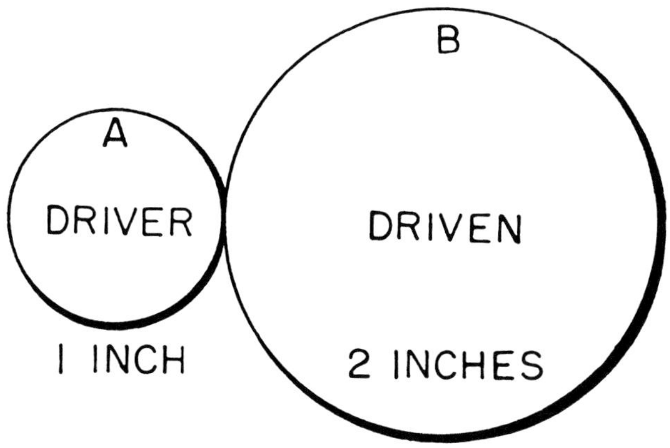

To prevent slipping and lost motion, it becomes necessary to gear the rollers with proper fitting teeth. The shape of these teeth is an important factor in obtaining a uniformity of motion, a sliding action between the teeth of two wheels engaged with one another, and a minimum amount of friction between the meshed gears.

GEARING TERMS DEFINED

The cogs on a wheel are known as teeth and those on a pinion are called leaves. Gears of 20 teeth or more are termed wheels and gears having less than 20 teeth are referred to as pinions. In Fig. 4 are illustrated the gear parts and their terms which apply to large and small gears alike.

Fig. 4

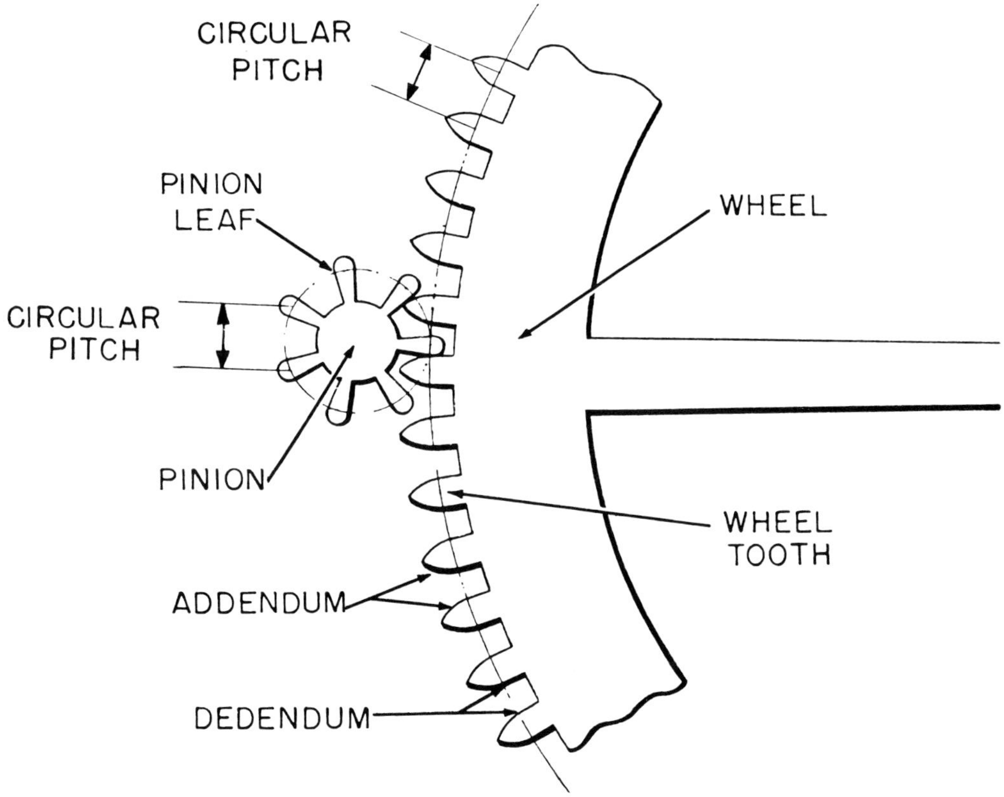

PITCH CIRCLE is an imaginary circle described from the center of a wheel to a point on its teeth which will be tangent to a corresponding circle of the meshing pinion.

ADDENDUM is the portion of the gear tooth above the pitch circle.

DEDENDUM is the portion of the gear tooth below the pitch circle.

CIRCULAR PITCH is the measurement of tooth and space. Wheels in gear must be of equal circular pitch.

PITCH DIAMETER is the diameter of the pitch circle of a wheel or pinion.

DIAMETRIC PITCH is the number of teeth for each inch of pitch diameter. For example, a gear of ½ inch diameter and 60 teeth would be a 120 pitch gear.

MODULE (used in European metric system) is the pitch diameter in millimeters divided by the number of teeth. For example, a gear of 60 teeth and 12 millimeter pitch diameter would have a module of .20.

To continue with the theory of rollers, it can now be stated that if these rollers have been geared with proper fitting teeth, then the ratio between the geared rollers or gears will also depend on the number of teeth in each. Since wheels in gear must be of equal pitch, gear ratios based on the number of teeth in each is merely using the principle of ratios between circumferences of rollers wherein the circumferences have been divided into many component parts known as teeth.

Thus as in Fig. 5, a wheel of 100 teeth will drive a pinion of 10 leaves exactly 10 times as fast.

Fig. 5

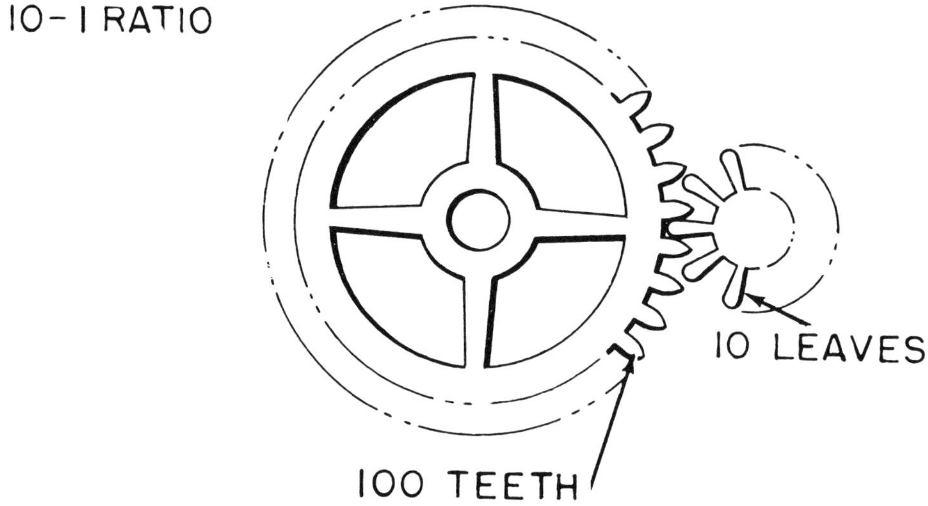

And vice versa, a pinion of 10 leaves will cause a wheel of 100 teeth to rotate at 1/10 of the revolutions of the pinion for the same period.

Fig. 6

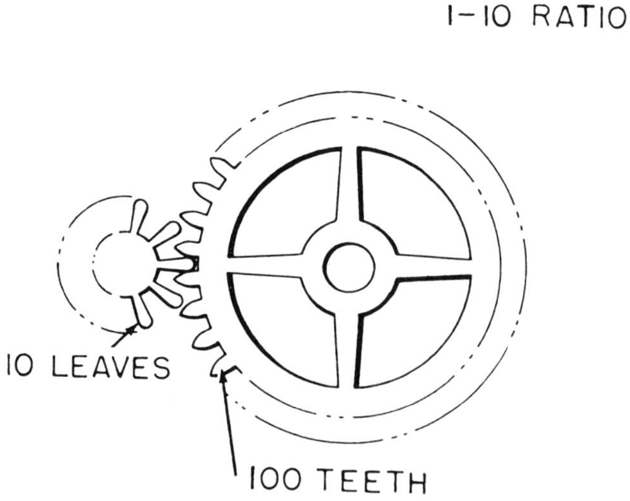

HOW THE WHEEL TRAIN FUNCTIONS

Fig. 7

DIAL UP

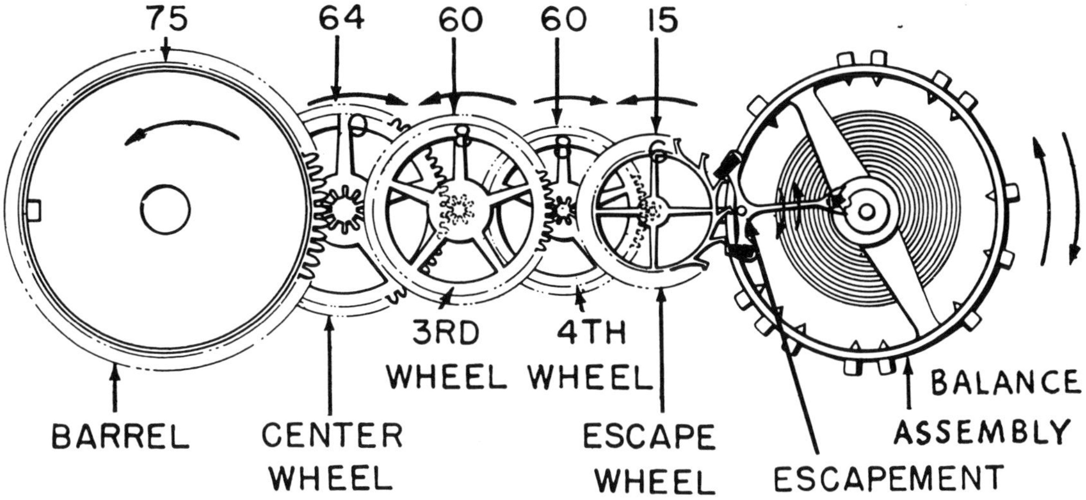

Fig. 7 represents a schematic of the barrel, wheel train and escapement of a 10 AK Bulova watch, giving the number of teeth for each wheel and pinion. It will be a simple matter to show how the ratio between these wheels and pinions was planned to obtain the proper speeds of the train wheels for indicating the passing of time. Actually, the drive is coming from the mainspring barrel, and the speed of the last unit of the train which is the escape wheel is controlled by the balance and hairspring through the escapement.

Because of this control, the center wheel which carries the minute hand makes one revolution per hour. It will be noticed in Fig. 7 that there are 75 teeth on the barrel and only 10 leaves on the center wheel pinion. This would indicate that the speed of the center wheel is 7½ times faster than that of the mainspring barrel because a ratio of 7½ to 1 has been established between the barrel and center wheel.

Similarly, because there are 64 teeth on the perimeter of the center wheel and only 8 leaves on the 3rd wheel pinion, this would set up an 8 to 1 gear ratio between these two members of the wheel train. In other words, the 3rd wheel will rotate 8 times faster than the center wheel in a counter-clockwise direction.

NOTE: Whenever two gears are engaged together, they will drive in opposite directions. For example, if in Fig. 8, A rotates clockwise, then B will rotate in a counter-clockwise direction.

Fig. 8

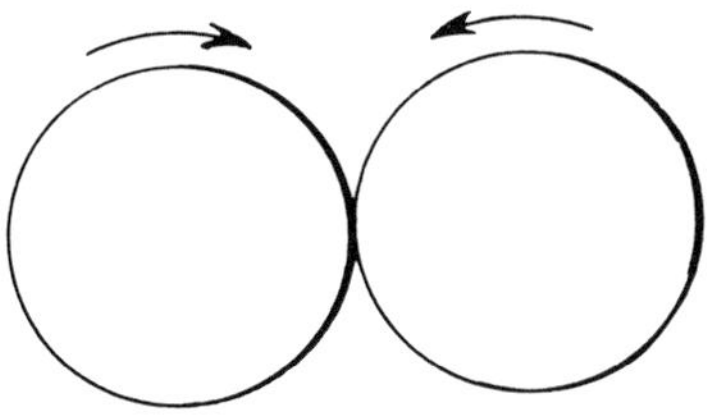

A B

However, if a ratio of 60 to 1 can be worked out between the center wheel and some other clockwise gear in the wheel train, it would then be possible to utilize this wheel by mounting a second hand on it to indicate time in seconds. This is accomplished by setting up a ratio of 7½ to 1 between the 3rd wheel and the 4th wheel. Sixty teeth have been cut into the 3rd wheel, and with only 8 leaves in the 4th wheel pinion, the 4th wheel will turn 7½ times faster than the 3rd wheel which has been shown to be making 8 revolutions for each one of the center wheel. In other words, the 4th wheel will rotate 60 times faster than the center wheel or make 60 clockwise revolutions in an hour.

$$\frac{64}{8} \times \frac{60}{8} = 60$$

The 4th wheel therefore is constructed with an elongated pivot which extends through the pillar plate and dial, and on this pivot is mounted the second hand.

On the perimeter of the 4th wheel there are 60 teeth and this against the 6 leaves of the escape wheel pinion makes a ratio of 10 to 1. The escape wheel will therefore rotate 10 times faster than the 4th wheel, and in an hour period will make 600 counter-clockwise revolutions.

Since the escape wheel is divided into 15 teeth, and it has just been shown that the escape wheel makes 600 revolutions in an hour, then it can be said that 15 x 600 or 9000 escape wheel teeth will pass a line of centers in one hour. However, each tooth of the escape wheel delivers two impulses to the pallet during one revolution of the escape wheel. It therefore follows that the pallet will receive 18000 impulses and impart 18000 vibrations to the balance wheel in a one hour period. For this reason this type of train is known as an 18000 train.

$$\frac{64 \times 60 \times 60 \times 15}{8 \times 8 \times 6} \times 2 = 18000$$

VELOCITIES OF 10 AK BULOVA WHEEL TRAIN

Barrel	*Center Wheel*	*3rd Wheel*	*4th Wheel*	*Escape Wheel*
.133 rev. per hour	1 rev. per hour	8 rev. per hour	60 rev. per hour	600 rev. per hour
.00222 rev. per min.	1/60 rev. per min.	2/15 rev. per min.	1 rev. per min.	10 rev. per min.

In examining the table above, it can be seen that the speeds of this train are designed for a 60 to 1 ratio between center wheel and 4th wheel. Although this table pertains to the Bulova 10 AK watch model alone, there are many other gear ratios on the market, but basically they all accomplish the same result with respect to the 60 to 1 ratio between center wheel and 4th wheel. Only when a watch model is built without a second hand, then of course the 60 to 1 ratio between center and 4th wheels need not necessarily be maintained.

While the 18000 train is the most used by the many watch manufacturers, there are others, known as odd beat trains which have been designed by various Swiss and American watch companies for wrist watch models. These fast odd beat trains tend to effect a more satisfactory unlocking of the escapement due to increased strength of the hairspring and a more rapid motion of the balance wheel. A greater tolerance in the lock with less tendency to remain locked is therefore possible in the small wrist watch. The most used of these odd beat trains as listed by the Watchmaster Handbook are:

19800	20994	21000
20160	20222	21600

Taking the 19800 train as an example, it can be broken down into its gear ratios as follows:

$$\frac{64 \times 66 \times 60 \times 15}{8 \times 8 \times 6} \times 2 = 19800$$

Here as well as in an 18000 train, the number of vibrations are computed by the simple formula of:

$$\frac{\text{center wheel} \times \text{3rd wheel} \times \text{4th wheel} \times \text{escape wheel}}{\text{3rd wheel pinion} \times \text{4th wheel pinion} \times \text{escape pinion}} \times 2 = \text{no. vibrations}$$

ASSEMBLING THE WHEEL TRAIN AND PALLET OF THE 10 AK BULOVA WATCH MODEL

1. Place the pillar plate in the movement ring, top side up. (Dial Down).
2. Insert escape wheel in pillar plate so that its bottom pivot rests in its jewel. The escape wheel pinion will denote the upper side of the wheel.
3. Insert 4th wheel in pillar plate so that its long pivot rests in the jewel. The 4th wheel will mesh with the pinion of the escape wheel.
4. Insert 3rd wheel in pillar plate so that its lower pivot rests in the jewel and the wheel meshes with the 4th wheel pinion.
5. Insert center wheel in pillar plate so that its long shaft will extend through the plate. The center wheel will mesh with the 3rd wheel pinion and the center wheel pinion will mesh with the teeth of the mainspring barrel (already inserted).
6. Place the train and escape wheel bridges in position. Carefully insert the top pivots of the wheels into jewels of the bridges.
7. Screw down the 3 screws in the two bridges.
8. Check the endshake and sideshake of the four wheels. If the jewels have been properly fitted, the sideshake should be right. Endshake can be adjusted by raising or lowering the jewels in the pillar plate.
9. Check for freedom of train. Also see that the wheels run true.

Continuing with Assembly:—

10. Insert pallet in bottom hole jewel with the guard pin on lower side.
11. Place pallet bridge in position.
12. Holding the bridge jewel over the top pivot with piece of pegwood, screw down pallet bridge.
13. Check pallet for endshake and sideshake. The pallet stones must make contact with the escape wheel.

Listed below are the proper endshakes and sideshakes for the parts assembled up to this point:

Barrel Arbor — in barrel
Endshake — .015 to .02 m.m.
Sideshake — .015 m.m.

Barrel arbor — in plates
Endshake — .015 m.m.
Sideshake — .015 m. m.

Center wheel
Endshake — .02 m.m.
Sideshake — .015 m.m.

Third Wheel
Same as center wheel

Fourth wheel
Same as center wheel

Escape wheel
Endshake — .02 m.m.
Sideshake — .015 m.m.

Pallet arbor
Endshake — .015 m.m.
Sideshake — .0075 m.m.

ASSEMBLY OF 10 AK BULOVA WHEEL TRAIN AND BARREL BEFORE BRIDGES HAVE BEEN INSTALLED

Fig. 9

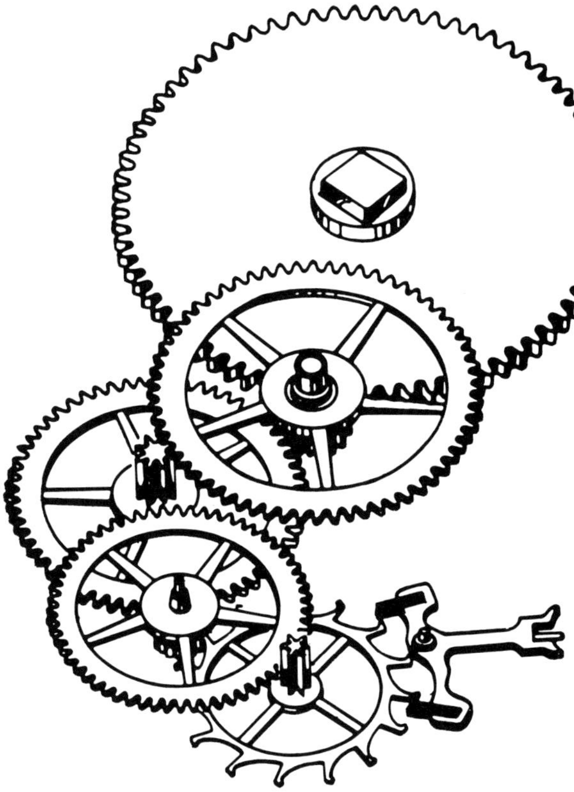

RECOIL ACTION OF THE ESCAPE WHEEL

After the wheel train has been completely assembled and the endshakes and sideshakes of the wheels have been carefully checked, the freedom of the train may be further ascertained by observing the recoil action of the escape wheel. In a well constructed and serviced watch in which the pivots are well polished, the jewel holes are round and polished, and the depthing is right, the escape wheel will momentarily reverse its direction immediately after the mainspring has become unwound. This recoil action should give positive assurance that the wheel train is free, and will therefore fully perform its function of transmitting the mainspring's driving power.

TRUING A TRAIN WHEEL IN THE FLAT

The rotation of the wheel train also makes it possible to observe whether the wheels are true in the flat. Obviously, a wheel that is not true in the flat will appear to wobble as it is turning. When such a condition is detected, the untrue wheel must be straightened to insure best results.

When a small wheel requires flat truing, it can be removed from the movement and inserted in a balance truing caliper. The truing procedure shown in Fig. 10 is similar to that of balance truing except that tweezers may also be used here.

Fig. 10

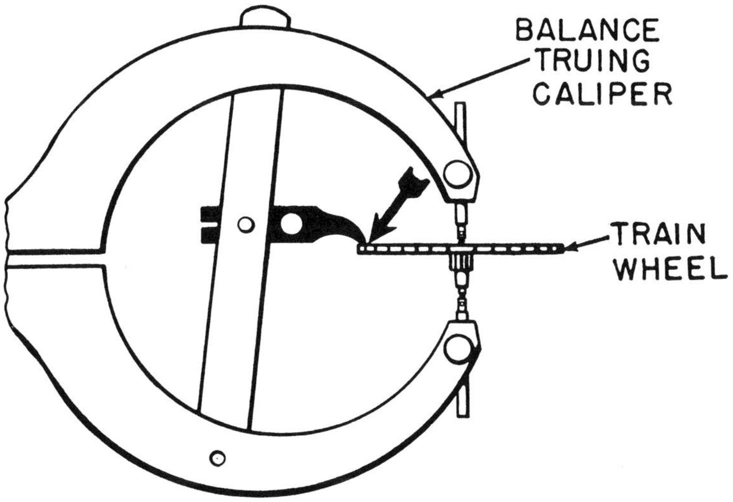

If the pivots are too large for a balance truing caliper, the wheel may be trued between the bridge and pillar plate as the wheel train is being rotated. See Fig. 11.

Fig. 11

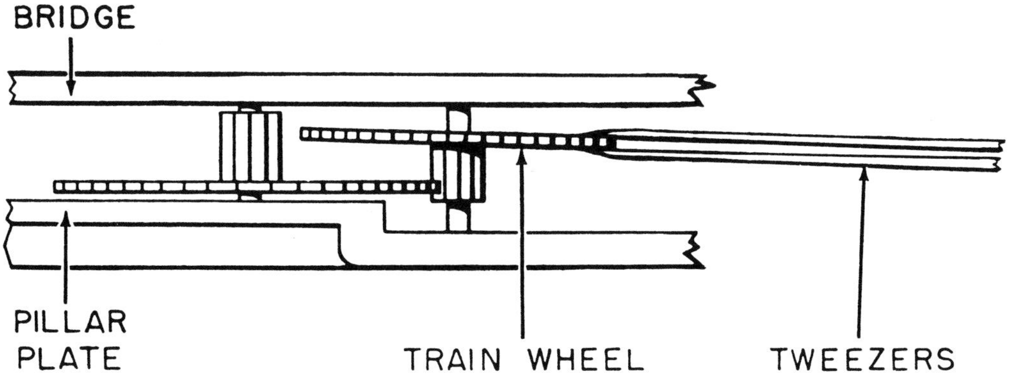

STAKING WHEEL AND PINION

If a wheel comes off its pinion, it can be restaked to its pinion. In this operation, shown in Fig. 12, the pinion is positioned on its shoulder in the staking tool. After carefully centering the wheel over the pinion, a flat punch with hole slightly larger than the wheel shoulder of the pinion is used to tap the wheel flush against the seat of the pinion.

Fig. 12

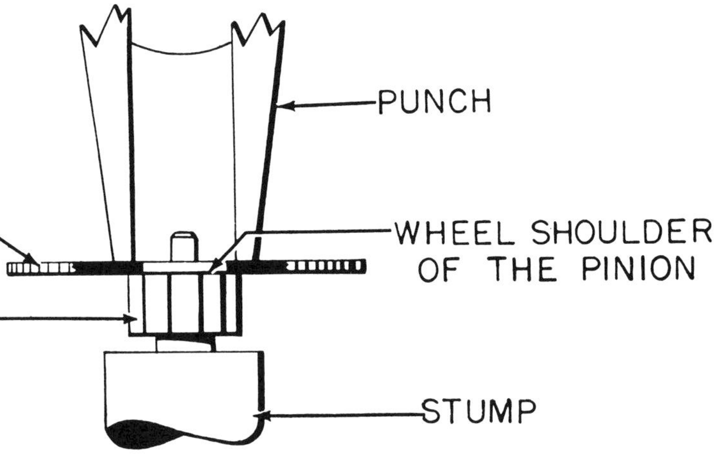

If the wheel hole is too large for the pinion, its diameter may be somewhat reduced by using an inside taper punch which is slightly larger in diameter than the wheel hole. Fig. 13 shows the wheel centered under the punch on a flat polished stump.

Fig. 13

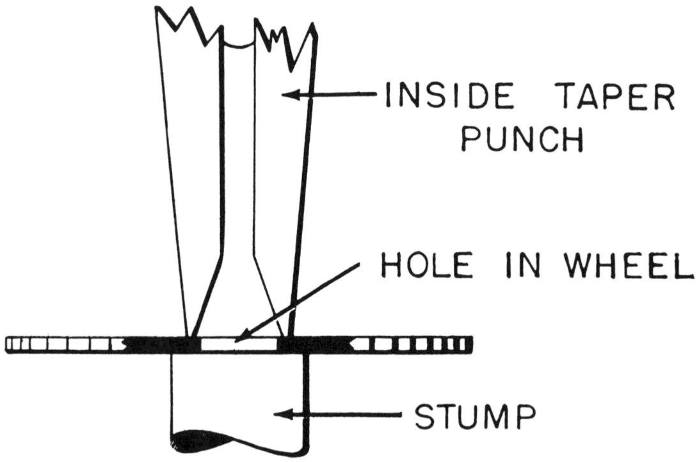

After the swaging operation, it should now be possible to stake the wheel to the pinion.

THE SAFETY PINION

When a mainspring breaks, a recoil action occurs in which the barrel is suddenly snapped back in a reverse direction. This will often result in serious damage to the teeth of the barrel, especially in the case of a large watch with a strong mainspring. As a safeguard against such possibility, many of the old 18 size American watches were equipped with a safety pinion threaded to the center wheel shaft. See Fig. 14.

Fig. 14

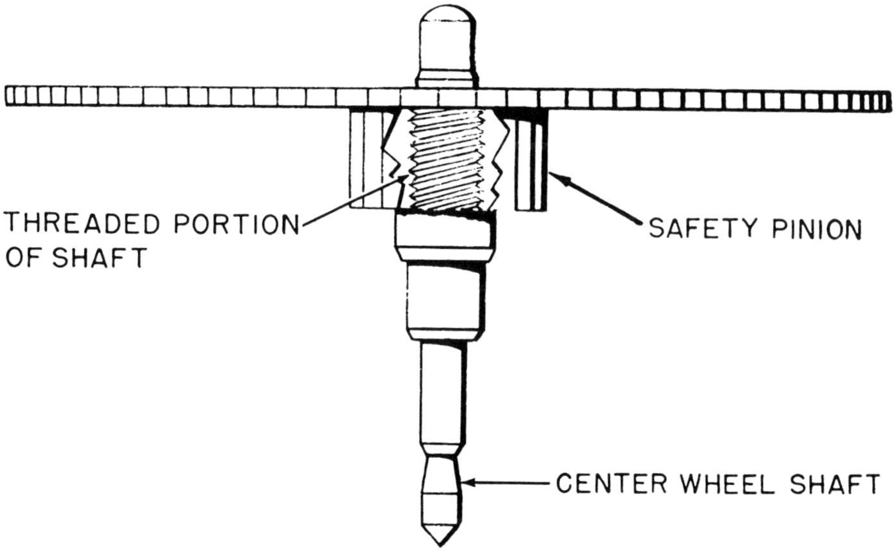

As the mainspring barrel drives the center wheel pinion and wheel train, a torque is applied to the pinion in the direction of its threads, causing it to remain fixed to the center wheel shaft. However, if the barrel drive should suddenly reverse its direction as when a mainspring suddenly breaks, the torque on the safety pinion will likewise be reversed. In this way, any strain on the barrel teeth due to this momentary shock will be absorbed in unscrewing the pinion from its shaft.

BULOVA SCHOOL *of* WATCHMAKING

TRAINING UNIT
NUMBER 10

★

ESCAPEMENT

ESCAPEMENT

The escapement is that unit of the watch which serves as the connecting link between the wheel train assembly and the control unit or balance wheel assembly.

Although the balance wheel and hairspring acting together constitute the timekeeping unit of the watch, the accuracy of this time keeping is also dependent upon the condition of the escapement. For this reason, it is first necessary to have an intimate knowledge of all the actions involved in an escapement before attempting to make adjustments.

BULOVA SCHOOL *of* WATCHMAKING

SUBJECT:

Escapement

OBJECTIVES:

1 To understand how the escapement functions.

2. To check a properly matched escapement.

The type of escapement that will be studied in this unit is known as the Detached Lever Escapement. Its three basic components are:

1. The escape wheel.
2. The pallet and fork.
3. The impulse and safety rollers.

HOW THE ESCAPEMENT FUNCTIONS

When energy is stored in the mainspring, by winding, the train wheels begin to turn. This energy is carried from the barrel through the center, third and fourth wheels to the escape wheel, turning the escape wheel in the direction of the arrow shown in Fig. 1.

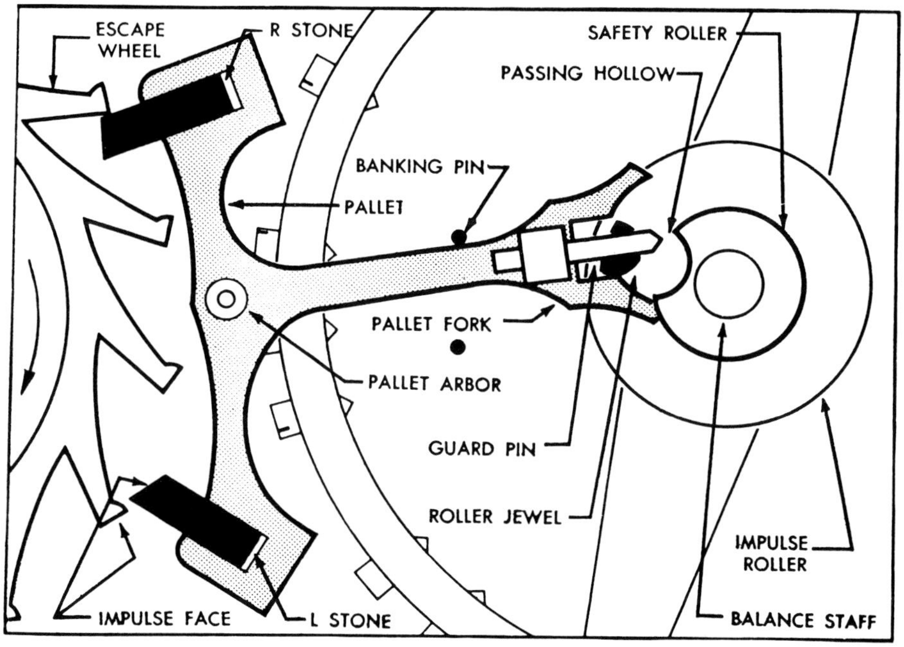

Fig. 1

Considering the escapement as a unit, it can be seen that the escape wheel supplies the power, whereas the pallet transfers the rotary motion of the escape wheel into a back and forth motion of the pallet.

The pallet in turn imparts motion to the jewel pin (roller jewel) which causes a vibrating motion of the balance wheel.

DETAILED SEQUENCE OF ESCAPEMENT ACTIONS

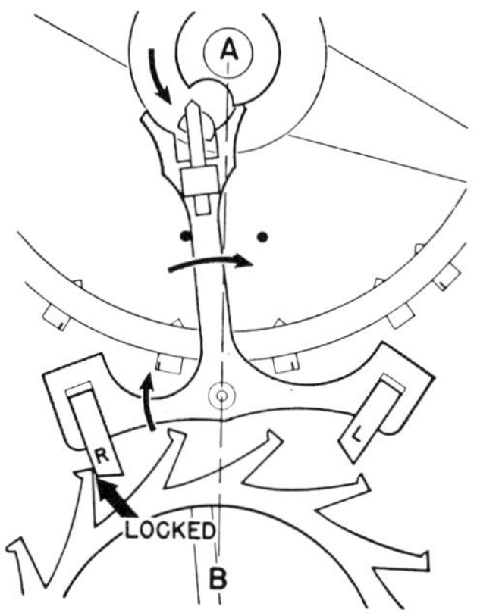

Fig. 2A

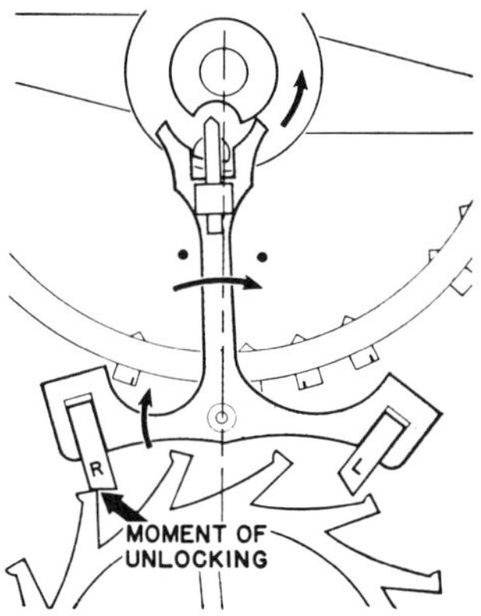

Fig. 2B

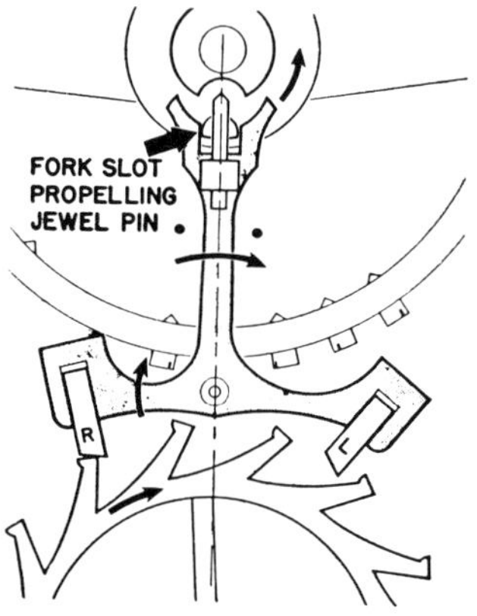

Fig. 2C

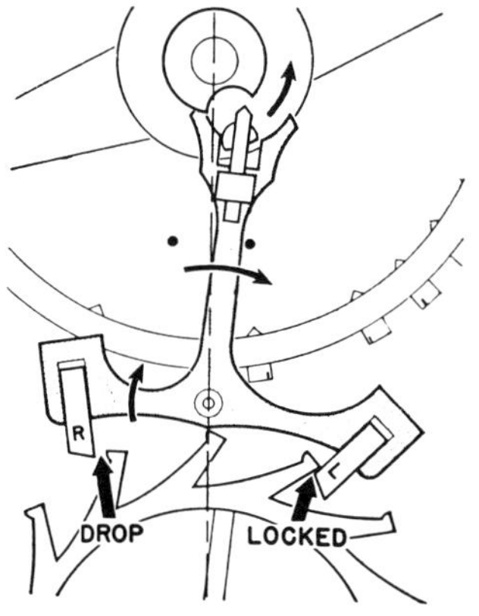

Fig. 2D

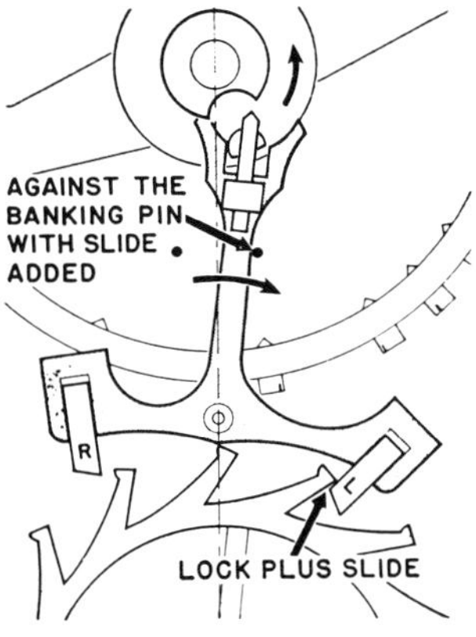

Fig. 2E

As the balance wheel swings, the jewel pin is carried back and forth across the center line of the escape wheel, pallet and balance wheel. Line AB in Fig. 2A.

As the jewel pin approaches the center line and when in the position shown in Fig. 2A, it enters and strikes the inside wall of the fork slot.

During the passage of the jewel pin over the center line, the fork is moved in the direction of the arrow (Fig. 2B), causing the R pallet stone to unlock an escape wheel tooth.

At the instant of unlocking, the escape wheel tooth is released, which creates an impulse on the R pallet stone. This impulse forces the pallet stone out of the path of the tooth. (Fig. 2C.)

The impulse begins at the instant of unlocking. Instead of the jewel pin driving the fork, as was the case in the unlocking period, the fork imparts a propelling force against the jewel pin. This impulse action supplies the energy to maintain the balance wheel in motion.

After the escape wheel tooth slides off the pallet stone, it travels a short distance, known as drop, until another tooth locks on the opposite pallet stone. (Fig. 2D.)

The pallet is caused to move slightly farther than the locking point because of the draw. Draw may be described as a force which tends to hold the pallet against the banking pin. It is the result of the force exerted by the escape wheel tooth on the locking face of the pallet stone. The amount of additional movement of the stone on the tooth is called slide. (Fig. 2E.)

The balance wheel turns until the energy stored in it from the impulse is overcome by the tension in the hairspring, pivot friction and air resistance.

This entire sequence of movements will be repeated when the balance wheel and jewel pin under the influence of the hairspring return to the fork slot.

Each time the jewel pin enters the fork slot, one vibration of the balance wheel has been completed. Two vibrations of the balance constitute one oscillation.

In the Bulova movement, the escape wheel rotates 10 revolutions per minute or 600 revolutions per hour. Since the escape wheel has 15 teeth, it is obvious that 15 X 600 or 9000 escape wheel teeth will pass a given point in an hour. During a revolution of the escape wheel, each of its teeth delivers two impulses to the pallet. Thus in one hour, 9000 escape wheel teeth will deliver 9000 X 2 or 18000 impulses to the balance wheel. Since each impulse causes one vibration of the balance wheel and there are 18000 impulses in a one hour period, the balance wheel will vibrate 18000 times an hour, 300 times per minute, or 5 times per second. Because of this, the power of the mainspring is arrested and released every 1/5 second by the locking and unlocking action of a pallet stone with an escape wheel tooth.

EXAMINATION OF THE FEATURES WHICH ALLOW AN ESCAPEMENT TO FUNCTION

TYPE I: ESCAPEMENTS WITH FIXED BANKING PINS

This method of examination is based upon the assumption that the jewel pin and both pallet stones are correctly located. A well adjusted watch movement, preferably a new one, with fixed banking pins should therefore be used for making this examination.

It is also arranged so as to use the fundamental principles already described, in a manner whereby the condition of the jewel pin shake will provide us with all of the knowledge necessary to make observations relative to the lock, drop, and correct or incorrect position of the pallet stones.

In the detailed examination of an escapement, the hairspring may remain attached to the balance wheel, eliminating the necessity for any disassembly of the movement. This greatly simplifies the work involved in passing judgment on the safe or unsafe condition of any escapement.

Note: In all the sketches shown in this unit, the fork and roller are shown upside down to show the impulse action more clearly.

Training Unit Number 10

Before proceeding with any further discussion of the escapement, it should be mentioned that there are three distinct and different terms used in connection with the jewel pin, namely:

1. Jewel Pin Freedom

Which is the difference between the width of the fork slot and the width of the jewel pin. (Fig. 3.)

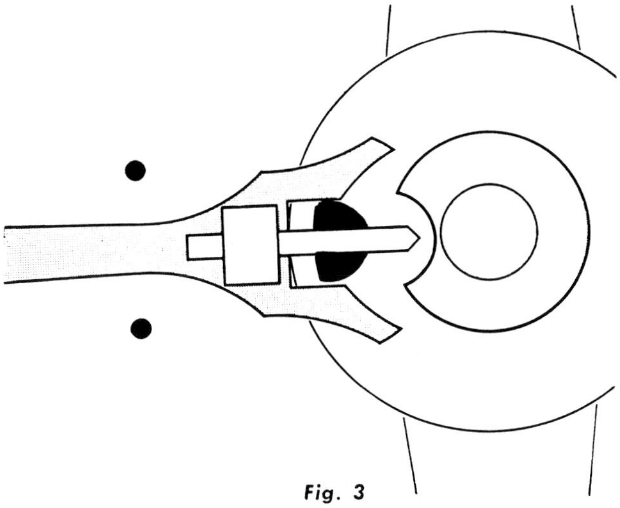

Fig. 3

2. Jewel Pin Shake

Which is the space arrangement between the inside corners of the pallet fork and the front and back of the jewel pin at the instant of "drop." (Fig. 4) Also refer to Fig. 10.

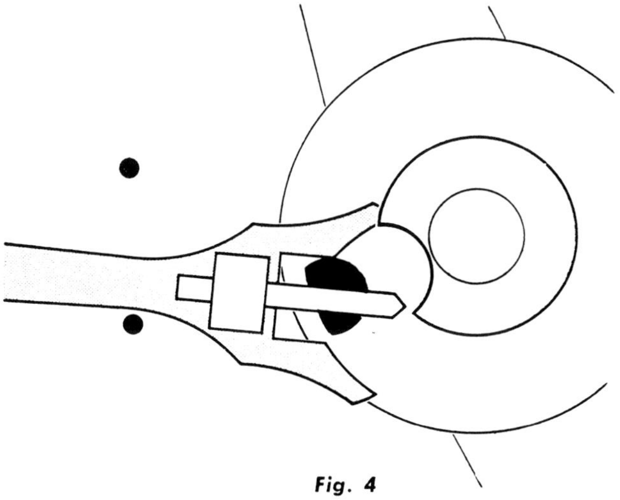

Fig. 4

3. Jewel Pin Clearance

Sometimes called fork horn clearance. This is the clearance for the jewel pin in passing out and past the fork horns. (See Fig. 5.)

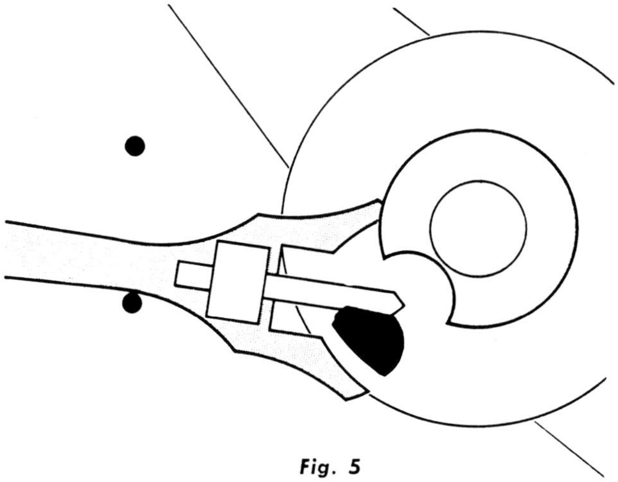

Fig. 5

The three terms relative to the jewel pin must always be used correctly.

EXAMINATION OF A CORRECT ESCAPEMENT

(Based upon the assumption that the jewel pin and both pallet stones are correctly located or set.) See Fig. 6.

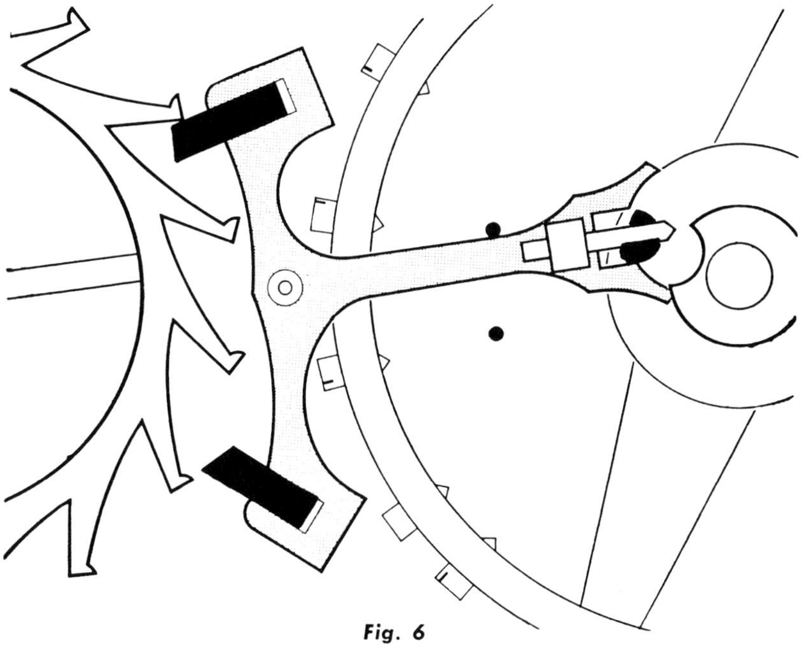

Fig. 6

Move the balance wheel slowly with the first finger, causing the jewel pin to enter the fork slot.

Observe carefully with a double eye loupe that this operation unlocks the escape wheel tooth. (Begin from the R stone side.) At the instant the tooth is unlocked, the impulse action begins. (See Fig. 7.)

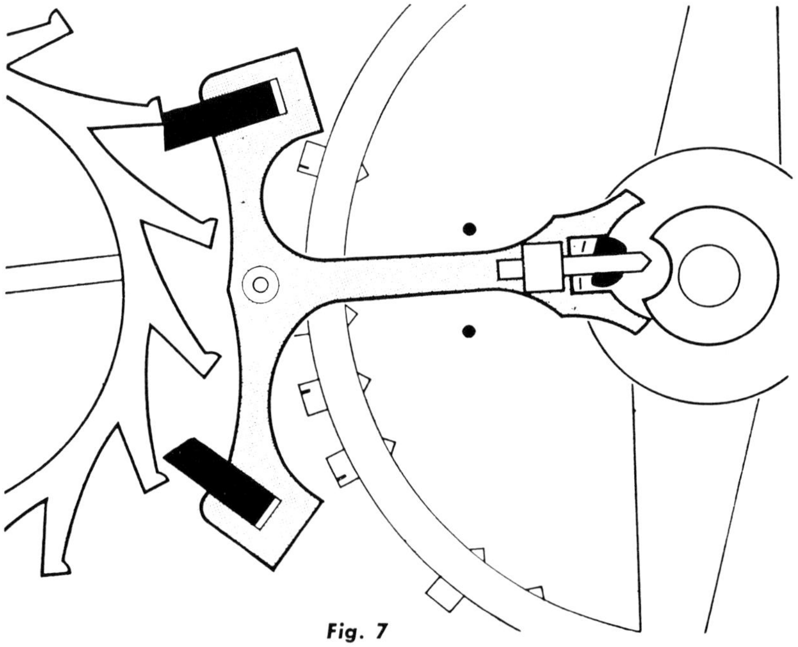

Fig. 7

The balance wheel should be turned slowly enough so that the travel of the escape wheel tooth across the pallet stone can be clearly observed. (See Fig. 8.)

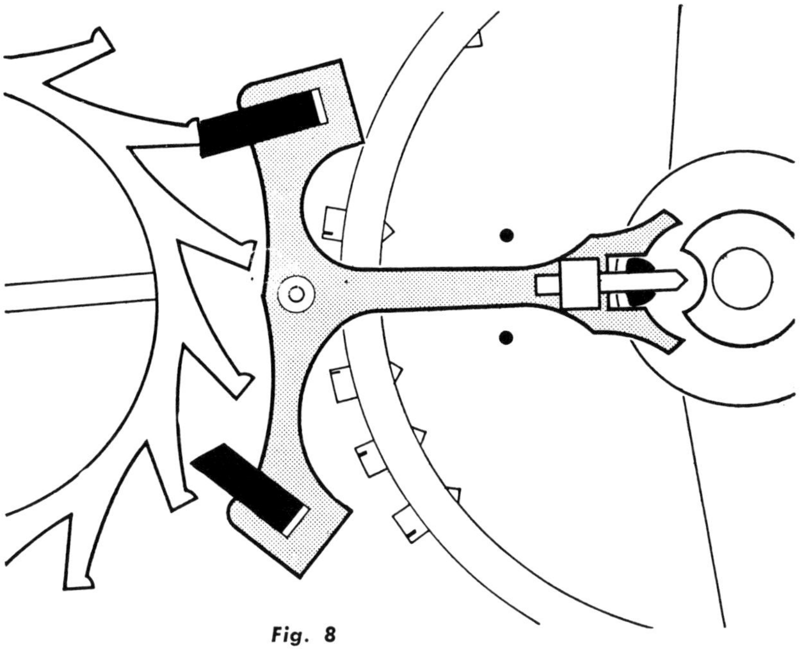

Fig. 8

JEWEL PIN SHAKE

At the exact instant the escape tooth "drops" from the pallet stone, another tooth will "lock" itself on the other pallet stone. The motion of the balance wheel and jewel pin must be stopped at the exact instant of drop and lock, so that the jewel pin shake can be tried or tested. (See Fig. 9.)

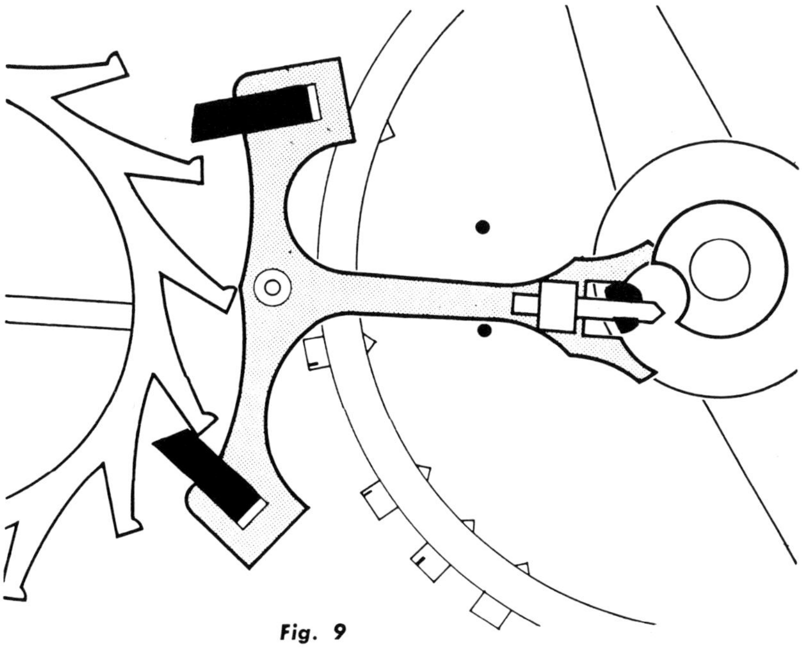

Fig. 9

The jewel pin shake is checked, tested or tried, by moving the fork back and forth. (An escapement trying tool or pivot broach may be used.) This will cause corner A of the fork slot (shown in Fig. 10) to touch the back side of the jewel pin, and corner B to touch the flat face of the jewel pin. During this test for jewel pin shake, it should be carefully noted that the fork has not yet touched the banking pin.

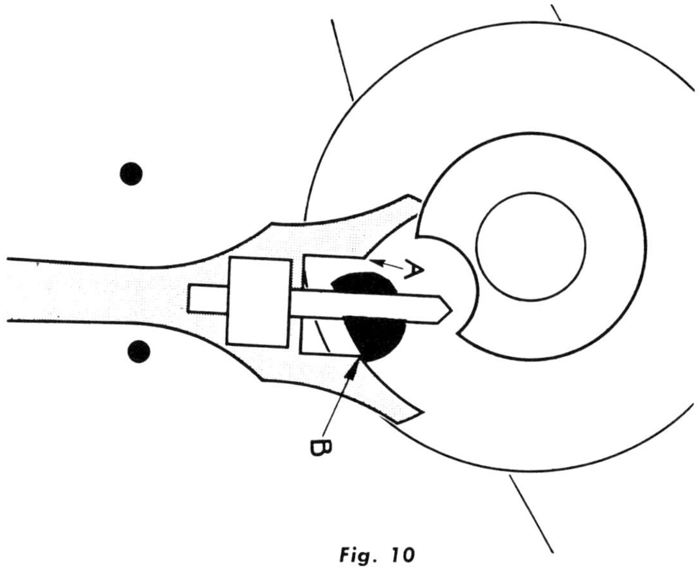

Fig. 10

Now reverse the motion of the balance wheel. The jewel pin will move the fork enough to unlock the escape wheel tooth, the impulse will begin its action. And when the impulse is completed, the drop will occur. Almost simultaneously with and immediately following the drop, the escape wheel will again come to lock. (See Figs. 11a, 11b, 11c)

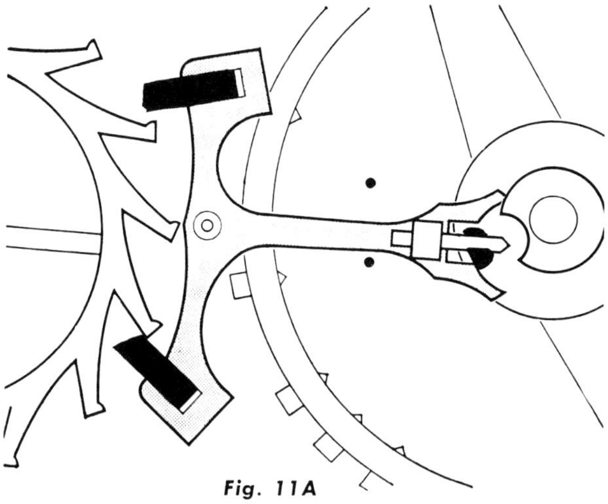

Fig. 11A

Fig. 11B

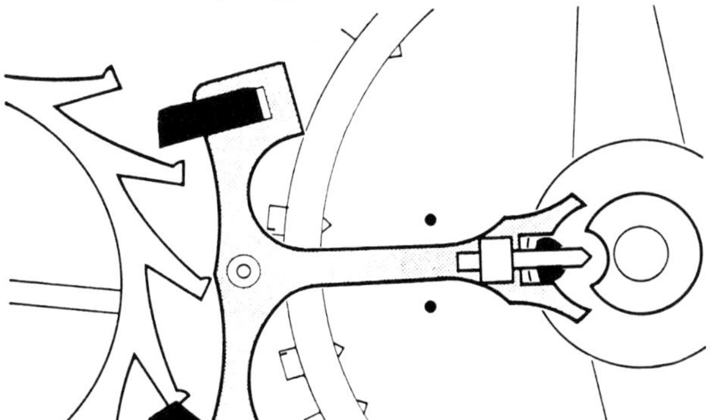

Fig. 11C

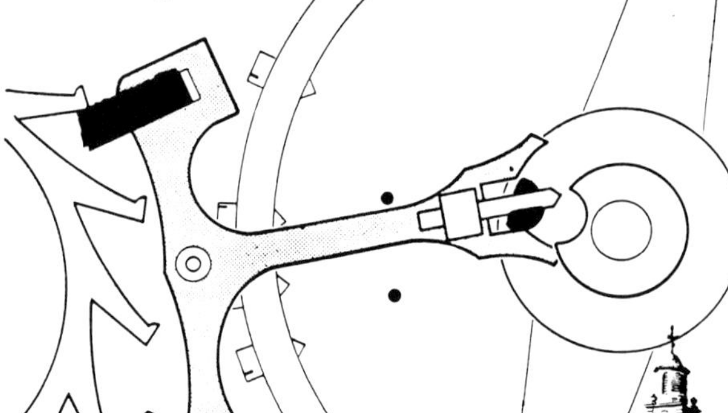

At this exact instant or in this exact position, the jewel pin shake is again tested.

Observe that the jewel pin shake is equal on both sides. This indicates that the fork moves as far on one side of the center line (line of centers) as it does on the other side.

JEWEL PIN CLEARANCE

After the jewel pin shake has been tested, we are ready to check the CLEARANCE of the jewel pin in its complete passage out of the fork slot. This is accomplished by turning the balance wheel slowly, causing the jewel pin gradually to pass out of the fork slot and past the fork horn. During this operation it should be noted that the fork is now permitted to come in actual contact with the banking pin.

This jewel pin clearance is tested several times in its (the jewel pin's) movement past the fork horns. In this test, the fork will strike or touch the banking pin at A (Fig. 12) and the fork horn will clear the flat face of the jewel pin at B.

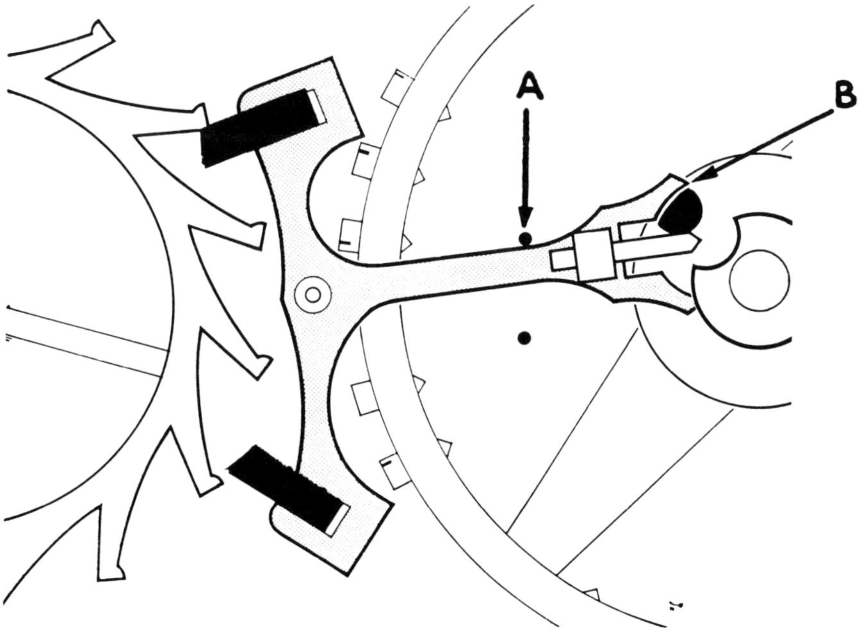

Fig. 12

GUARD PIN SHAKE

After the jewel pin has passed out of the fork slot and beyond the fork horn, the GUARD PIN SHAKE is tested.

The guard pin shake, which is the distance between the guard pin and safety roller when the fork rests against the banking pin, should be slightly LESS than the lock. This is determined by noting that when the guard point is pressed against the safety roller, the escape wheel tooth is still safely locked on the pallet stone. (See Fig. 13.)

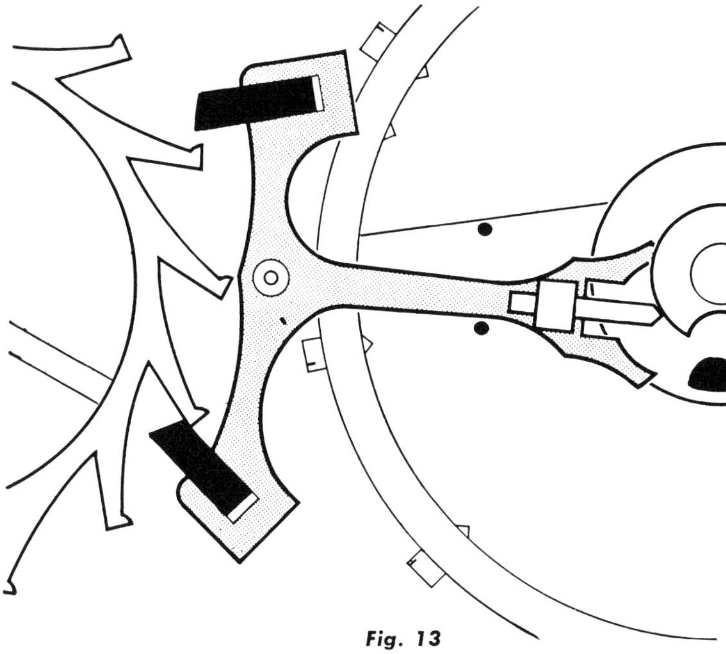

Fig. 13

DRAW

The DRAW or DRAFT feature can be tested at the same time and along with the test for guard pin shake.

It would be appropriate to remark here that the draft feature or draw on the pallet stone and escape wheel teeth are for the special purpose of holding the fork securely against the banking pin during the detachment of the jewel pin from the fork slot.

When the guard pin is touching the safety roller and the lock has been observed, quickly release the fork and the draw will function. It will at once pull the guard pin away from the roller and hold the fork against the banking pin.

SLIDE

Slide is the space a pallet stone travels downwards on an escape wheel tooth after the lock occurs.

To observe the pallet stone *sliding* on the escape tooth, move the jewel pin quickly away from the position in which the jewel pin shake is tried and in a direction AWAY from the line of centers.

REVIEW OF ESCAPEMENT EXAMINATION PROCEDURE

The watchmaker first moves the balance wheel, causing the jewel pin to engage the fork slot. He then continues this motion until:

1. The escape tooth unlocks (R stone).
2. The impulse is completed.
3. The drop occurs.
4. The lock is established.

(see Figs. 6, 7, 8 and 9.)

The passage of the jewel pin into the fork slot, the completion of the impulse, and the final lock must be done slowly and with absolute accuracy. The entire escapement must be brought under observation at the EXACT INSTANT OF THE DROP; the jewel pin must be held in that position without the slightest move back or forth so as to observe and check the condition of:

Lock
Drop
Jewel pin shake

The balance wheel is then turned farther away from the line of centers to check the:

Jewel pin clearance
Guard pin shake
Slide
Draw

Now repeat the procedure, starting with the unlocking of the L stone.

Since the watch used in this examination was known to be in good working order, all clearances and safety actions will be found to be satisfactory.

However this checking procedure should be used on any watch with fixed banking pins whenever any doubt exists as to the proper functioning of its escapement.

EXAMINATION OF THE FEATURES WHICH ALLOW AN ESCAPEMENT TO FUNCTION

TYPE 2: ESCAPEMENTS WITH MOVABLE BANKING PINS

BANKING TO A DROP

Another method of checking the escapement is known as "banking to a drop". This approach to the study of escapement matching is based entirely on the movable banking pins with which some watches were provided before the development of more accurate escapement making machines.

Because millions of these watches are still in use today, an explanation of their adjustment and repair should be in order. For this reason, "banking to a drop" will be briefly discussed.

Here again, the discussion is based upon a correct escapement, and a properly adjusted watch (preferably a new one) with movable banking pins should be used in the procedure to follow.

In a watch with movable banking pins, "banking to a drop" can be considered the basic reference for investigating the lock, drop, jewel pin shake, and guard pin shake.

After removing the balance wheel, turn the L banking pin so that it will be as close as possible to the line of centers of the escape wheel, pallet arbor and balance wheel. This is possible to do because the banking pins are located off center on the screws which support them. The fork will now rest against the L banking pin, being held there by the power of the mainspring transmitted through the train wheels to the impulse faces of the escape wheel teeth and the pallet stones. (See Fig. 14.)

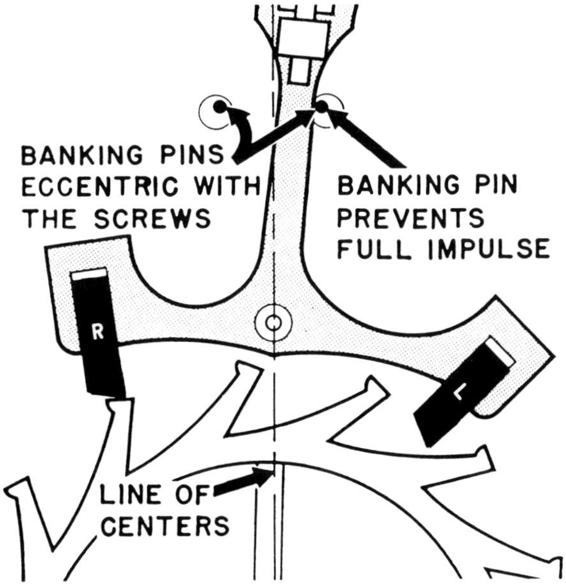

Fig. 14

Hold the watch in the left hand, dial down, and with the screw driver in the right hand, approach the watch from the dial side. Place the screw driver in the screw against which the fork rests, and turn this screw so that the eccentric pin will move away from the line of centers. It is necessary in this operation to look at the escape wheel with a loupe and observe the movement of the impulse face of the escape wheel tooth across the impulse face of the pallet stone. As soon as the impulse face of the escape wheel tooth drops off the R pallet stone, another tooth will lock itself on the L stone. The lock will be about ¼ the width of the impulse face of the pallet stone. It is important to remember that at this point we are interested in the amount of lock just after an escape wheel tooth has lost contact with a pallet stone. The eccentric banking pin screw should be turned just enough to allow a tooth to escape a pallet stone and no farther. (See Fig. 15.)

Fig. 15

Also understand that the location of the banking pins is determined by the position of the pallet stones. The farther out the pallet stones are, the farther away from the line of centers must the eccentric banking pin be moved before a tooth will escape. Now turn the R banking pin so that it will be as close as possible to the line of centers.

The fork must be moved so that it will rest against the R banking pin. (See Fig. 16.)

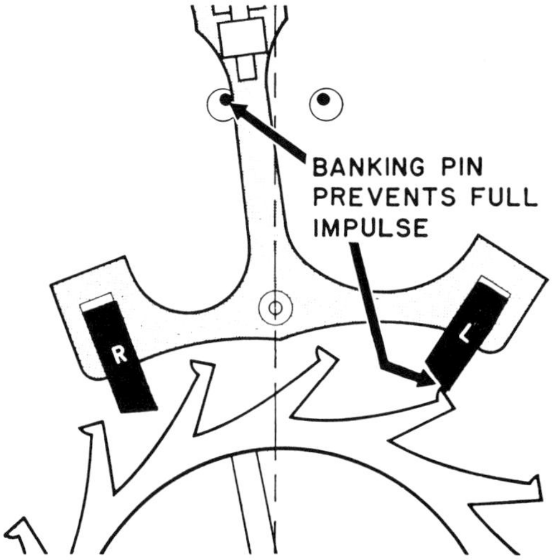

Fig. 16

The screw for this pin must now be turned just enough so that the escape wheel tooth drops off the L pallet stone. (See Fig. 17.)

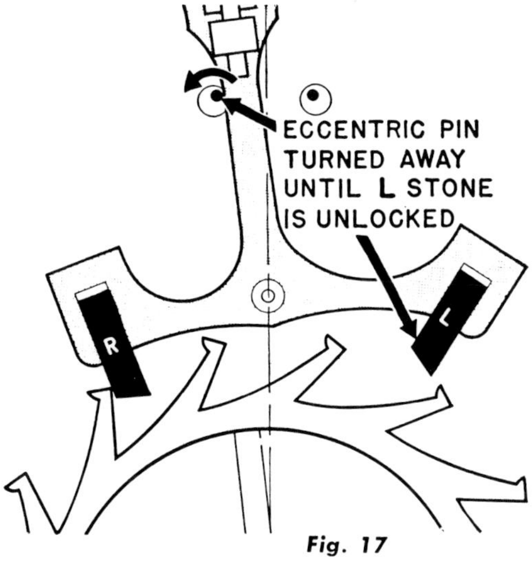

Fig. 17

The fork must now be moved back and forth until the escape wheel has made one complete turn.

The pallet may now be considered as banked to the drop. When this is so, the fork should move an equal distance from each side of the line of centers. (Fig. 18.)

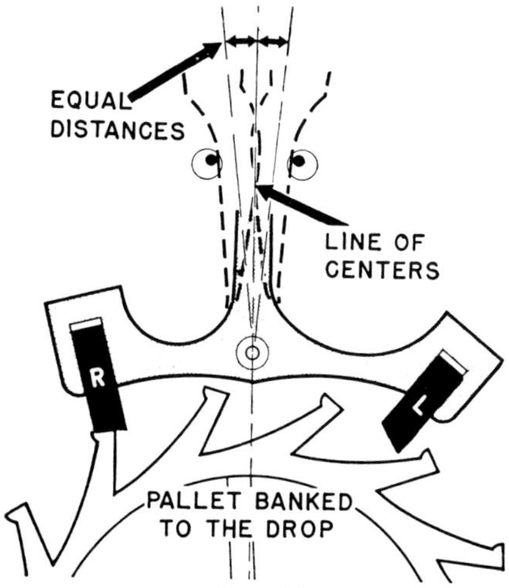

Fig. 18

Replace the balance wheel and hairspring unit in the watch. From here on, the checking of lock, drop, jewel pin shake and guard pin shake proceeds as in the fixed banking pin method.

As the balance is turned away from the line of centers, guard pin shake is tested exactly as in the fixed banking pin procedure. Guard pin shake at this point will be just barely noticeable, not more than ½ degree in a light lock. As the lock is increased, more guard pin shake may be permitted, as much as 1 degree.

After the lock, drop, jewel pin and guard pin shakes have been found to be satisfactory, slide can now be added. This is accomplished by turning both banking pins away from the line of centers as indicated by the arrows in Fig. 19.

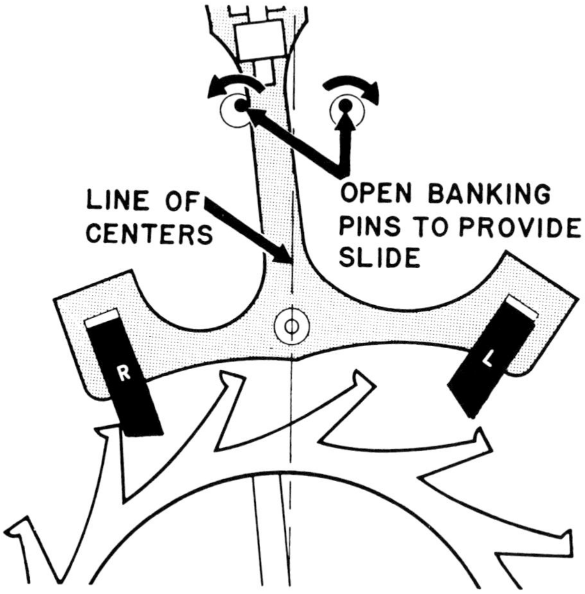

Fig. 19

The pins should be turned just enough to permit an increase of about 1/3 the lock established when banking to a drop. Guard pin shake will now also be found to have been slightly increased on both sides.

The slide may be observed by carefully turning the balance with the finger. Immediately after the drop, as the balance is turned farther away from the line of centers, the pallet jewel will be seen to move up or slide a short distance on the escape wheel tooth. Slide will be present on each tooth of the escape wheel.

Since the watch used to make these observations was properly adjusted to begin with, it will now be found to function satisfactorily once more.

TRAINING UNIT

NUMBER 10A

★

HOW TO REPAIR PARTS OF THE ESCAPEMENT

REPAIRING THE ESCAPEMENT

THE escapement parts consist of the escape wheel, the pallet, and the roller.

The occasion frequently arises wherein one or more of these parts may require repairs or replacement. For this reason, a correct procedure will be given on how to make these repairs on the major escapement parts. Also, the principles involved in positioning pallet stones will be explained, and a simple rule will be given to determine which stone must be moved, when adjustment is necessary.

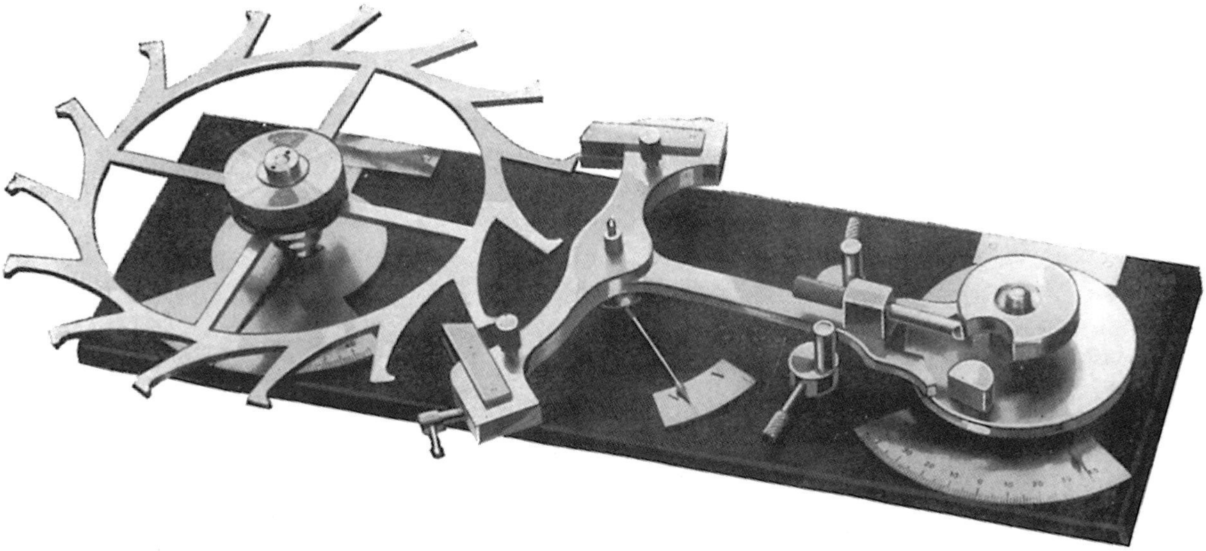

BULOVA SCHOOL *of* WATCHMAKING

TRAINING UNIT NUMBER 10A

SUBJECT:

Repairing Parts of the Escapement

Objectives:

To learn how to repair escapement parts

To learn how to move pallet stones.

New Tools Required:

1 Pallet Warmer

2 Pallet Stone Pusher

3 Escapement Tool

4 Jewel Pin Warmer

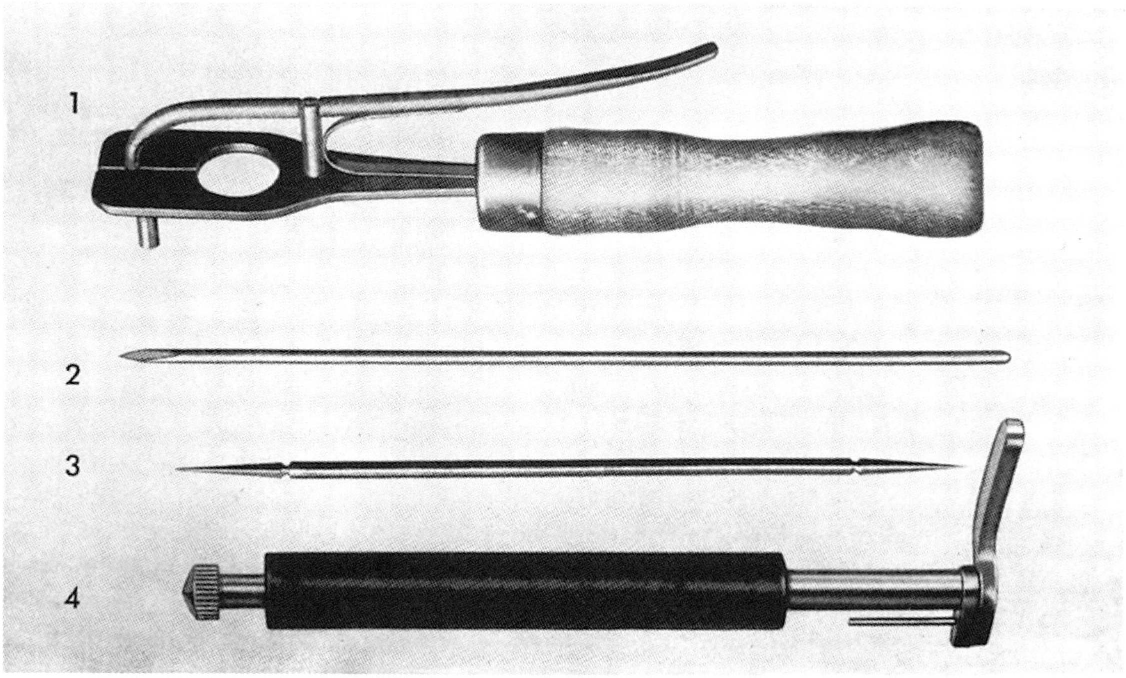

HOW TO REPAIR THE PARTS OF THE ESCAPEMENT

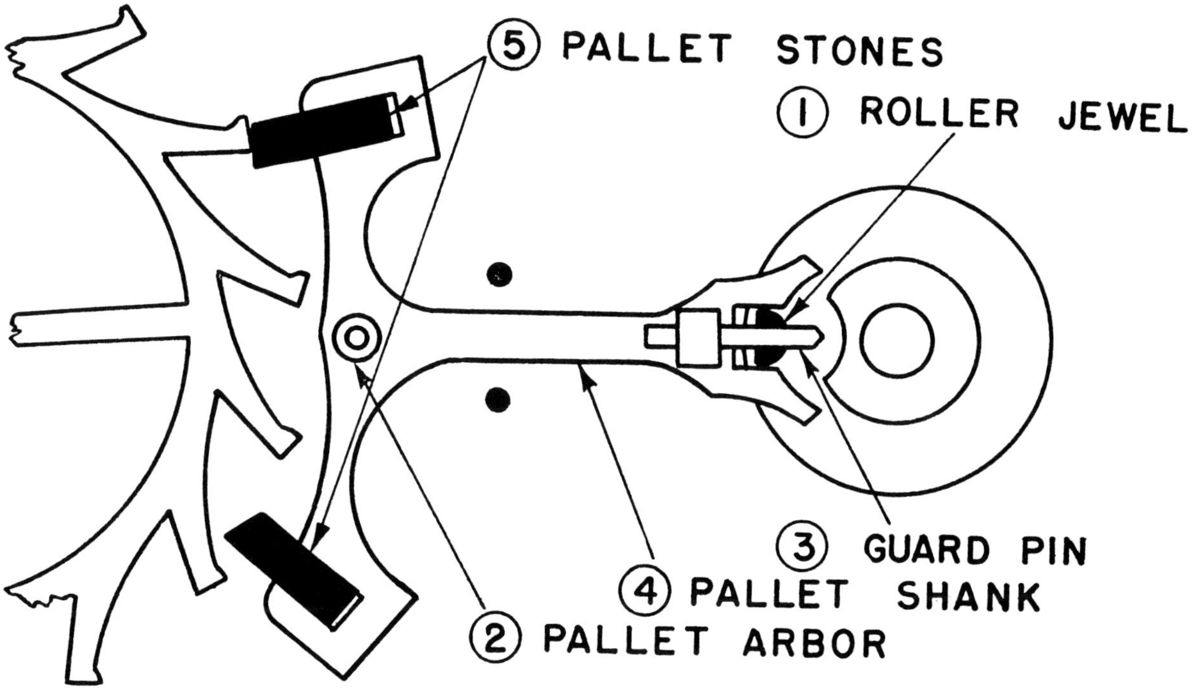

Repairing of these parts will be treated in numerical sequence in the pages which follow.

No. 1: HOW TO SET A NEW ROLLER JEWEL

If a new roller jewel is to be inserted in the roller, it should first be tried in the fork of the pallet. Obviously, the jewel cannot be larger than the fork slot. On the other hand, a jewel that is too small for the slot will cause lost motion of the pallet. Jewel pin freedom of .01 mm. for small watches to .03 mm. for 16 size watches should permit a safe impulse action.

Now thoroughly clean the roller jewel opening and properly fitted jewel in alcohol to insure permanent adhesion by the shellac.

Place the jewel into the roller, being careful not to apply too much pressure on the tweezers. If the tweezers are held too tightly, the jewel might snap away.

The roller should now be placed in a jewel pin warmer (the balance wheel may or may not be attached to the roller). As the jewel tool is heated over an alcohol lamp, a piece of string shellac is applied on top of the impulse roller, directly over the jewel. See Fig. 1.

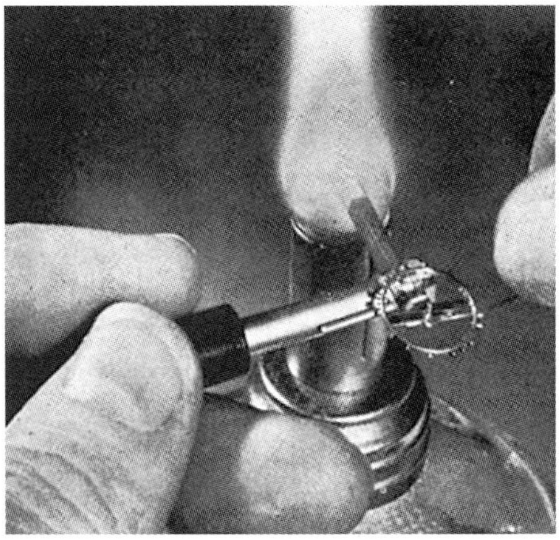

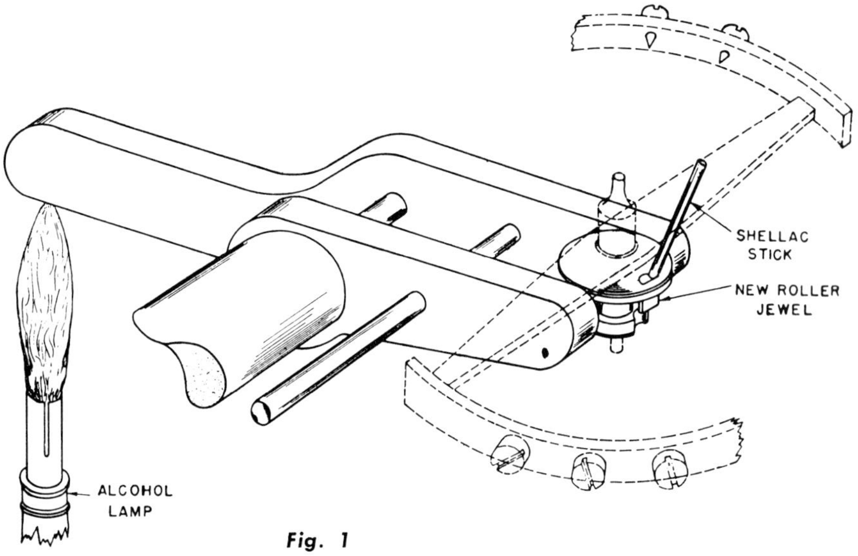

Fig. 1

Care must be taken to insure that the jewel is perpendicular to the roller and also that it will be of proper length. In Fig. 2 are shown a roller jewel which is of proper length, a roller jewel which is too short, and one that is too long.

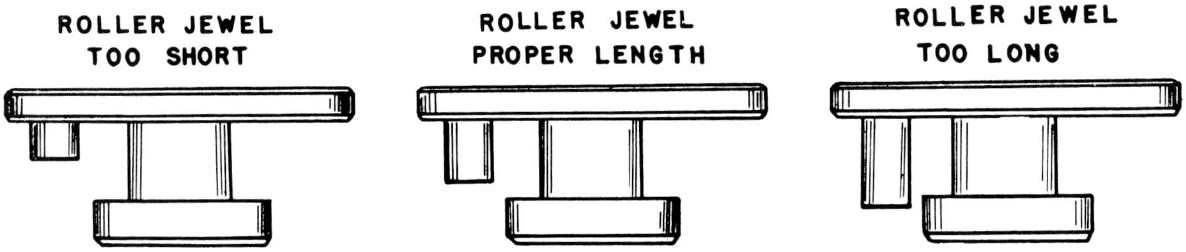

Fig. 2

When the jewel has been set, the balance assembly should be tried in the watch to check its action with the fork of the pallet.

Remove any excess shellac on the jewel with an edged piece of pegwood which has been dipped in alcohol.

No. 2: PALLET ARBORS

Pallet arbors are constructed in two types: the screw arbor and friction arbor.

The friction type shown in insert of Fig. 3 is inserted or removed from the pallet by the staking method. Fig. 3 shows a punch being tapped against the friction arbor so that it will be inserted in the pallet. The punch which is flat-faced and with hole diameter slightly larger than that of the pivot rests on the shoulder of the arbor. The pallet is supported by a small stump which must not interfere with the pallet jewels.

The pallet is usually positioned slightly below the upper pivot of the arbor but can be shifted to any desirable height. Removal of the friction pallet is accomplished by reversing the process; that is, by driving out instead of driving in the arbor.

Fig. 3

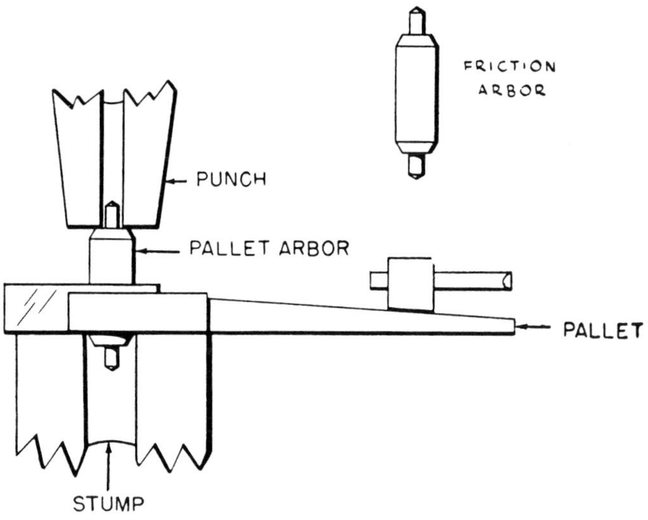

The screw type, illustrated in Fig. 4, is threaded on its upper shoulder so that it will screw into the pallet. Insertion or removal of the arbor can be easily performed with the aid of a pin vise which is tightened on the lower shoulder of the arbor.

Fig. 4

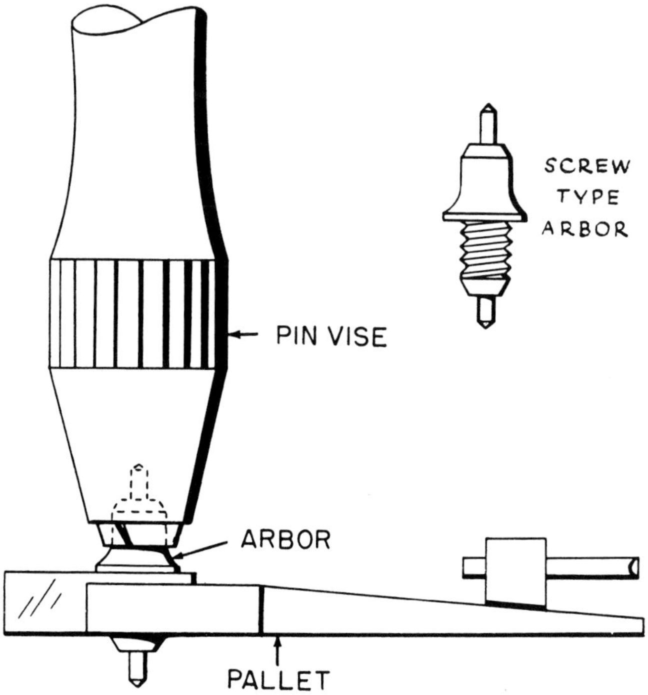

No. 3: HOW TO FIT A NEW GUARD PIN

Before a guard pin can be replaced, it may be necessary to remove the old one. This can best be accomplished by first shortening the pin and forcing it out from the tail end towards the pallet arbor with pliers or tweezers. See Fig. 5. If the pin is broken off at the hole, the stump should be driven out with a pin pushed in the same way that a hairspring stud is cleared of the stud pin.

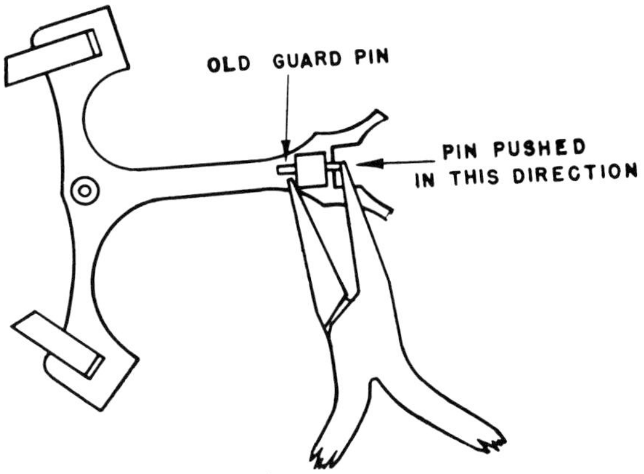

Fig. 5

The new guard pin is made by inserting a piece of thin brass wire into a pin vise and roll-filing it on a box-wood slip to a fine taper. After burnishing, the wire is forced friction tight into the hole, so that the small end of the taper pin is away from the pallet arbor. See Fig. 6. In small watches, the tapered pins used in hairspring work will be found suitable.

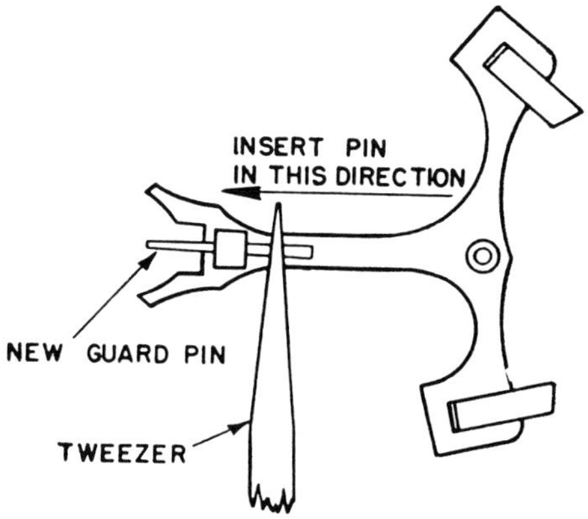

Fig. 6

The new guard pin can now be clipped with cutting pliers, allowing just enough length for finishing. Finish off the tip with an oilstone slip, so that the point makes an angle of about 90 degrees. See Fig. 7.

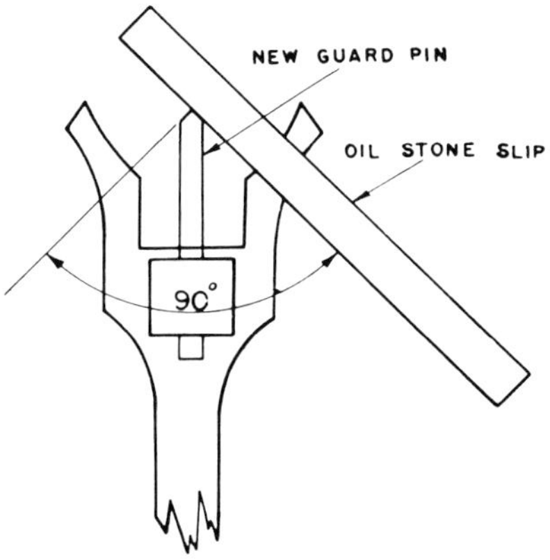

Fig. 7

From here on, the pallet must be tested in the watch. If the guard pin shake is excessive, the pin may be slightly lengthened by pressing the thick end of the pin closer to the fork. If the pin is found to be too long, shortening is accomplished in the same way as finishing, by using an oilstone slip.

No. 4: HOW TO STRAIGHTEN THE PALLET LEVER

Frequently, a watch that has been improperly repaired will come in with the pallet bent out of the level to accommodate a poor fitting staff. When a genuine or proper fitting staff has been replaced in the watch, it will then be necessary to bend the pallet again back to its original form.

In a case such as this, the pallet should be supported on a boxwood slip or on the wooden handle of a brush so that the pallet arbor is within a small hole already prepared in the wood.

The pallet can now be stroked with tweezers as shown in Fig. 8 until the bent shank is returned to its original level condition.

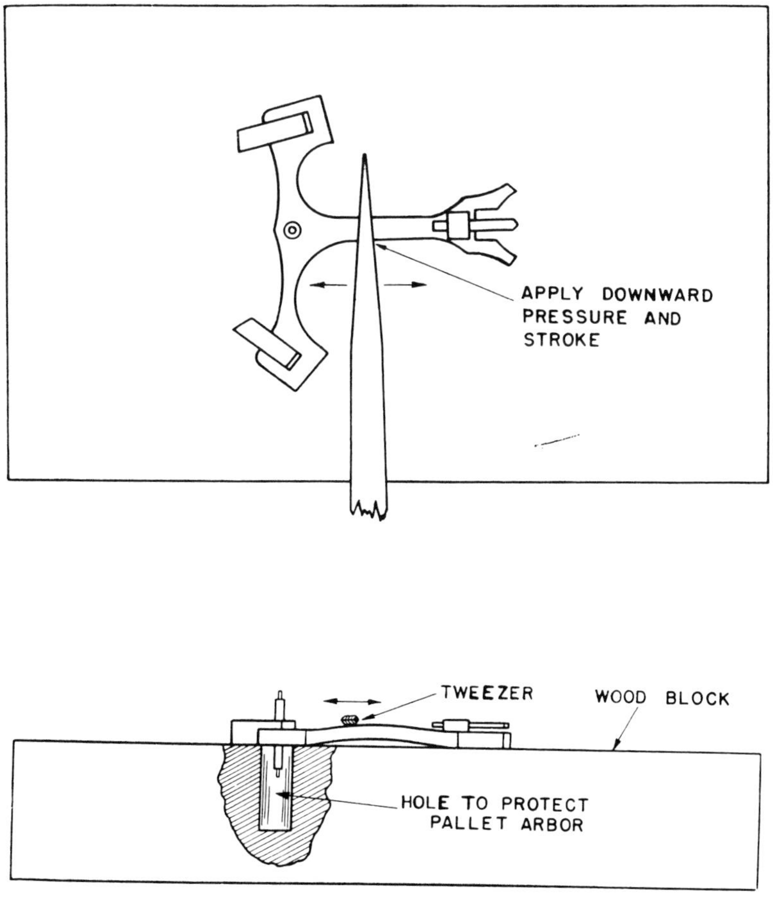

Fig. 8

No. 5: HOW TO MOVE PALLET STONES

In order to move a pallet stone, the pallet must be removed from the watch and placed upside down on a pallet warmer, with the lower pivot placed in the hole provided in the movable arm.

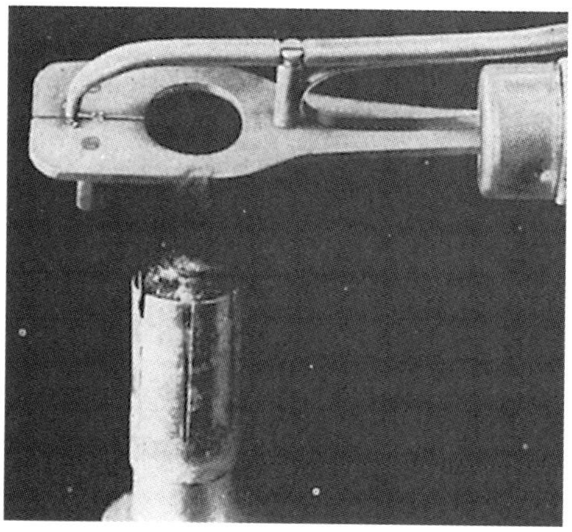

Fig. 9

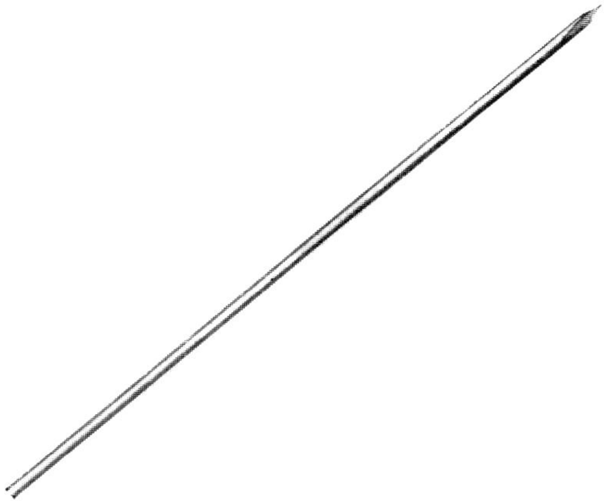

Fig. 10

The shellac holding the pallet stones is softened by heating the pallet warmer over an alcohol lamp. See Fig. 9. Care must be taken that the shellac does not become too hot and spoil. The pallet warmer is now removed from the lamp.

One or both stones may be moved in or out with a nickel or brass tool. (A convenient shape is shown in Fig. 10 and may be made from brass or nickel wire approximately of 1½ mm. diameter.) The pallet is then removed from the warmer by lifting straight up so that the stones will remain level with the body of the pallet. Excess shellac on the top (polished) surface of the pallet may be removed with a piece of pegwood or a buff dipped in alcohol. The pallet may now be replaced in the watch.

THE PRINCIPLES OF MOVING PALLET STONES

When a pallet stone is moved out (towards the escape wheel tooth), this will cause the pallet fork to swing a greater distance from the line of centers before the lock occurs on the opposite stone.

When the pallet stone is moved in (away from the escape wheel tooth), this fork angle (the distance the pallet swings from the line of centers before locking with the other stone) is reduced.

Because the movement of a pallet stone affects the distance that the pallet will move, it can be stated that if one stone is moved out or in (causing the lock to be increased or decreased), the lock on the opposite stone will be increased or decreased by the same amount.

Let us suppose that the R stone has been pulled out a small amount. Careful inspection of the escapement will show that because the pallet now swings farther from the line of centers before locking on the L stone, that:

1. The lock has been increased the same amount on the L stone.
2. The pallet has moved closer to the L banking pin at the instant of lock, thereby permitting less slide. (See Fig. 11.)

R STONE MOVED OUT — L STONE OK

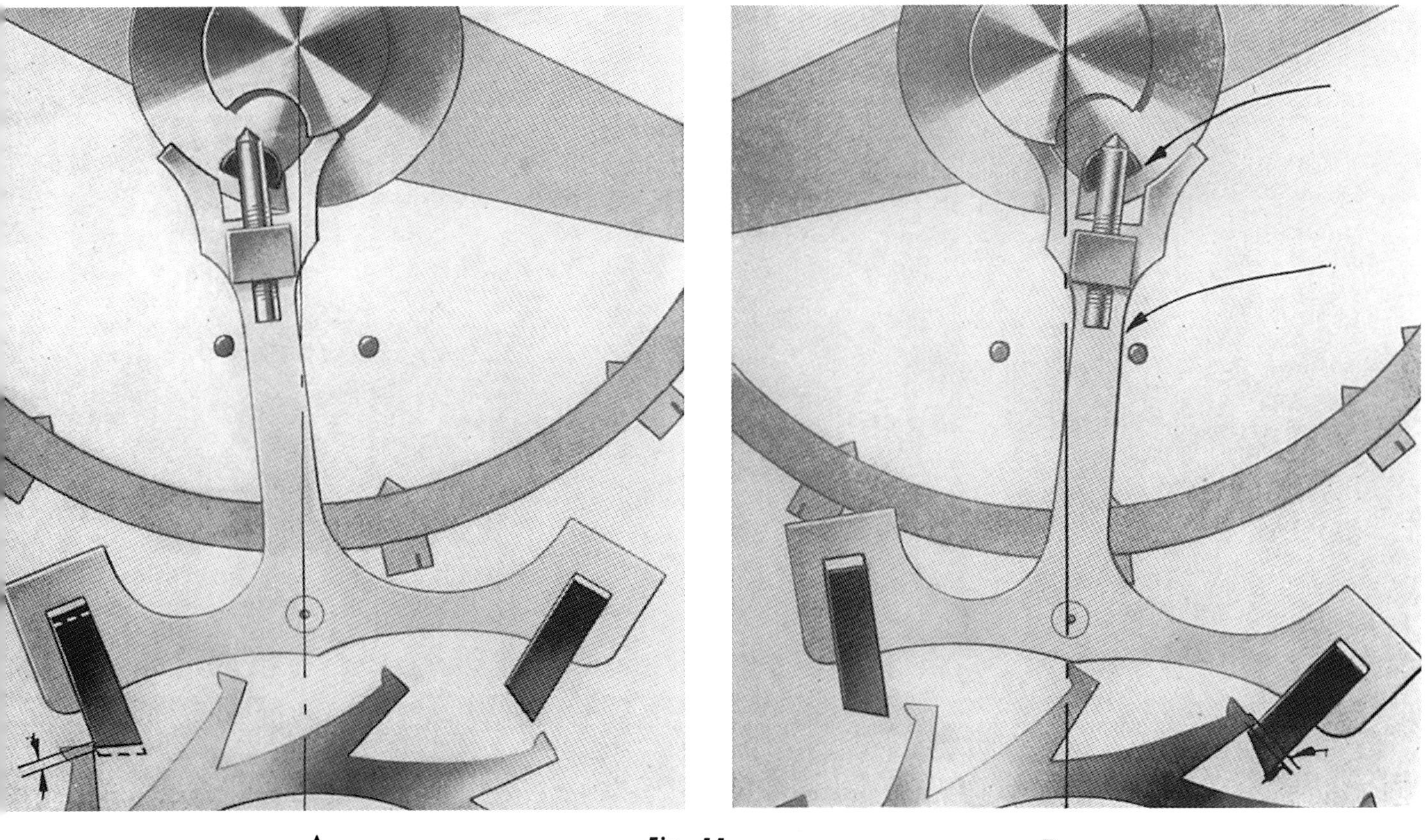

A   *Fig. 11*   B

Basically, this is the key to what takes place if either stone or both stones should be shifted in any way. Had the R stone been pushed in, so that the pallet now would swing a shorter distance from the line of centers to the point of drop, the lock would have been decreased an equal amount on the L stone and the slide would now be greater on the L stone side. If in place of the R stone, the L stone had been moved, the same conditions would have resulted on the R side.

With the balance wheel in the watch, the amount of travel of the pallet from the line of centers can be determined by trying jewel pin shake. If the pallet is farther away from the line of centers at the instant of lock, then the jewel pin shake will be greater. If the pallet is closer to the line of centers, then the jewel pin shake will be less.

PALLET STONE RULE

If the lock is too light, pull out the stone opposite the LESSER jewel pin shake. If the lock is too strong, push in the stone opposite the GREATER jewel pin shake.

Remember this simple procedure and you master every situation involving the movement of pallet stones.

ESCAPEMENT HEIGHT RELATIONSHIP CHECKS

A. ESCAPE WHEEL:

End shake of the escape wheel should be from .02 to .04MM. Look out for the following:

1. Escape wheel too close to the pillar plate.
2. If the lower escape pivot is capped, make sure the lower cap screw does not extend up to the escape wheel.
3. Cracked or out of upright escape wheel hole jewels. If there is a cap jewel, check to see if it is seated on the pillar plate or train bridge. If it is not, check for the presence of foreign material between the cap and its seat. The cap screw should be snugly in place and upright. The jewel in the cap should be even with the side of the cap assembly that faces the pivot. The cap jewel should be concentric with the hole jewel. Finally, the crown or top of the hole jewel should be .02MM below the cap seat to allow proper space for oil.

B. PALLET ARBOR:

Pallet arbor end shake should be from .02 to .04MM. In most watches the upper pallet hole jewel is flush with the underneath side of the pallet bridge. When this is so, then the pallet arbor end shake can be adjusted by raising or lowering the lower pallet hole jewel in the pillar plate.

C. BALANCE WHEEL ASSEMBLY:

Balance staff end shake should be from .02 to .04MM. In case the upper or the lower or both the pallet arbor pivots are capped, the same checks should be made, as outlined in paragraph "C" below, relative to the balance staff end shake. A good rule here would be to have the end shake of the pallet arbor equal to that of the balance staff. This would avoid rubbing between the fork horns and the impulse roller. Also the guard pin would stay clear of the end of the roller jewel when changing the watch from one dial position to the other. This, of course, presupposes that the fork is centered between the two rollers to begin with. The key points to be checked concerning the staff end shake are as follows:

1. Balance bridge bent up or down.
2. Burrs on the pillar plate where the balance bridge seats or on the underneath side of the balance bridge itself.
3. Loose or non-flush jewels in either the upper or lower cap. Replace pitted cap jewels.
4. Out of upright or loose cap screws.
5. Foreign material under the caps or burrs on the cap seats.
6. Out of upright balance hole jewels that would cause friction by crowding out the necessary side shake of the pivots in their holes.

C. BALANCE WHEEL ASSEMBLY Cont'd.:

7. The proper space of .02MM should exist between the seat of the cap jewel and the top or crown of the hole jewel for the deposit of oil.
8. Check to see if the roller is seated against the hub of the balance staff.
9. Be sure the roller jewel is tightly cemented with the cement free of cracks and chips. Also the roller jewel should be upright and free of cracks and chips. There should be no cement on the sides of the roller jewel. Besides cutting down or eliminating the jewel pin freedom, particles of the cement might chip off and cause untold interference in the train, the hairspring or elsewhere.

D. CHECKS CONCERNING THE PALLET:

1. The pallet stones should be free of cracks or chips on the locking faces, locking corners, impulse faces or let-off corners. The working surfaces of the pallet stones should be clean and free of shellac.
2. The stones should be flush with the top of the pallet and free of cement. See if the stones are properly cemented on the underside.
3. The pallet arbor should be checked for tightness in its hole.
4. Friction pallet arbors are chamfered on each end. The position of the pallet on its arbor should not be above the bottom edge of a chamfer. If this limit is exceeded a loose arbor may result. Also the top of the pallet would get too close to the underneath side of the pallet bridge and may contact it.
5. The guard pin should be friction tight and properly shaped with a 90 degree end.
6. Finally, check to see if the pallet shank is bent up or down. This error would destroy the ideal fork-roller relationship as shown in the figure 12.

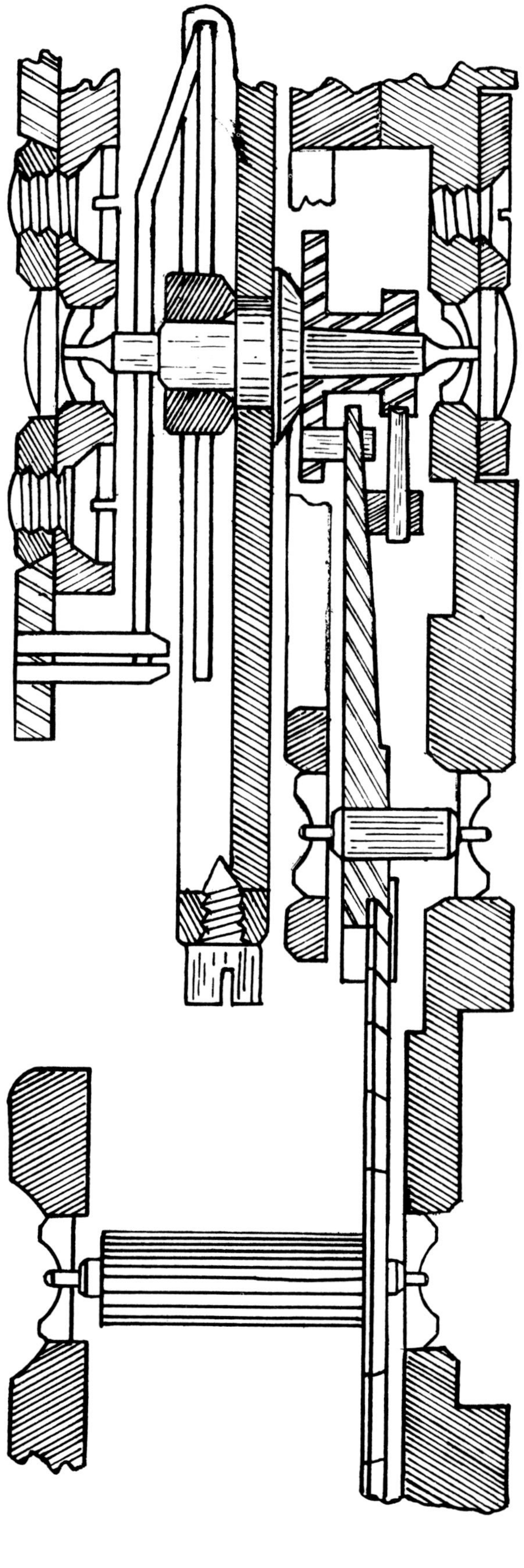

SIDE VIEW OF BALANCE AND ESCAPEMENT

Fig. 12

CHECKS YOU MUST MAKE ON EVERY ESCAPEMENT TO ASSURE MINIMUM SAFE FUNCTIONING IN A WATCH

☆

1

CHECK JEWEL PIN FREEDOM IN FORK SLOT

Take off all power and allow balance to move to center position. Jewel pin will be approximately on line of centers. Hold balance in this position with finger and shake pallet to observe shake between fork slot and jewel pin. There must be some freedom but not enough to cause significant lost motion between fork and jewel pin at impulse. See Fig. 1.

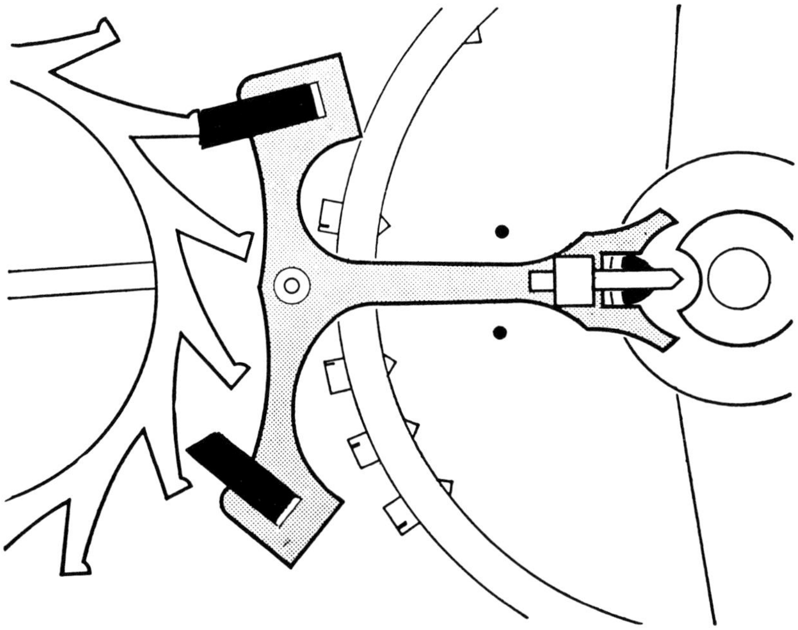

Fig. 1

2

CHECK LOCK AND SLIDE

With some power on train, move balance to exact instant of drop and hold carefully at this point. Note whether lock is safe. Observe whether there is *any* slide by noting whether there is any clearance between fork and nearest banking pin. Repeat these checks for minimum lock and slide on both stones and for one revolution of the escape wheel. See Fig. 2.

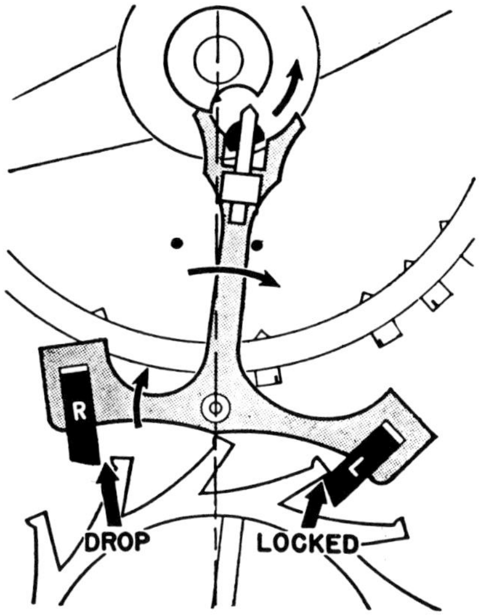

Fig. 2

With balance wheel held with jewel pin out of fork and fork against banking pin, note lock plus slide. Lock plus slide may not exceed ⅓ of width of impulse face of pallet stone. Check both stones, but it is not necessary to check all escape teeth. See Fig. 3.

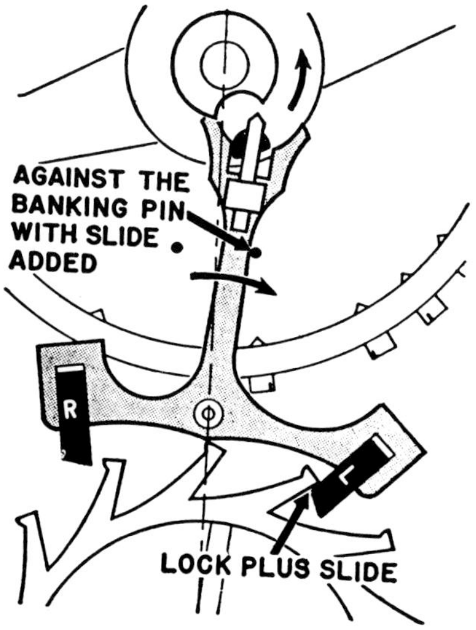

Fig. 3

3

CHECK JEWEL PIN CLEARANCE

With some power on the train, hold pallet against banking pin and allow balance wheel to turn under hairspring action until jewel pin lies against inside side of fork slot. Hold balance carefully in this position and note shake of pallet between jewel pin and banking pin. With pallet pushed against the jewel pin, lock must remain safe. Also, there must be some shake of fork between banking pin and jewel pin. See Fig. 4.

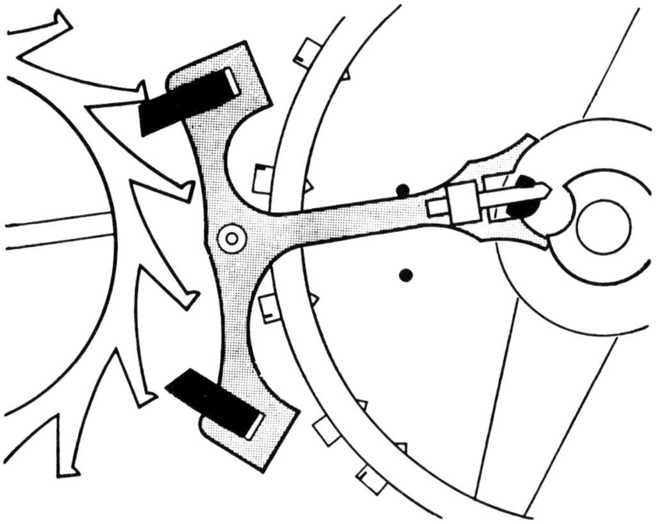

Fig. 4

Move balance slightly in a direction to disengage jewel pin from fork and repeat check for minimum and maximum jewel pin clearance. Repeat until jewel pin no longer lies within fork horn. Repeat check on opposite side. In cases where lock is barely safe with fork horn against jewel pin in above check, it is well to repeat test on all teeth, particularly if escape wheel teeth are not all of equal length as found in test Number 2. See Fig. 5.

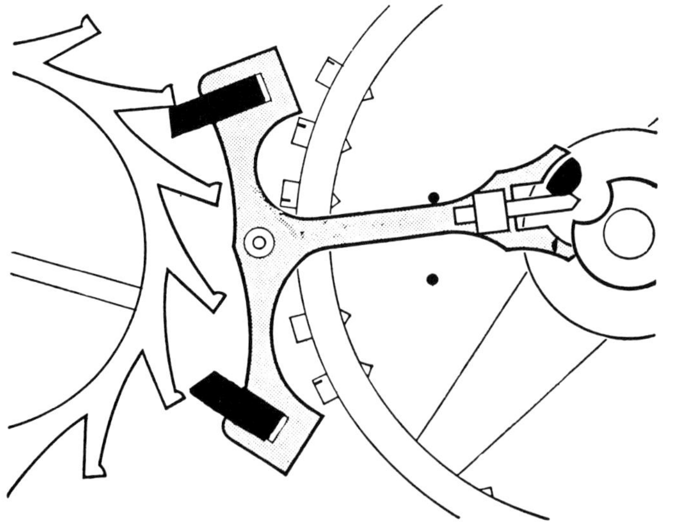

Fig. 5

4

CHECK GUARD PIN SHAKE

Check shake of fork between banking pin and safety roller for a number of positions of the balance wheel, beginning at balance wheel position where the jewel pin has just left the fork horn. There must be guard pin shake for all positions of the balance wheel. Repeat check with escape wheel locked on other pallet stone. See Fig. 6.

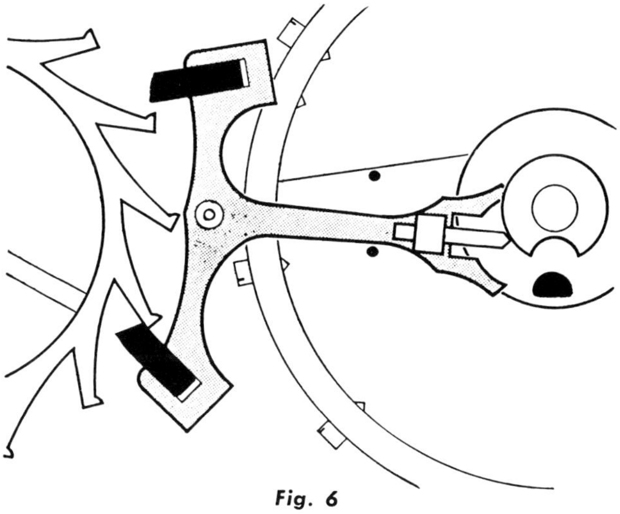

Fig. 6

Now reverse train direction by pushing on 3rd or 4th wheel. This pushes the guard pin against the safety roller. With the finger rotate the balance wheel back and forth to bring the roller jewel into engagement with the pallet, to make certain it does not catch on the tip of fork horn on either side. If the jewel pin catches on the tip of the fork horn, this indicates excessive guard pin shake. See Fig. 7.

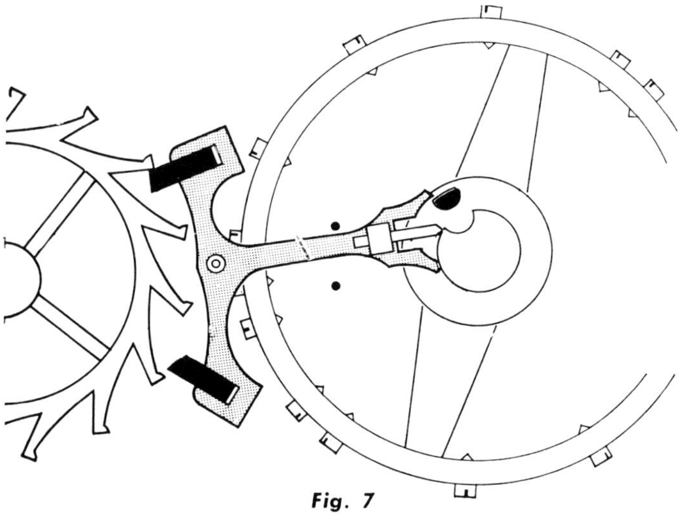

Fig. 7

• • • • • • •

NOTE—While making this check with the train reversed, it is well to make sure that the jewel pin can be pushed smoothly into the fork slot. Any catching of the jewel pin on the inside of the fork horns is an indication of roughness or burrs on this portion of the pallet. See Fig. 8.

Fig. 8

BULOVA SCHOOL *of* WATCHMAKING

SUPPLEMENT TO TRAINING

UNITS 10 & 10A

★

ESCAPEMENT TERMINOLOGY

1. **Function of an Pallet** is to change the rotary motion of the escape wheel into the vibrating motion of the balance wheel assembly.

2. **Sequence of Events** in an escapement during one vibration of the balance wheel assembly. Starting from a condition of total lock, the sequence is as follows:
    1. Jewel pin enters the fork slot.
    2. Escape wheel tooth unlocks.
    3. Impulse takes place.
    4. Drop.
    5. Lock.
    6. Slide.

3. **R or Receiving Stone** is that pallet stone which first receives or locks with a given escape tooth in an escapement action.

4. **L or Let-off Stone** is that pallet stone which last makes contact with the given escape tooth.

5. **Pallets** designate the assembly of R and L stones together with the metal arm which joins them.

6. **Pallet lever** is the metal rod that joins the pallet assembly with the pallet fork. The pallet lever is sometimes called the Shank.

7. **Fork** is that part located at the tail of the pallet lever. It consists of two horns and a slot. It is the sides of the fork slot which alternately receives the unlocking blow from the roller jewel and delivers the impulse to the roller.

8. **Horns** are projections on each side of the fork slot which provide safety action from the movement of let-off until the guard pin is along side of the rounded portion of the safety roller.

9. **Jewel Pin.** Also called roller jewel. It is vertically supported by the impulse roller. The jewel pin is the connecting link between the pallet and balance wheel and its purpose is to deliver the unlocking blow and to receive the impulse.

10. **Impulse Roller** is a flat circular disk to which the roller jewel is attached.

11. **Safety Roller** is also a flat circular disk. A portion of it is cut out which is called the crescent. The safety roller together with the guard pin provides additional safety action.

12. **Single Roller** is a roller which consists of a single metal disk serving both as the safety and impulse roller.

13. **Double Roller** is a roller consisting of two metal disks. The upper and larger disk supports the roller jewel and is called the impulse roller. The lower and smaller disk contains the crescent and is called the safety roller.

14. **Crescent** is the notch in the safety roller which permits the guard pin to pass freely across the line of centers.

15. **Guard Pin.** The function of the guard pin is to prevent the escapement from unlocking when the watch receives a jar. It is a small brass pin located below the fork. Because of the safety action that it provides the pallet fork will be on the correct side of the line of centers ready to receive the jewel pin on its return trip.

16. **Guard Pin Shake.** This check is made when the jewel pin is turned completely away from the fork. This shake is the freedom of movement existing between the banking pin and the safety roller. The pallet shank stops on the banking pin and the guard pin stops on the safety roller.

17. **Line of Centers.** This is the imaginary straight line that joins the balance staff, pallet arbor, and escape wheel staff. It is conveniently used as a basic reference line in escapement adjusting.

18. **Toe** of an escape wheel tooth is that part that engages the pallet stone when the escapement is locked.

19. **Heel** of an escape wheel tooth is the corner of the tooth that last makes contact with a pallet stone.

20. **Impulse Face** of an escape wheel tooth is that part between the toe and the heel.

21. **Locking Face** of a pallet stone is that part which engages with an escape wheel tooth when the escapement is locked.

22. **Locking corner** of a pallet stone is that corner separating the locking face and the impulse face of a pallet stone.

23. **Let Off Corner** of a pallet stone is that part which last makes contact with an escape wheel tooth in an escapement action.

24. **Impulse Face** of a pallet stone is the inclined plane between the locking and let-off corner. When an escape wheel tooth sweeps across this face it delivers an impulse that throws the pallet lever to the opposite side of the line of centers. Because of these impulse faces the rotary motion of the escape wheel is changed into the vibrating motion of the balance wheel assembly.

25. **Impulse.** The impulse begins right after tooth and stone are unlocked. As the tooth sweeps across the impulse face of the stone, the entire pallet turns about its arbor. This causes a side of the fork to strike the jewel pin thus providing it with the force that causes the balance wheel assembly to vibrate.

26. **Banking Pins** limit the amount of angular swing of the pallet lever from each side of the line of centers.

27. **Draw.** Sometimes called draft. It is a force exerted by an escape wheel tooth against the locking face of a pallet stone together with the draft angle of the stone. This draws or pulls the pallet lever against the banking pins and holds it there.

28. **Slide.** Slide is caused by the draw. It is the distance a pallet stone travels downward on an escape wheel tooth immediately after lock occurs. Sliding stops when the pallet lever comes up against a banking pin.

29. **Drop** is the free motion of the escape wheel from let-off to lock. Or it may be spoken of as the distance from the let off corner of a pallet stone to the heel of the tooth that has just let off.

30. **Close Outside.** Always refers to the L stone. This condition exists when the distance between the L stone and the heel of the escape tooth just let off is too small to permit the R stone to unlock.

31. **Close Inside.** Always refers to the R stone. This condition exists when the distance between the R stone and the heel of the escape tooth just let off is too small to permit the L stone to unlock.

32. **Jewel Pin Freedom.** The difference between the diameter of the jewel pin and the width of the fork slot.

33. **Lock.** (DROP LOCK) This is the amount of overlap that exists between tooth and stone right after drop. Sometimes referred to as drop lock or lock at the instant of drop.

34. **Total Lock.** This lock consists of the lock at the instant of drop plus slide.

35. **Jewel Pin Shake.** The escapement is checked for jewel pin shake at the exact instant of drop. It is the amount of angular movement of the pallet lever permitted at this point. Jewel pin shake is used to determine if the swing of the pallet from the line of centers up to the bank to the drop point is the same on both sides or not.

36. **Jewel Pin Clearance.** The jewel pin must be permitted to pass freely toward and away from the fork slot. Without some distance (clearance) between the jewel pin and the fork horn the watch would stop. If the jewel pin clearance is too much, due to a short fork condition, premature unlocking would result.

37. **Long Fork.** This condition exists when the jewel pin cannot pass out of the fork slot when the escapement is banked to the drop. Caused by the jewel pin being tilted away from the balance staff or by the pallet shank being too long.

38. **Short Fork.** If the jewel pin shake is so great as to allow the pallet stone to unlock when the escapement is banked to a drop then the condition of short fork is present. Caused by a short shank or by the jewel pin being tilted toward the balance staff.

39. **Banking to a Drop.**
   1. Fixed pins. By manually moving the pallet lever with or without the aid of the balance wheel until an escape wheel tooth just lets off a pallet stone - we bank to a drop.
   2. Movable pins. An escapement may also be banked to the drop by first turning the pins in toward the line of centers then away again until an escape wheel tooth can just let off a pallet stone.
      It is when the escapement is banked to the drop that we examine for lock and jewel pin shake.

40. **Overbanked.** (out of action) This fault is due to a short or a bent guard pin or possibly a low fork. Under this condition a tooth will unlock from a stone prematurely and depending on the amount of the error the pallet lever may move over to the opposite banking pin relative to the return of the jewel pin. Can also be caused by a short roller jewel or excessive end shake in either the balance staff or pallet arbor.

41. **Banking** because of too strong a mainspring, the balance wheel develops too much motion which causes the jewel pin to hit on the back of the fork horns. This is also called knocking, racing and rebanking.

42. **Train Reversal.** There is a slight train reversal which occurs during the unlocking stage.

43. **Two basic classifications of the escapement.**
   1. Frictional
      a. Cylindrical
      b. Duplex
   2. Detached
      a. Lever
         1. Circular
         2. Equidistant
         3. Semitangental
      b. Chronometer

44. **Circular pallet escapement.** In this type of escapement the center of the impulse face of each pallet stone is at an equal distance from the pallet arbor.

45. **Equidistant escapement.** In this type of escapement the locking corner of each pallet stone is the same distance from the pallet arbor.

Effects Of Moving Pallet Stones
(Assuming a perfectly adjusted escapement)

1. **Pull R Only**

R Lock increases	R JPS remains the same	R slide same
L Lock increases	L JPS increases	L slide decreases

2. **Push R Only**

R Lock decreases	R JPS remains the same	R slide remains the same
L Lock decreases	L JPS decreases	L slide increases

3. **Pull L Only**

R Lock increases	R JPS increases	R slide decreases
L Lock increases	L JPS remains the same	L slide remains the same

4. **Push L Only**

R Lock decreases	R JPS decreases	R slide increases
L Lock decreases	L JPS remains the same	L slide remains the same

5. **Pull Both Stones Same Amount**

R Lock increases	R JPS increases	R slide decreases
L Lock increases	L JPS increases	L slide decreases

6. **Push Both Stones Same Amount**

R Lock decreases	R JPS decreases	R slide increases
L Lock decreases	L JPS decreases	L slide increases

7. **Pull R and Push L Same Amount**

R Lock stays same	R JPS decreases	R slide increases
L Lock stays same	L JPS increases	L slide decreases

8. **Push R and Pul L Same Amount**

R Lock stays same	R JPS increases	R slide decreases
L Lock stays same	L JPS decreases	L slide increases

TYPES OF PALLET STONE ERRORS AND THEIR SOLUTION

Object: To obtain a satisfactory lock and equal jewel pin shakes

1. If the locks are light and the jewel pin shakes are unequal?
   Solution: Pull the stone opposite the lesser jewel pin shake.

2. If the locks are heavy and the jewel pin shakes are unequal?
   Solution: Push the stone opposite the greater jewel pin shake.

3. If the locks are light and the jewel pin shakes are equal?
   Solution: Pull both stones the same amount.

4. If the locks are heavy and the jewel pin shakes are equal?
   Solution: Push both stones the same amount.

5. If the locks are satisfactory and the jewel pin shakes are unequal?
   Solution: Pull the stone opposite the lesser jewel pin shake, and push the stone opposite the greater jewel pin shake by the same amount.

Training Unit
Number 11

★

FINISHING

FINISHING

Finishing is the term applied to the assembly of the complete balance and hairspring unit into the watch movement. It requires the combined skills and knowledge given in earlier training units. After a watch is properly finished, it is ready for the final timing and special adjusting.

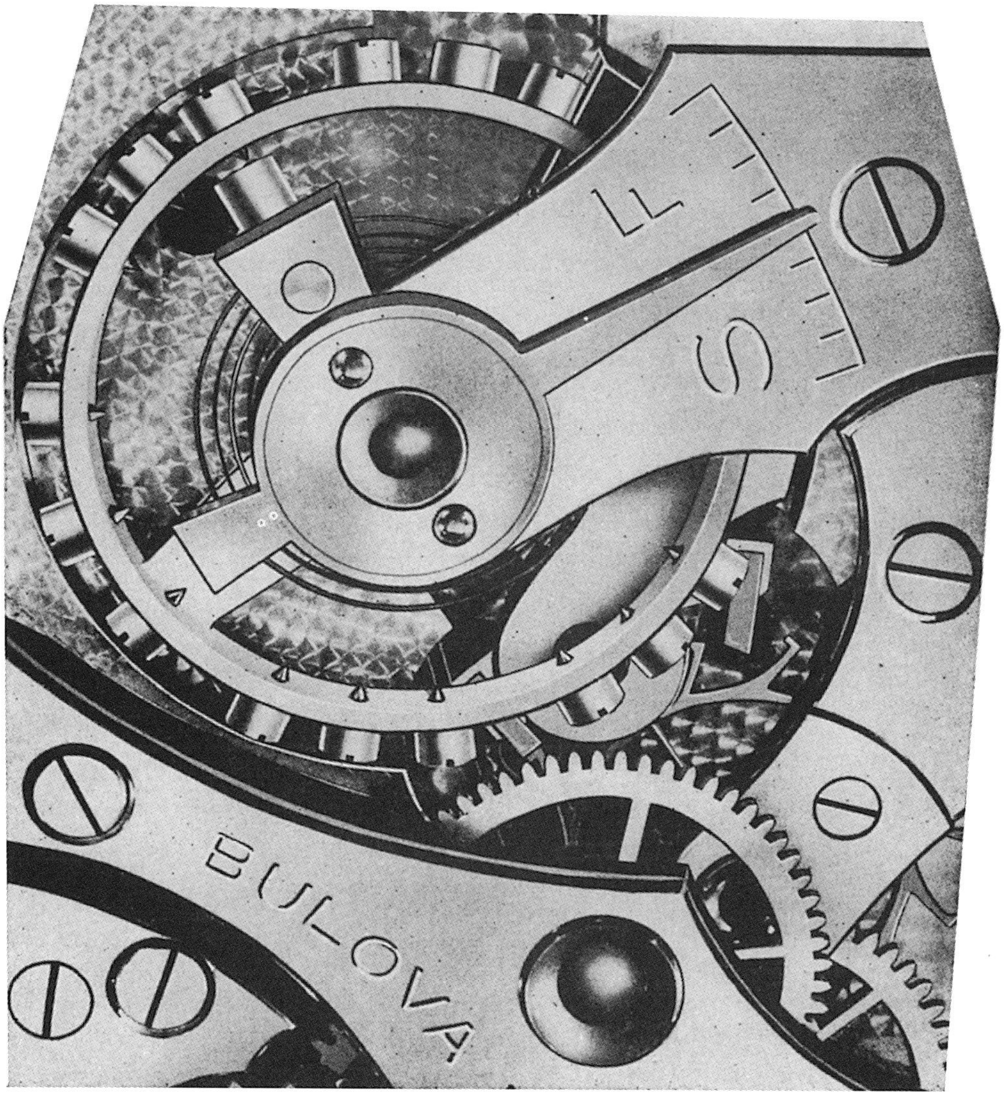
F
S
BULOVA

BULOVA SCHOOL of WATCHMAKING

SUBJECT:

Finishing.

OBJECTIVE:

To present the basic requirements for the location of the overcoil and hairspring stud — to the end that hairsprings can be fitted easily into a watch movement.

TOOLS NEEDED:

1. Hairspring Leveler (Double End).
2. Regulator Pin Opener.
3. Hairspring Beat Tool and Hairspring Removing Tool.
4. Escapement Trying Tool.

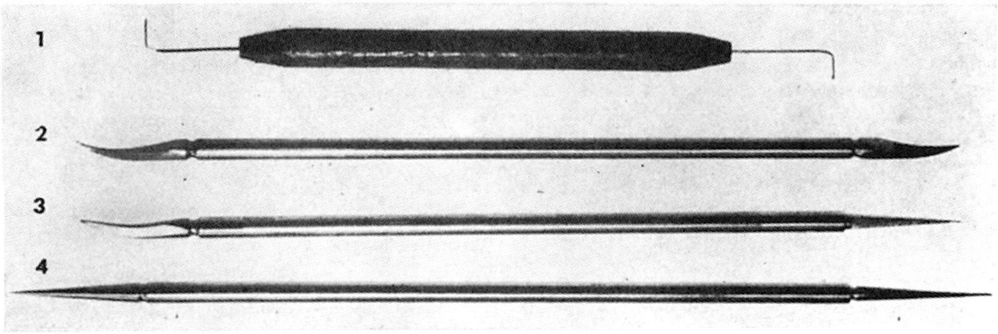

FINISHING

The art of finishing will be presented under the following main subdivisions:

- Finishing
- Common Terms Used in Finishing
- Correct Conditions for a Properly Fitted Hairspring
- Circling the Overcoil — Analysis of Overcoil Errors
- Vibrating the Spring between Regulator Pins and Circling the Overcoil
- Centering the Spiral Portion of the Spring
- Leveling the Spiral Portion of the Spring
- Errors Which Will Result in Out of Level Conditions
- Combining Out of Center and Out of Level Errors
- How to Put a Watch in Beat

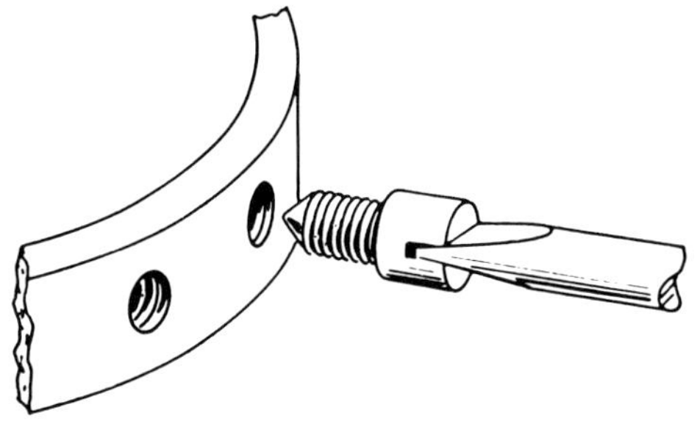

Balance screws must be tight in the balance wheel.

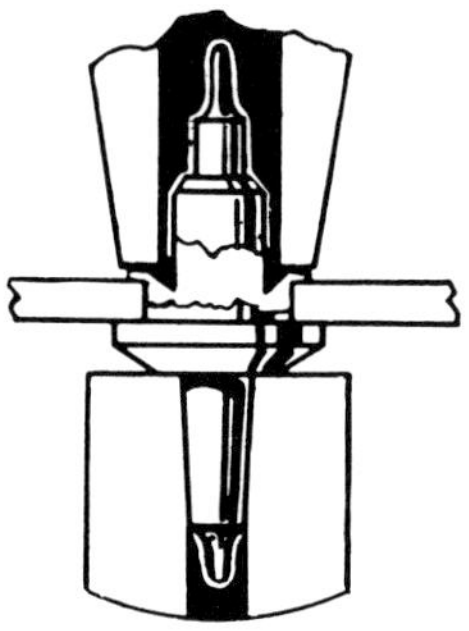

Balance staff must be staked tight on the balance wheel.

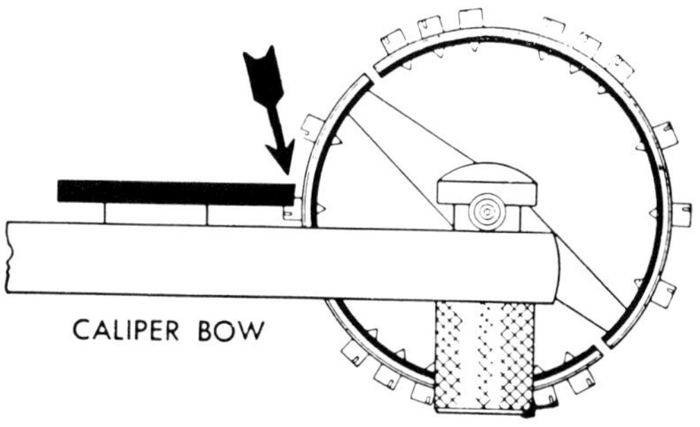

The balance wheel must be true in the round and flat.

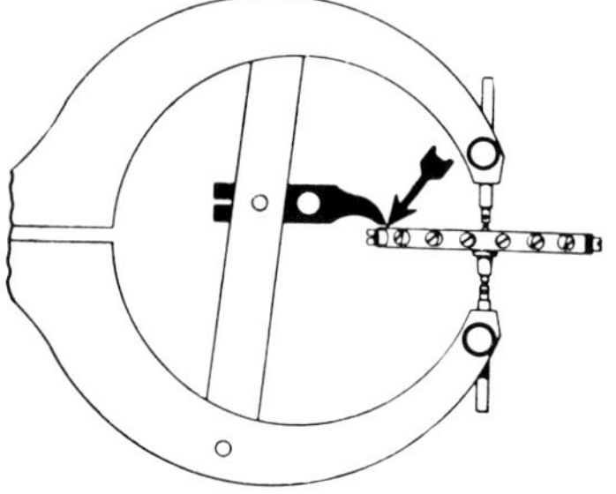

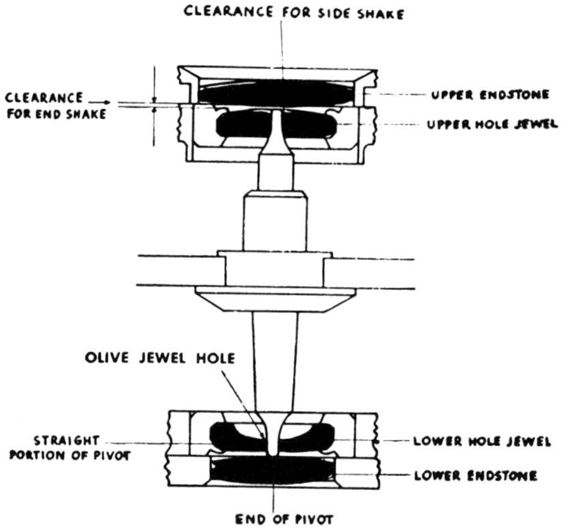

Pivots must be correctly shaped, properly burnished and accurately fitted to the jewels.

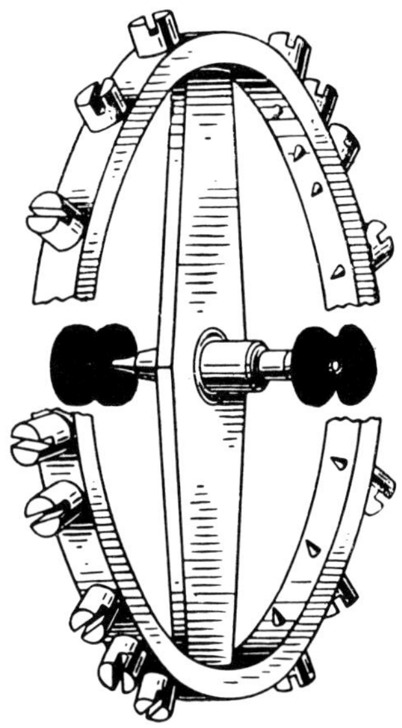

The balance wheel must be poised.

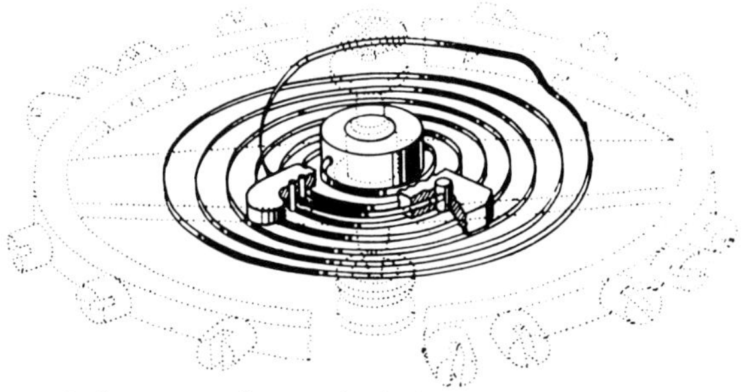

Collet must be staked firmly to the staff, the spring must be true in the round and flat.

FINISHING

Finishing should be approached with the assumption that the following conditions exist:

1. That the Balance wheel is tight on the balance staff. (See Training Unit No. 1).
2. That the Balance wheel is true in the round and flat. (See Training Unit No. 2).
3. That the Balance Pivots are correctly shaped, properly burnished and accurately fitted to the jewels. (See Training Unit No. 4.).
4. That the Balance Wheel is Poised. (See Training Unit No. 5.).
5. That the Collet is correctly staked and the spring is true in the round and flat. (See Training Unit No. 6).
6. That the Hairspring stud is tight and properly attached to the balance cock. (See training Unit No. 7).
7. That the Overcoil is the correct height and the correct shape. (See Training Unit No. 8).
8. That the Barrel Assembly and complete Train are free. (See Training Unit No. 9).
9. That the Escape Wheel and Pallet are properly matched. (See Training Unit No. 10).
10. Every moving part from the Barrel to the Balance Wheel should be free and properly fitted.

COMMON TERMS USED IN FINISHING

Hairspring OUT OF CENTER.

Hairspring OUT OF LEVEL.

Overcoil OUT OF LEVEL.

CIRCLE OVERCOIL.

VIBRATE OVERCOIL.

CLOSE REGULATOR PINS.

OPEN REGULATOR PINS.

MOTION. (Of the balance wheel)

PUT ESCAPEMENT IN BEAT.

CORRECT CONDITIONS FOR A PROPERLY FITTED HAIRSPRING

When the Overcoil is the correct height and shape and the stud is properly attached, the conditions shown in Fig. 1 will be established.

Here it is shown that the distance from the balance staff center to the stud and the points F-E-D-C. are the same as the distance from the balance jewel hole to these same points. Therefore the stud will go into the stud hole, and the overcoil will fit between the regulator pins, permitting all of the coils of the spiral to be equally spaced and centered around the balance center.

Fig. 1 shows that the overcoil is the proper distance above the main body of the spiral. This insures that the spiral portion will be level. When the overcoil is too high, the spiral will be low. When the overcoil is too low, then the spiral will be too high.

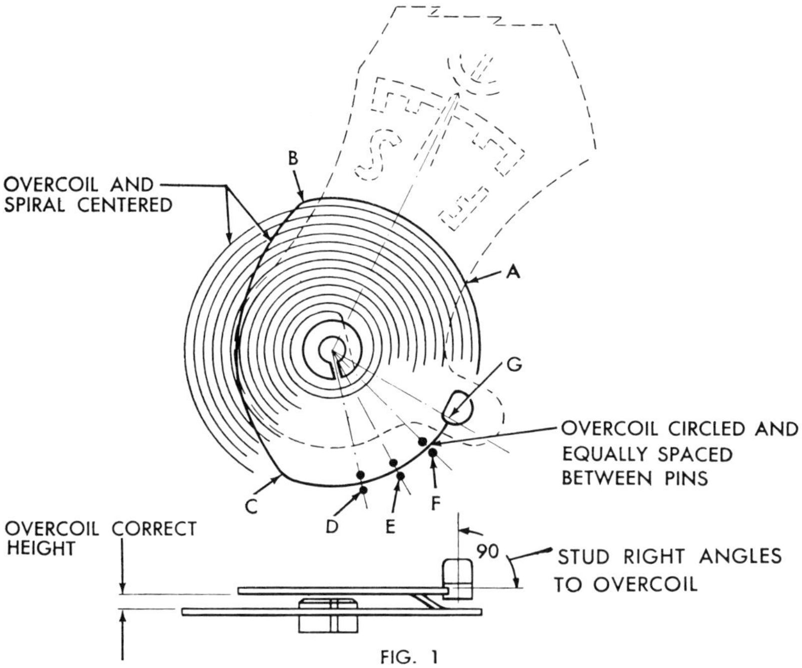

FIG. 1

E, REGULATOR PINS, WITH REGULATOR ON CENTER.

F, REGULATOR PINS, WITH REGULATOR TO SLOW.

D, REGULATOR PINS, WITH REGULATOR TO FAST.

CIRCLING THE OVERCOIL—ANALYSIS OF OVERCOIL ERRORS

Figure 2 shows the overcoil bent towards the balance staff. This is a direct result of a bend at point G. When the stud is placed in the balance cock, the overcoil should conform to the conditions shown in Fig. 1, but because of the bend at G, the overcoil will gradually curve inward towards the balance staff. The correction of this condition is known as "Circling the Overcoil."

Inspection of the watch which reveals this condition proceeds as follows:

1. Place regulator at the "slow" side of the balance cock, which moves the regulator pins to point *F*.

2. Move regulator from "slow" to "fast," causing regulator pins to move from *F* to *D*.

As the regulator pins move from *F* to *D*, the overcoil will move away from the balance staff. This shows that the overcoil is not circled. To correct this condition, move the regulator so the regulator pins will be as close to the stud as possible, then bend the overcoil in the direction of the arrow. Repeat the operation of moving the regulator from "slow" to "fast" and again observe the overcoil to determine if it remains motionless, or if it moves either towards the staff or away from the staff.

Figure 3 shows the overcoil bent away from the balance staff. From the explanation covering the condition in Fig. 2, it is possible to make the following statements in reference to circling the overcoil.

1. When the overcoil moves away from the balance staff, as the regulator pins are moved away from the stud, it will be necessary to bend the overcoil away from the staff to correct the error. (Fig. 2.)

2. When the overcoil moves towards the staff, as the regulator pins are moved away from the stud, it will be necessary to bend the overcoil toward the staff. (Fig. 3.)

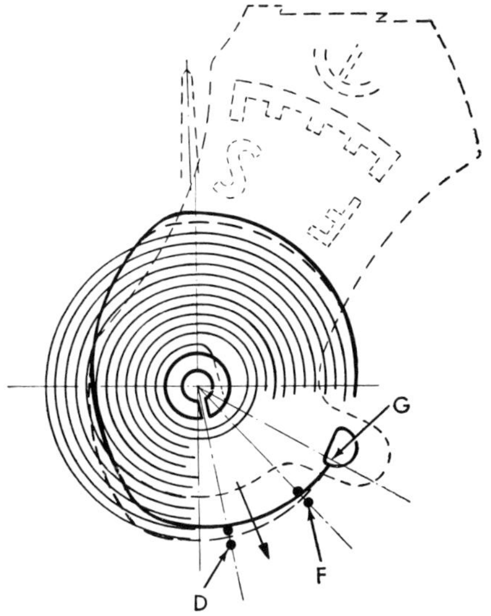

FIG. 2

CHART SHOWING OVERCOIL BENT TOWARD BALANCE STAFF. ARROW SHOWS DIRECTION OVERCOIL MUST BE BENT TO CORRECT.

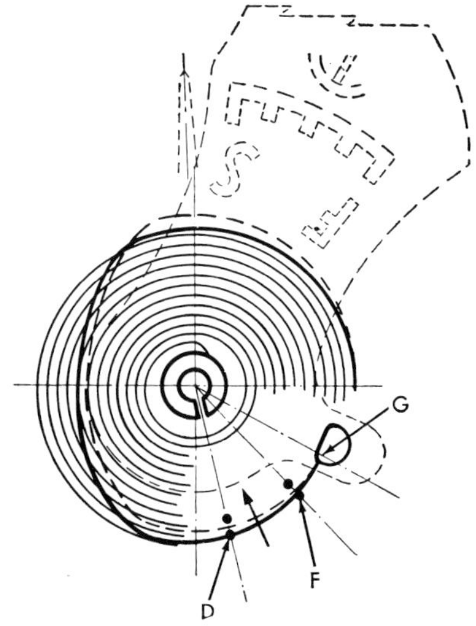

FIG. 3

CHART SHOWING OVERCOIL BENT AWAY FROM BALANCE STAFF. ARROW SHOWS DIRECTION OVERCOIL MUST BE BENT TO CORRECT.

VIBRATING THE SPRING BETWEEN REGULATOR PINS AND CIRCLING THE OVERCOIL

After the overcoil has been bent away from the balance staff as explained in Fig. 2, the regulator pins should be moved to position *F*. Look at the position of the overcoil between the pins; it will probably conform to one of the following conditions:

1. Equally spaced and pins properly closed. (Fig. 4.)
2. Equally spaced and pins slightly open. (Fig. 5.)
3. Touching inside pin. (Fig. 6.)
4. Touching outside pin. (Fig. 7.)
5. Pins too far apart and overcoil touching bottom of pins only. (Fig. 8.)
6. Pins too far apart but outside pin bent so end of pin will be parallel with pin. (Fig. 9.)

If the overcoil touches either the inside or outside pin (Fig. 6-7), the overcoil should be bent so as to bring it more central between the pins.

Let us now assume that the overcoil is directly in the center of the two pins (Fig. 5), make sure of this condition by stopping the balance wheel with the jewel pin in the fork slot and observe the location of the overcoil with a double eyeglass. Now move the regulator so the pins will be located at *D*, look at the overcoil again and decide if it still remains in the same central position between the pins. If it does, it can be seen that when the regulator pins are closed up to the spring that it will be possible to move the regulator from "slow" to "fast," carrying the pins from *F* to *D* without disturbing the location of the overcoil.

This method of locating the overcoil in the center of the regulator pins for all points from *F* to *D* accomplishes two operations at once; that is:

1. Vibrates the spring between the regulator pins.
2. Circles the overcoil.

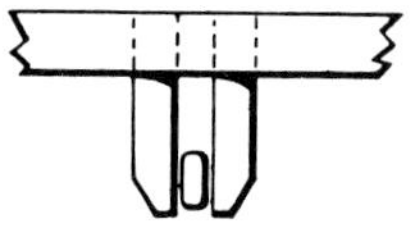

FIG. 4

NO. 1. EQUALLY SPACED AND PINS PROPERLY CLOSED. FREEDOM FOR OVERCOIL SHOULD BE .0003" OR .0076 MM.

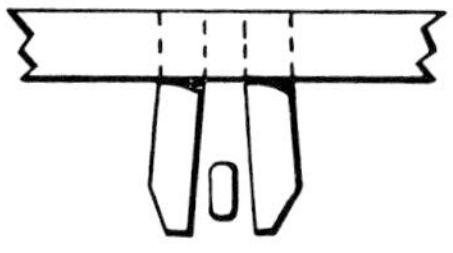

FIG. 5

NO. 2. EQUALLY SPACED AND PINS SLIGHTLY OPEN.

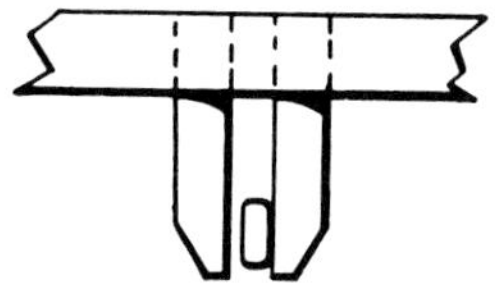

FIG. 6

NO. 3. SPRING TOUCHING INSIDE PIN.

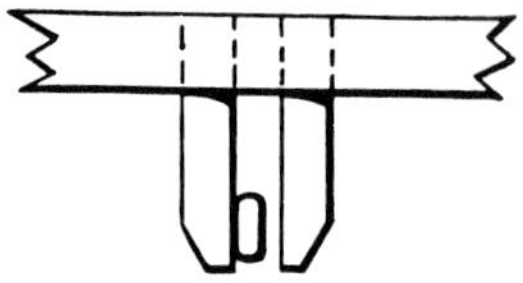

FIG. 7

NO. 4. SPRING TOUCHING OUTSIDE PIN

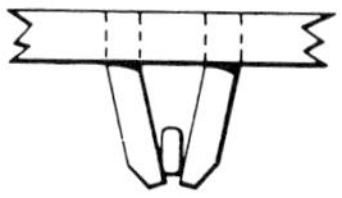

FIG. 8

NO. 5. PINS LOCATED TOO FAR APART IN REGULATOR, PINS AT ANGLE, SPRING TOUCHES PINS AT BOTTOM. THIS WILL CAUSE ERROR IN RATE BETWEEN THE DIAL UP AND DIAL DOWN POSITIONS.

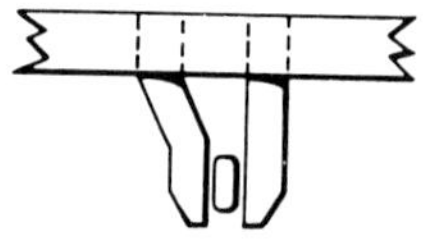

FIG. 9

NO. 6. PINS LOCATED TOO FAR APART IN REGULATO BUT OUTSIDE PIN IS BENT SO THAT END OF PIN WILL BE PARALLEL WITH INSIDE PIN.

CENTERING THE SPIRAL PORTION OF THE SPRING

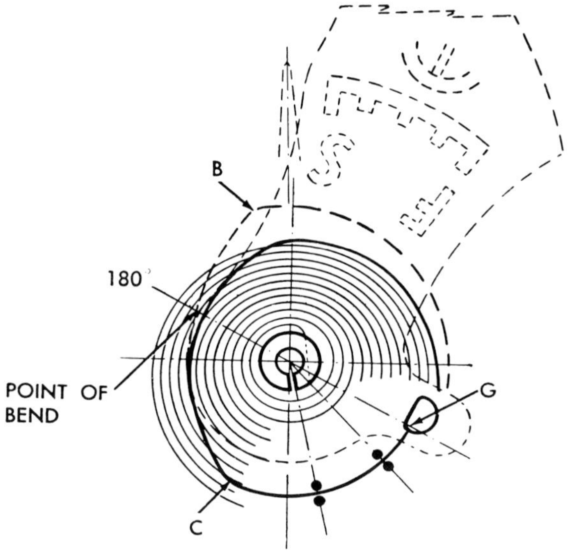

FIG. 10

CHART SHOWING OVERCOIL BENT 180 FROM STUD, BENDING POINT B TOWARD STAFF.

Figure 10 shows the overcoil bent 180° from the stud, bending point *B* towards the balance staff. The overcoil C-G under this condition will no longer lie over the top of the second coil as shown in Fig. 1, but depending upon the amount of bending at the point 180° from the stud, the overcoil may lie over the third or fourth coil. All of the coils underneath the regulator pins will be wider apart than the coils underneath the balance cock and the spring will be "out of center."

The remedy for this error is to grasp the spring with a pair of good tweezers 180° from the stud and bend the overcoil in such a manner as to carry point *B* away from the balance staff.

Figure 11 is just the reverse from Fig. 10 and all of the errors and the remedy stated above will be reversed.

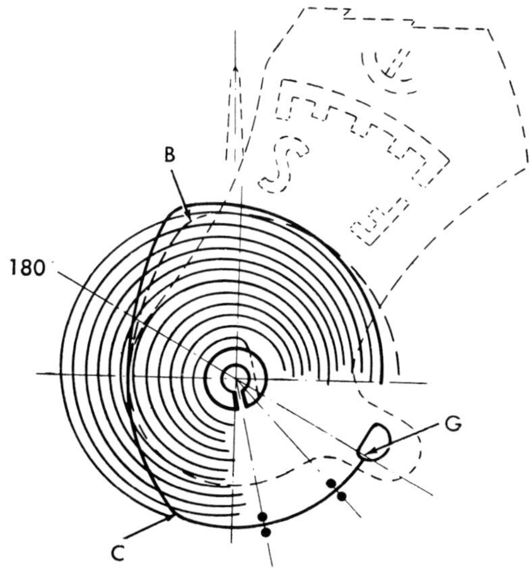

FIG. 11

CHART SHOWING OVERCOIL BENT 180° FROM STUD, BENDING POINT B AWAY FROM STAFF.

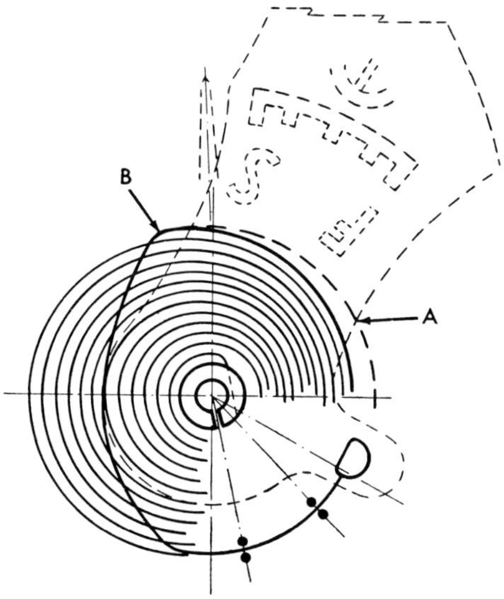

FIG. 12

CHART SHOWING OVERCOIL BENT AT POINT B, TAKING POINT A TOWARD STAFF.

Figure 12 shows the overcoil bent at point *B* carrying point *A* towards the staff. This will cause the spring to be "out of center." The remedy is to bend the spring at point *B* so as to carry point *A* away from the staff.

CENTERING THE SPIRAL PORTION OF THE SPRING

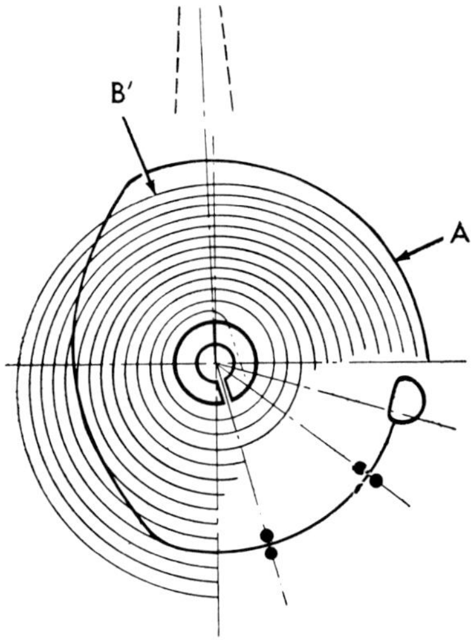

Fig. 13

CHART SHOWING OVERCOIL BENT AT POINT A, TAKING POINT B′ TOWARD STAFF.

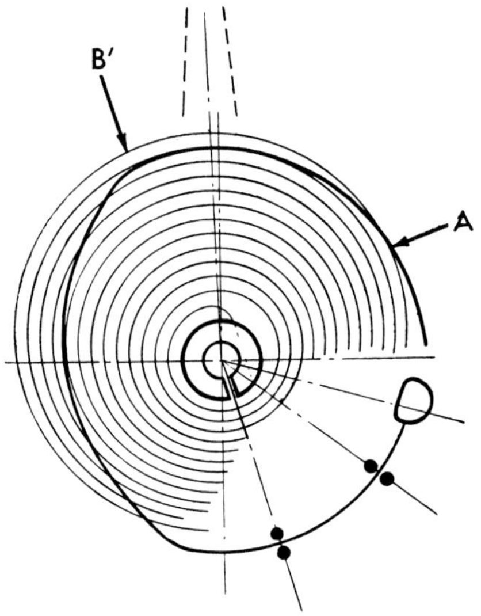

Fig. 14

CHART SHOWING OVERCOIL BENT AT POINT A, TAKING POINT B′ AWAY FROM STAFF.

If all the figures from 1 to 14 are studied carefully, keeping in mind the conditions shown in Fig. 1, one begins to form a mental picture of all the fundamental operations of finishing.

It also can be seen that in each case, the error caused by a bend at any point always throws the spring out of center in the same manner as described in Figs. 6-7-8-Training Unit No. 6.

This provides a means of readily locating the point to grasp the spring for centering.

LEVELING THE SPIRAL PORTION OF THE SPRING

In order that the spiral portion of the spring be level when placed in a watch, the overcoil must be the correct height above the spiral.

This height must be equal at all points of the overcoil. The stud must be attached perpendicular to the plane of the overcoil, observed from any position. (Fig. 15.)

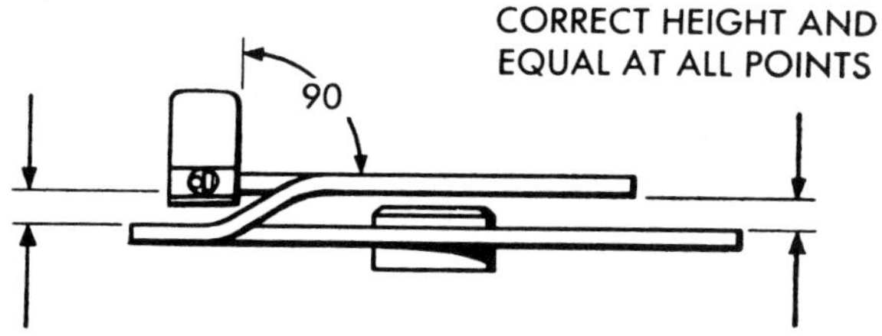

FIG. 15

ERRORS WHICH WILL RESULT IN OUT OF LEVEL CONDITIONS

OVERCOIL TOO HIGH. (Fig. 16.)

RESULT —Spiral part of spring will be too low when placed in the watch.

CORRECTION —Remove the balance and spring, then lower the overcoil to its correct height.

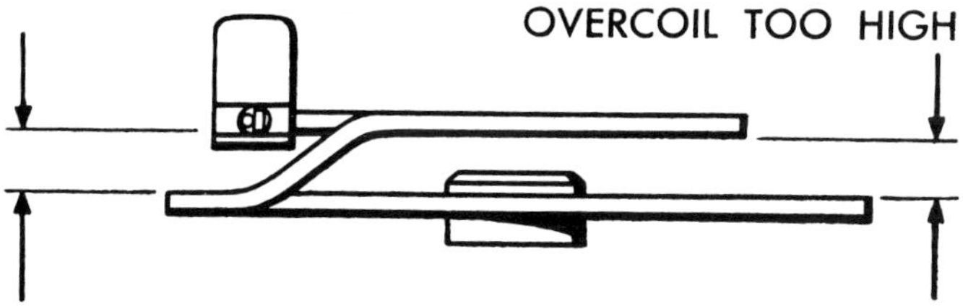

FIG. 16

OVERCOIL TOO LOW. (Fig. 17.)

RESULT —Spiral part of spring will be too high when placed in the watch.

CORRECTION —Remove the balance and spring, then raise the overcoil to its correct height.

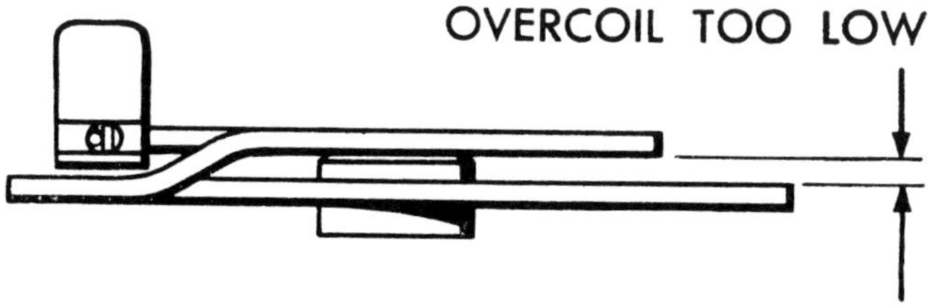

FIG. 17

STUD TOO HIGH. (Fig. 18.)

RESULT —Spiral part of the spring will be too low under the stud.

CORRECTION —Grasp the overcoil opposite the stud, at point A and bend the body of the spiral upwards, under the stud.

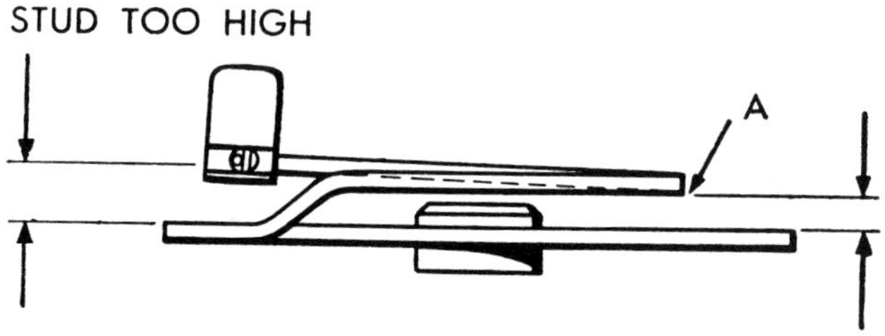

FIG. 18

STUD TOO LOW. (Fig. 19.)

RESULT —Spiral part of spring will be high under the stud.

CORRECTION —Grasp the overcoil opposite the stud, at point A and bend the body of the spiral downwards, under the stud.

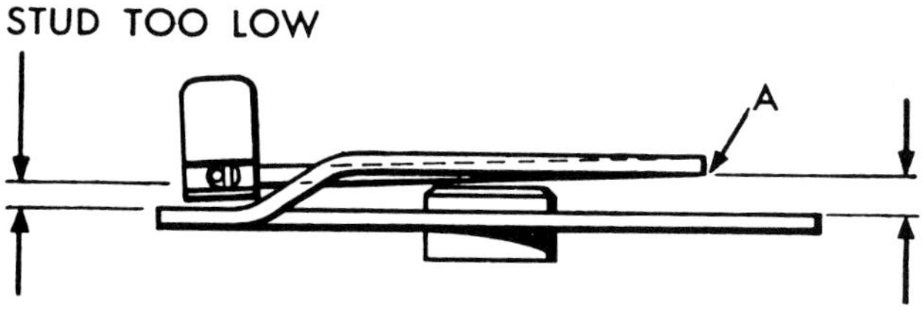

FIG. 19

LEVELING THE SPIRAL PORTION OF THE SPRING

OVERCOIL HIGH OPPOSITE THE STUD. (Fig. 20.)

RESULT — The spiral part of the spring will be too low opposite the stud.

CORRECTION — Grasp first coil of spiral under the stud at point B, and bend spiral up. Or, remove the balance, place on block, and bend the overcoil down, with a pair of tweezers.

OVERCOIL HIGH OPPOSITE STUD

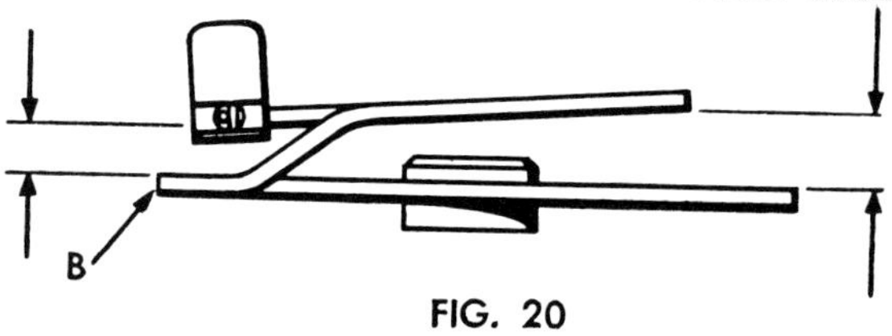

FIG. 20

OVERCOIL LOW OPPOSITE STUD. (Fig. 21.)

RESULT — The spiral part of the spring will be high opposite the stud.

CORRECTION — Grasp the first coil of spiral, under the stud at point B, and bend spiral down. Or remove the balance, place on block and bend the overcoil up.

OVERCOIL LOW OPPOSITE STUD

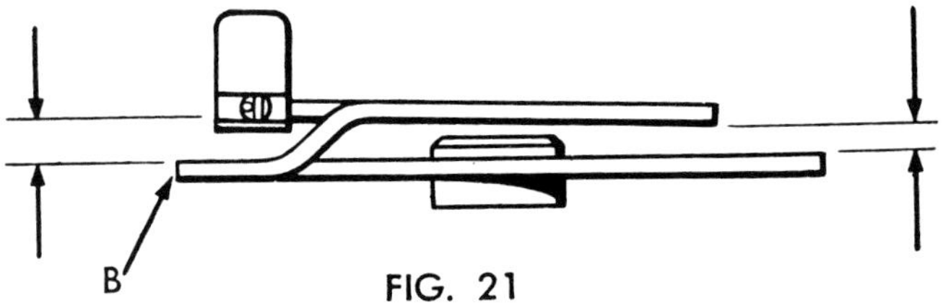

FIG. 21

OVERCOIL HIGH 90° TO LEFT OF STUD. (Fig. 22.)

RESULT — Spiral part will be low 90° to left of stud.

CORRECTION — Note that the bend will probably be at point C under the balance cock, which makes it difficult to grasp and bend. Safest method is to remove the balance, then correct overcoil by grasping spring at point C and bend point D down.

OVERCOIL HIGH 90° TO LEFT OF STUD

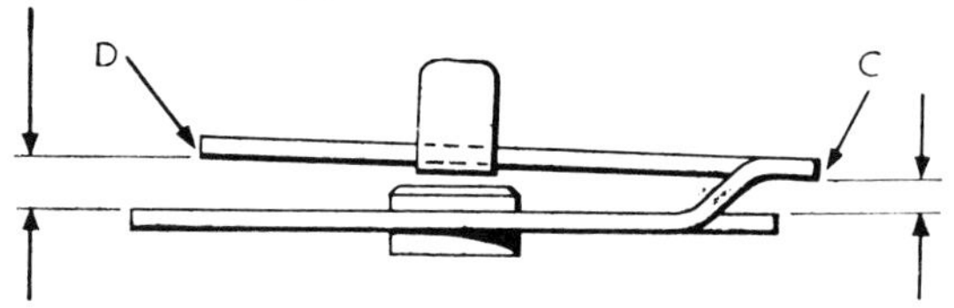

FIG. 22

OVERCOIL LOW 90° TO LEFT OF STUD. (Fig. 23.)

RESULT — Spiral part will be high 90° to left of stud.

CORRECTION — Remove the balance, place on block and raise overcoil, by grasping at point C and bend Point D up.

OVERCOIL LOW 90° TO LEFT OF STUD

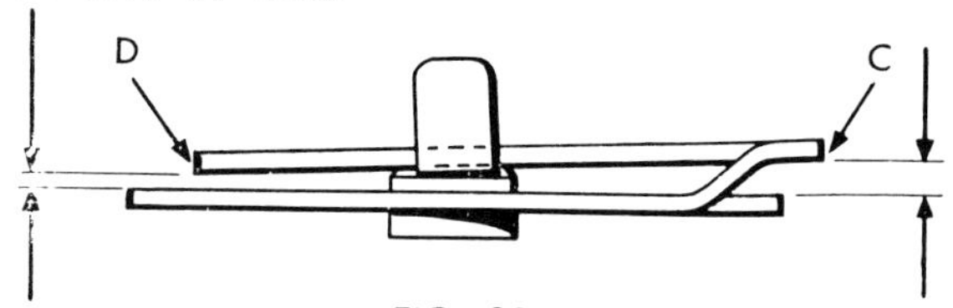

FIG. 23

OVERCOIL HIGH 90° TO RIGHT OF STUD. (Fig. 24.)

RESULT — Spiral will be low 90° to right of the stud.

CORRECTION — In this condition the spiral can be grasped at point E, and the outside coil of the spring bent down. NOTE—This error could have been the result of too great a rise at point F. Therefore overcoil could be lowered at bend up point F.

OVERCOIL HIGH 90° TO RIGHT OF STUD

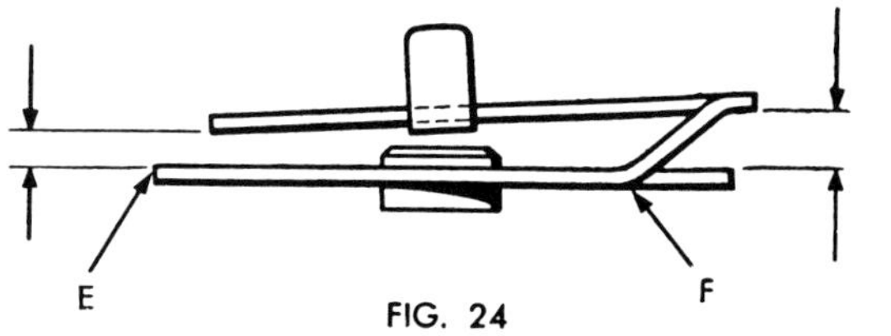

FIG. 24

OVERCOIL LOW 90° TO RIGHT OF STUD. (Fig. 25.)

RESULT — Spiral part of spring will be high 90° to the right of the stud.

CORRECTION — Grasp the spring at point E and bend the outside coil of the spiral upwards. Or raise overcoil at bend up point F.

OVERCOIL LOW 90° TO RIGHT OF STUD

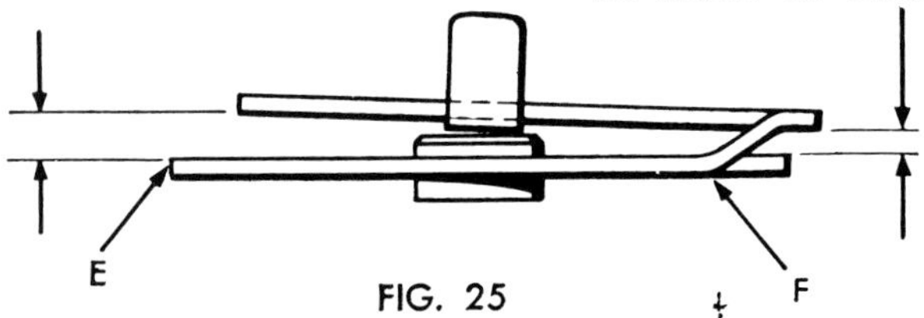

FIG. 25

LEVELING THE SPIRAL PORTION OF THE SPRING

STUDS NOT PERPENDICULAR TO THE PLANE OF THE OVERCOIL

ERROR — Stud bent away from the balance staff. (Fig. 26.)

RESULT— Overcoil and spiral low opposite the stud.

CORRECTION— Grasp spring at stud and bend overcoil up.

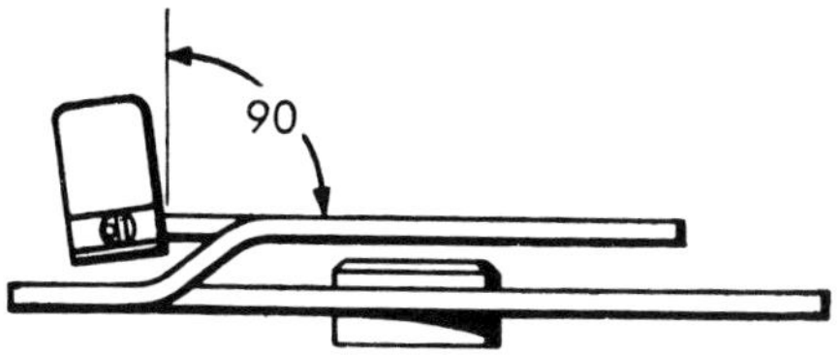

FIG. 26

STUD BENT AWAY FROM BALANCE STAFF.

ERROR — Stud bent toward the balance staff. (Fig. 27.)

RESULT— Overcoil and spiral high opposite the stud.

CORRECTION— Grasp spring at stud and bend overcoil down.

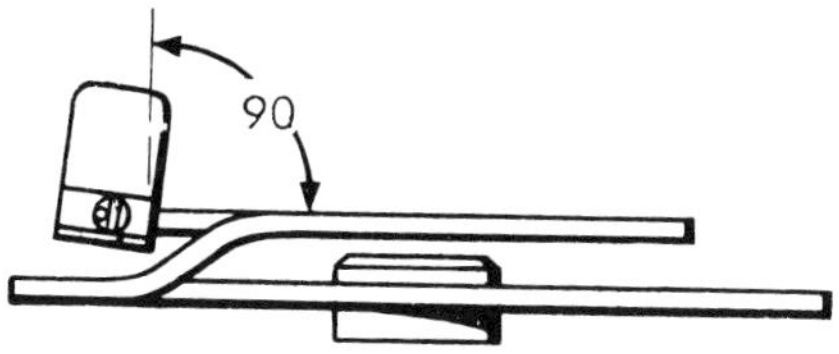

FIG. 27

STUD BENT TOWARD BALANCE STAFF.

ERROR — Stud bent to the right of the balance staff. (Fig. 28.)

RESULT— Overcoil and spring low to the left of staff.

CORRECTION— Place tweezers under overcoil at stud and bend up.

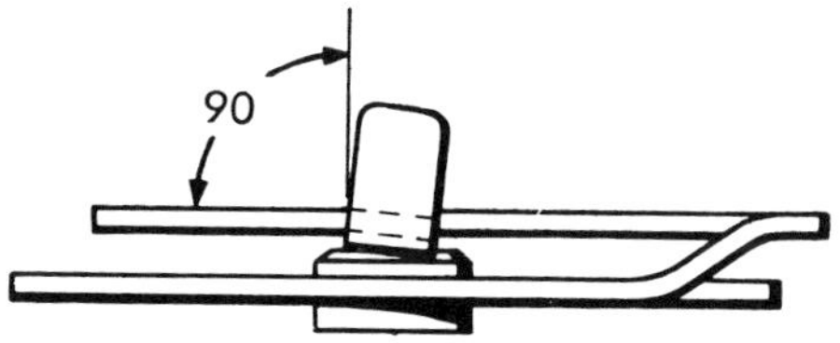

FIG. 28

STUD BENT TO RIGHT OF BALANCE STAFF.

ERROR — Stud bent to the left of the balance staff. (Fig. 29.)

RESULT— Overcoil and spring high to the left of staff.

CORRECTION— Place tweezers on top of the overcoil, at stud, and bend overcoil down.

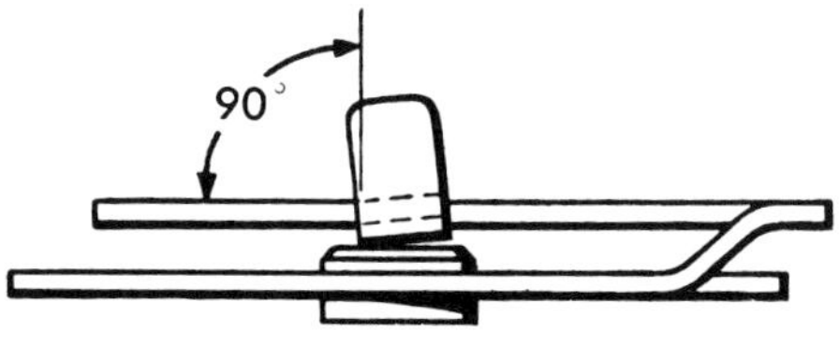

FIG. 29

STUD BENT TO LEFT OF BALANCE STAFF.

It must be obvious from the foregoing analysis of OUT OF LEVEL conditions, that finishing becomes necessary because of the slight errors in forming the overcoil, and attaching the stud. The actual bending operations necessary to correct these conditions can usually be determined from the error itself

COMBINING OUT OF CENTER AND OUT OF LEVEL ERRORS

Fig. 30 shows the spiral part of the spring out of center with the coils too wide apart directly under the regulator pins, and too close to the stud in the flat position.

Grasp the spring opposite the stud at point A, as explained in the manipulative operations, and turn the tweezers counter-clockwise, at the same time tipping the tweezers towards the balance staff. This will center and level the spring with one bend.

Double bends of this type can be used for practically all conditions where a double error exists.

It is not possible to illustrate correctly all of the various combinations of errors that do exist. However, it can be seen that finishing requires the ability to perform all of the bends explained under MANIPULATIVE OPERATIONS and to combine this ability with a knowledge of the errors which exist in either the OUT OF CENTER or OUT OF LEVEL condition of the overcoil or spiral portion of the spring.

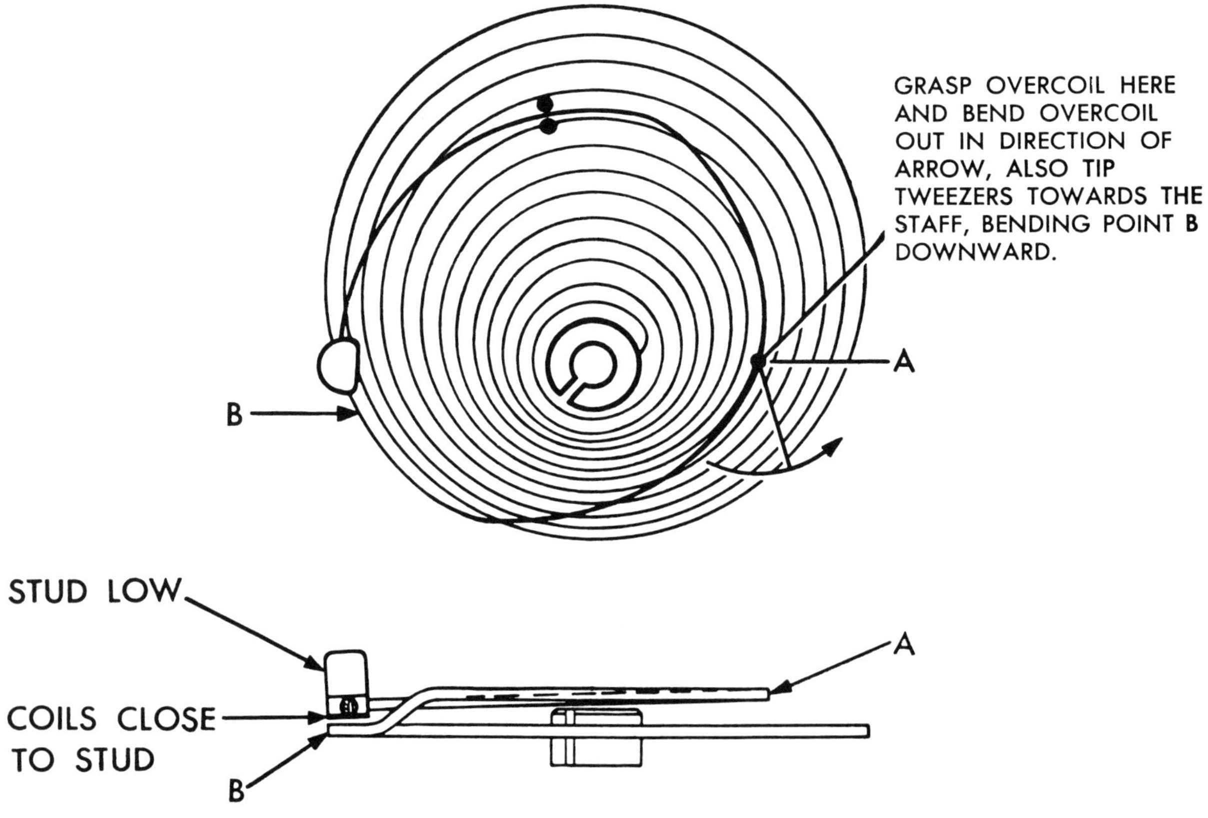

FIG. 30

CENTERING A FLAT HAIRSPRING

1. Remove hairspring from its balance wheel and insert stud into balance bridge. See that top of stud is flush with top of bridge and tighten stud screw.

2. Hairspring collet should lie directly over balance hole jewel. If necessary adjust hairspring to center collet over the jewel hole.

3. It is now important to examine the distance between stud and jewel hole and compare with the distance between regulator pins and jewel hole. We will refer to the stud distance as the stud radius and to the regulator pin distance as the regulator pin radius.

   A. When the radii are equal length as shown in Fig.31, it may be necessary to bend outer coil very slightly outward at point A. to prevent the second coil from touching the inside regulator pin or stud.

   B. If the stud radius is longer than the radius of the regulator pins, make a sharp bend in the hairspring near the stud as shown in Fig.32. A careful adjustment at this point can shorten the stud radius to be exactly equal to the radius of the regulator pins and so provide the proper regulator arc.

4. It should be emphasized that such adjustments can disturb the centering of the collet over the balance jewel so that this should be rechecked and corrected if necessary. Also be sure on completing the necessary adjustments that the regulator can be moved freely in either direction without changing the position of the hairspring on the balance bridge.

5. Stake hairspring to balance wheel and insert the balance assembly into the watch movement. Recheck the hairspring centering in the movement and make any necessary final adjustments.

TWO TYPES OF CENTERING FLAT HAIRSPRINGS

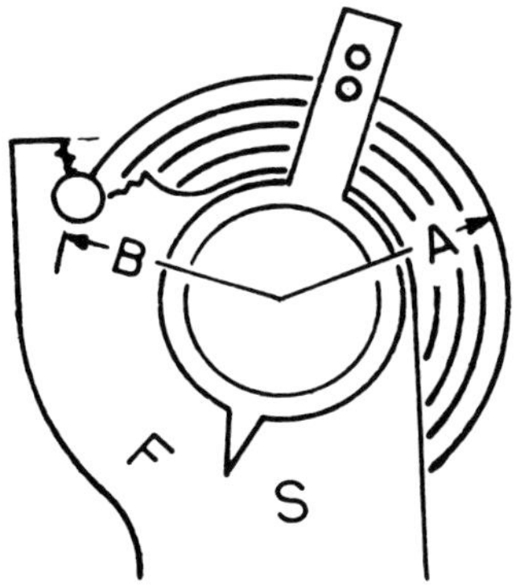

Reg. pin "A" and stud "B" radii the same.

Fig. 31

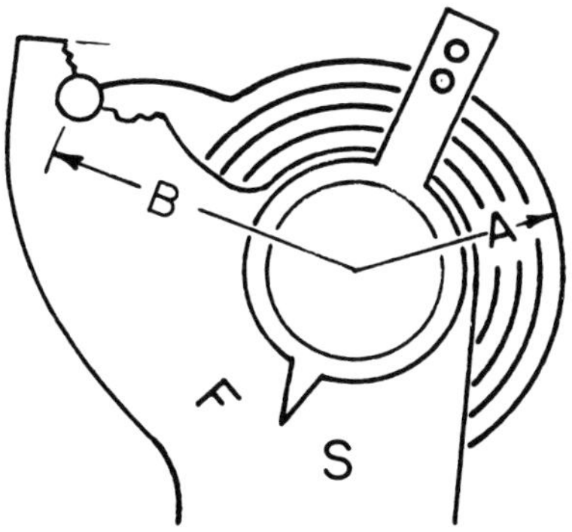

Reg. pin radius shorter "A" than stud radius "B"

Fig. 32

HOW TO PUT A WATCH *IN BEAT*

After the overcoil has been leveled, circled and vibrated and the spiral portion of the spring leveled and centered, the balance and hairspring must be put IN BEAT.

The function of the hairspring is to return the jewel pin to the line of centers of the escape, pallet and balance in a pre-determined time. From this can be developed the statement that a watch is considered to be in beat if the jewel pin lies on this line of centers when the hairspring is unstressed. See Fig. 33 Or it can be stated that a watch is in beat when the hairspring is unstressed and the jewel pin is midway betweeen the two banking pins, on condition that the banking pins are properly located.

A watch is considered to be out of beat when the hairspring is attached to the balance staff in such a position that when the hairspring is unstressed, it will hold the jewel pin some number of degrees away from the line of centers.

A simple way to determine whether or not a watch is in beat is to hold the watch in the dial down position. Then by use of a beat tool proceed to stop the balance, that is, try to get an escape wheel tooth to stay locked on either the R or the L Stone of the pallet. This operation must be done gently and care must be taken not to damage the balance pivots or scratch the balance wheel. It can best be done by allowing a balance screw to hit against the beat tool. If it is impossible to stop the motion of the balance, it may be considered that the watch is in beat.

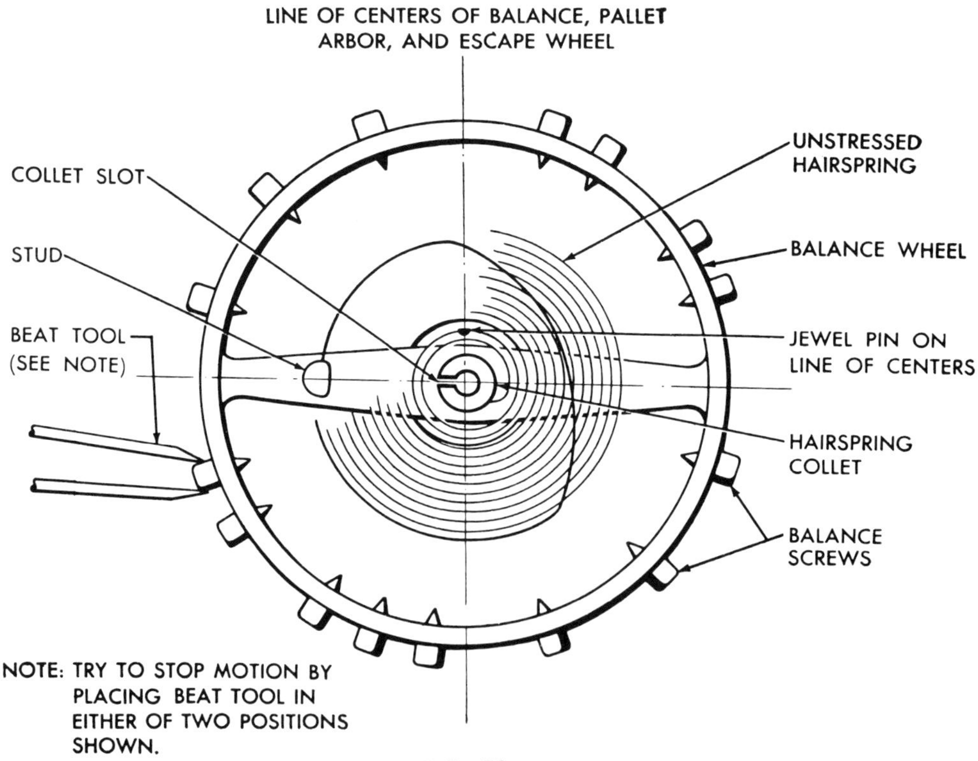

FIG. 33

WATCH IN DD POSITION.

HOW TO PUT A WATCH *IN BEAT*—*(Continued)*

If it is found that the balance can be stopped, the next step is to test for the direction of the starting push. This is done with the beat tool which is carefully brought into contact with a balance screw. When the slightest touch or push releases the escapement and starts the balance in motion, it can be considered to be the correct direction of the starting push. Had the beat tool been put on the opposite side of the same screw and pushed, the push would have been applied in the wrong direction and the escapement would not have been released. It is the starting push that causes the jewel pin to hit against the fork slot of the pallet which in turn unlocks the tooth of the escape wheel. We are thus able to determine on which side of the line of centers the jewel pin was when the motion of the balance wheel was arrested. This information tells us that when the hairspring is unstressed, the jewel pin will lie on that side.

To correct this error and thus bring the watch in beat, we may use either one of the following two methods:

1. Hold the collet by inserting the beat tool into the collet slot. Then proceed to turn the balance wheel in the same direction as the starting push.

2. Hold the balance wheel and with the beat tool move the collet in the opposite direction of the starting push.

Only experience will teach us how much to move the collet relative to the balance staff in order to put the escapement in beat.

Fig. 34 is a case where the balance wheel has been stopped with the beat tool, the pallet re-

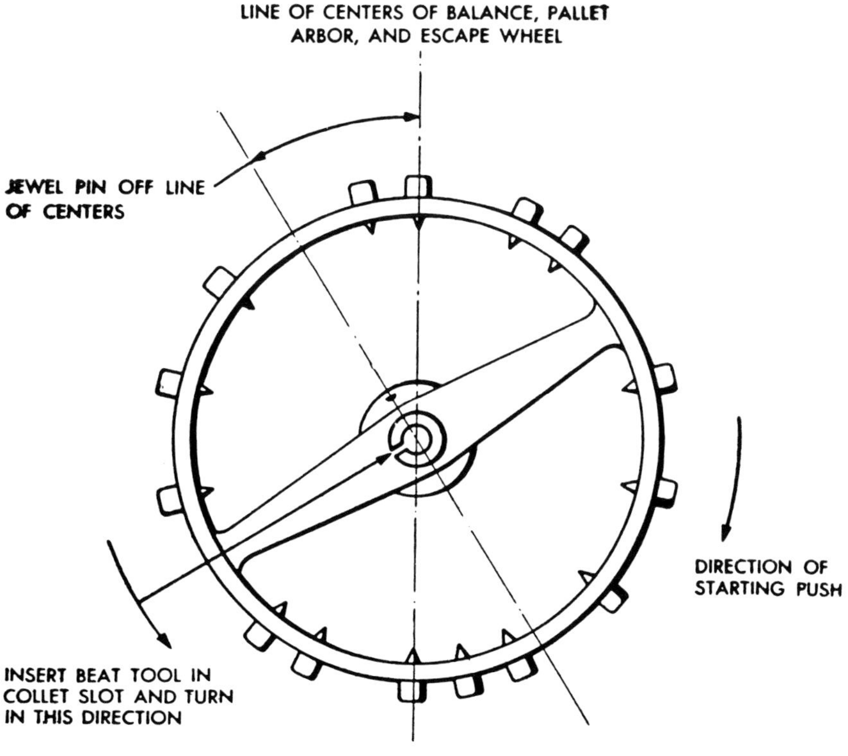

FIG. 34

HOW TO PUT A WATCH *IN BEAT*—*(Continued)*

maining locked on the let-off or L stone. The jewel pin, as indicated, is positioned to the left of the line of centers. (The balance wheel is nearest the observer while the escape wheel is the farthest away.) From what has been explained before, it is understood that the watch is out of beat and that the jewel pin would lie to the left of the line of centers when the hairspring was unstresed. Under this particular circumstance, it would be found that by pushing the balance with the beat tool in the direction of the arrow, the balance would continue to turn in that direction until it reached the end of a vibration. This proves that the jewel pin was to the left of the line of centers.

The remedy for this error is to employ any one of the two aforementioned methods. Either method achieves the same objective, namely, that of shifting the hairspring collet so that the jewel pin will be on the line of centers.

Fig. 35 shows a case where the balance wheel has been stopped, the pallet remaining locked on the receiving or R stone. The jewel pin, as indicated, is positioned to the right of the line of centers (balance nearest the observer-escape wheel farthest away). We have an indication here that the watch is out of beat and that the jewel pin would be to the right of the line of centers when the hairspring is unstressed.

By use of either of the two above mentioned methods, the collet can be shifted relative to

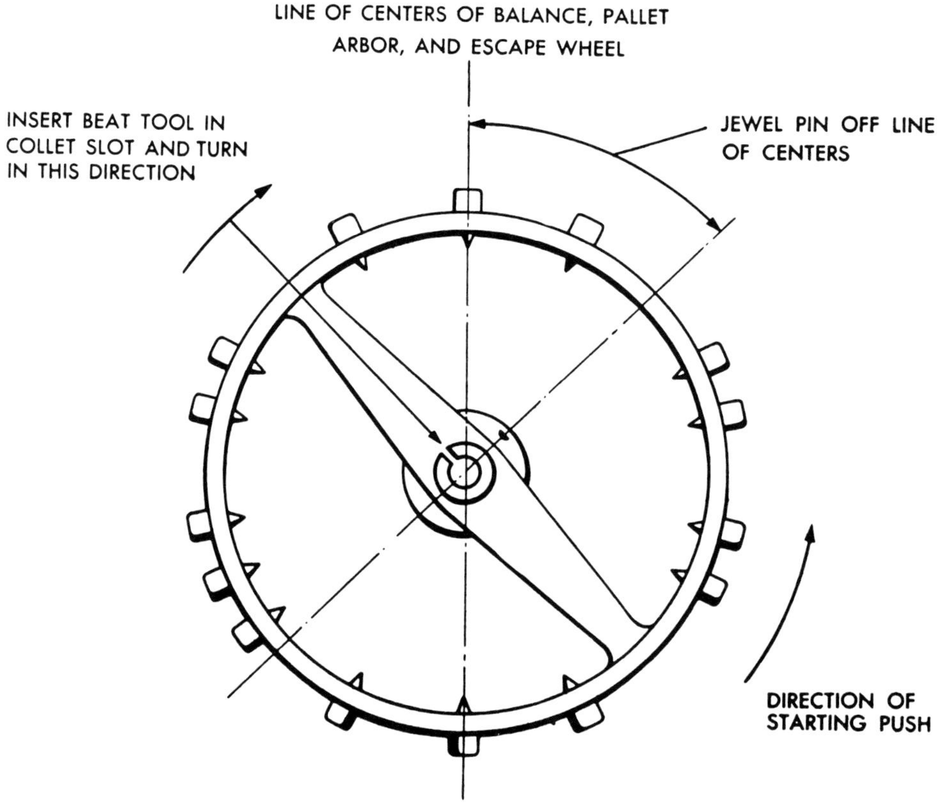

FIG. 35

the balance staff and thus the jewel pin will be brought to the line of centers, thereby putting the watch in beat.

When it is impossible to reach the collet while the balance wheel is in the watch, then it is advisable to remove the balance cock before attempting to shift the hairspring collet. This method will avoid the possibility of damage to the hairspring or pivots.

As shown in Fig. 36, the balance cock is placed on a tapered piece of pegwood or brass, resting in a hole of the bench anvil. The suspended balance wheel is then held in the fingers of the left hand as the collet is shifted to its proper position by carefully locating the beat tool held in the right hand through the coils of the hairspring into the collet slot.

The direction in which the collet is moved is exactly the same as explained when putting the watch in beat without removing the balance cock; namely, turn the collet clockwise when a counter-clockwise movement of the balance wheel is desired, and vice versa.

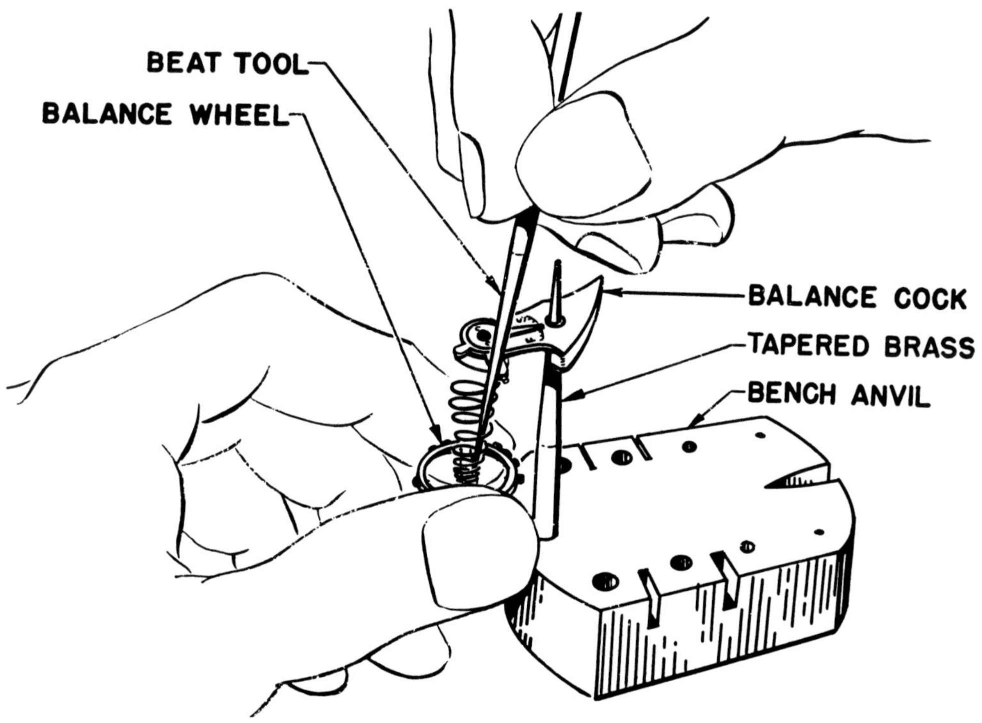

FIG. 36

JOSEPH BULOVA SCHOOL OF WATCHMAKING

NOTES ON TEMPERATURE COMPENSATION

Temperature compensation has to do with the rate of the balance wheel and hairspring combination when the temperature changes. If the combination is correctly designed and set up, then there will be little or no change in rate as the temperature varies between 45°F and 95°F. In the field of watchmaking there are two combinations of balance wheel and hairspring that are met with: 1. Bimetallic wheel using either a steel or a palladium hairspring. 2. Monometallic wheel using an alloy hairspring of which Elinvar is the most common.

The Bimetallic Wheel—Steel or Palladium Hairspring Combination

The bimetallic wheel is a self-compensating wheel. It is made up of two strips of metal fused together in a circular form. A strip of brass is on the outside and one of steel is on the inside. Since brass has the property of expanding more than steel when the temperature rises and of contracting more than steel when the temperature drops, we have a way of offsetting or compensating for the change in elasticity in the steel or palladium hairspring when the temperature changes.

A steel or palladium hairspring will vibrate at a slower rate when the temperature rises because its elasticity is reduced. The same hairspring will vibrate at a faster rate when the temperature drops because its elasticity is increased. In its broadest terms, elasticity is that property of a piece of material that causes it to return to its unstressed or relaxed condition after pressure or tension has been removed.

By combining a self-compensating wheel (bi-metallic) with a steel spring, the following things will happen when the temperature rises: The hairspring which now has less elasticity would of itself vibrate at a slower rate—but at the same time the rims of the bimetallic wheel would curve inward toward the staff and by so doing would offer a smaller load for the weaker hairspring to handle. The whole assembly thus would continue to vibrate at nearly the same rate as before the temperature changed.

A reduced temperature would increase the elasticity of the hairspring and make it tend to vibrate faster because its elasticity increased. But at the same time, the rims of the balance wheel would move away from the staff, thus increasing the load on the stronger hairspring. The net effect then would be for the balance wheel assembly to vibrate the same as before the temperature dropped.

From the foregoing it can be seen that the location of the screws on the balance wheel is important to the temperature compensating ability of the assembly. The manufacturers of a particular bimetallic hairspring assembly have, after proper research, located the screws so that the wheel will be neither over nor under-compensated. By running a series of tests at 45°F and 95°F they can finally locate the screws so that the rate of vibration will remain nearly constant regardless of the temperature. Thus it can be seen that, if a watchmaker alters the location of balance wheel screws on a bimetallic wheel, he will upset the temperature compensation. For example, if a pair of opposite screws are moved from a point near the arms to a point near the cut or free end, the wheel would vibrate faster in the heat and slower in the cold. The opposite effect would take place if screws were moved from the cut end toward the arm. The watch would lose in the heat and gain in the cold.

The Monometallic Wheel—Alloy Hairspring Combination

This combination is a vast improvement over the above-described combination. Around the turn of the century the French scientist, Guillaume, developed a type of alloy for hairsprings whose elasticity did not change when the temperature changed. This alloy is known as Elinvar, based on two words—*elasticity* and *invariable*. It does not rust or retain magnetism as does a steel hairspring. Because its elasticity does not change, a solid or monometallic wheel can be used with it. This type of solid wheel, which is uncut, retains its form in round and flat much better than the bimetallic wheel and it does not rust or retain magnetism as does the bimetallic wheel.

How to Differentiate between Hairspring Types

The best way to determine if a hairspring is made of steel, Elinvar or palladium, is by use of a compass and a small magnet.

Lay the hairspring on a piece of stiff paper and approach from beneath with the magnet. If the spring stays motionless as the nearby magnet is moved, then it is made of palladium. Palladium is not attracted by a magnet, neither does it retain magnetism. If the spring shows signs of being drawn by the moving magnet, then it is either steel or Elinvar. If it is steel it would retain its magnetism as evidenced by use of the compass. If it does not affect the compass, then the spring is made of Elinvar. When employing the compass technique, lay the spring to be tested on a flat, non-magnetic surface and set the compass down beside it. Push the compass with peg wood along a circular path around the outside of the spring. If the compass needle stays pointed towards the north throughout one circular path around the spring, then we know that there is no magnetism in the spring and that it is made of Elinvar. If the compass needle swings back and forth while being moved, then we know that the spring is magnetized and is made of steel.

Mean Time Screws

Mean time screws are used for regulating the rate of vibration of the balance wheel. If a wheel has two of them they are located near the ends of the balance arms. If there are four of them, the extra pair are located approximately 90 degrees away from the arms. By unscrewing an opposite pair of them, one must do so on equal amounts. This will slow down the rate in all positions. By screwing in an opposite pair, the rate of vibration will be increased in all positions. They are never to be used to poise the wheel.

HOW OLD IS THIS WATCH?

Watchmakers know when a watch has been in service for 20 or more years, it can no longer be considered a routine repair. Parts are worn and often unavilable.

To guide you in properly estimating repairs on older watches we have prepared this reference charts. The dates of manufacture are approximate.

Approximate Dates of
Manufacture of AMERICAN WATCHES, Year and Number

WALTHAM	ELGIN	HAMILTON	BALL	ILLINOIS
1942 30,750,000	1942 39,147,000	1942 C 2,100	1942 654,000	1940 5,650,000
1941 30,566,000	1941 38,777,000	1941 C 1,500	1941 652,000	1930 5,100,000
1940 30,560,000	1940 38,410,000	1940 C 1,001	1940 651,000	1920 4,500,000
1939 30,000,000	1939 38,400,000	1939 2,750,000	1939 650,000	1910 3,800,000
1938 29,630,000	1938 37,000,000	1938 2,650,000	1938 647,000	1900 3,200,000
1935 28,000,000	1935 36,000,000	1935 2,600,000	1935 641,000	See *
1930 27,500,000	1930 33,000,000	1930 2,500,000	1931 803,000	**HOWARD
1925 24,500,000	1925 28,000,000	1925 2,250,000	1930 801,000 637,000	1,500,000
1920 23,000,000	1920 23,000,000	1920 1,845,000	1928 800,000	1920 1,400,000
1915 21,000,000	1915 18,500,000	1915 1,450,000	1925 620,000 270,000	1910 1,010,000
1910 16,000,000	1910 15,000,000	1910 1,250,000	1920 610,000 260,000	1900 850,000
1905 11,850,000	1905 12,000,000	1905 1,125,000 See ***	1915 603,000 250,000	
1900 8,827,977	1900 9,000,000	1900 1,015,000	1910 216,201 600,000	

NOTE: During 1940 to 1942, same Walthams with numbers 28 to 30 Million were released using plates made but not previously used. Check factory for correct dates of these.

NOTE: During 1940 to 1942, some Elgins with numbers 29 to 38 Million were released using plates made but not previously used. Check factory for correct dates of these.

NOTE: A new watch 992B with Friction jeweling and improved Elenvar was made in 1940. **Elenvar was used after 1930. *** 18S #940 Model discontinued 1910.

NOTE: *Illinois Watches out of production since 1940. **Howard Watches out of production since 1930.

GENERAL REPAIR INFORMATION

When ordering a crown you need the following:

1. Tap
2. Diameter of crown
3. Size of tube if water proof
4. Post (Flush, medium, long)
5. Style
6. Color

TAP SIZE CHARTS

For Bracelet Crowns	
.85 to .95 mm. Thread	10
.95 to 1.05 mm. Thread	9
1.05 to 1.15 mm. Thread	8
1.15 to 1.25 mm. Thread	7
1.25 to 1.35 mm. Thread	6

For Pocket Watch Crowns	
1.15 to 1.25 mm. Thread	3
1.40 to 1.50 mm. Thread	2
1.80 to 1.90 mm. Thread	1
2.10 to 2.20 mm. Thread	0

EXAMINATION OF THE WATCH PRIOR TO DISASSEMBLY AND REPAIR

THE BAND

1. Does it need repairing or replacement?
2. If the old strap has a gold buckle use it on the new strap.
3. If the band is a piece of jewelry, check the stone settings and band links. Inform the customer of any defects.
4. Be sure the spring bars are right for the case and check their spring tension.

THE CASE

1. Is the case in proper condition to protect the movement? Is it free of holes and does it close properly? If not, can it be repaired by you or must it be sent to a case expert?
2. Are the case loops or lugs in good condition?
3. Does the bezel hold the crystal properly? Does it fit tightly on to the frame or back of the case?
4. Is a new crystal necessary? Since there are many kinds and qualities of crystals available, get the customer to agree to the type recommended.
5. If the case is jeweled, check the stones and their settings. Inform the customer if any stones are missing or broken.
6. Check the fit of the movement in the case.
7. Be sure that the right style of crown is on and properly fitted.

DIAL, HAND, WINDING AND SETTING MECHANISM

Does the dial need to be refinished? Are the hands mismatched, loose, bent, rusty, worn or in need of refinishing?

REMOVE MOVEMENT FROM THE CASE

Since there are several types of cases in use to-day, care must be taken to use the proper approach and opener appropriate to the type on hand. Be sure the opening is done over the skin drawer so that any loose parts will not be lost.

When watches are fitted with male and female stem and crown combinations, the crown may be snapped off by using cutting pliers or tweezers as a wedge. Always be sure of the proper procedure before attempting to pry or to press the case apart when opening it.

There are several types of case-locking devices in use today and it cannot be too strongly urged upon the student that he avail himself of all the literature that he can that will help him understand the various structural features that are used in case work. Special tools are used in opening some cases and if this is not known or the proper procedure followed, then damage may be done to the case and movement.

If it is an automatic watch, check for marks on inside of the back and on outside of the oscillating weight or movement which will indicate that the weight was rubbing these places.

When the stem must be removed in order to remove the movement from the case, first pull the stem into the setting position, then loosen the setting lever screw. Remove the stem. After removing the movement from the case, replace the stem in the movement and then check the fit of the stem bearing in the movement.

BALANCE WHEEL ASSEMBLY

1. If the Balance Assembly is free, check the wheel for truth in the flat and round. Are there any screws missing or loose?
2. Is the hairspring centered and level?

A general observation of the movement at this stage may save a lot of valuable time as some movements are beyond repair and often no further disassembly is needed to determine this.

DISASSEMBLY

Before proceeding with the disassembly, it is well to demagnetize the watch. This will prevent small screws from clinging to the tweezers and screw drivers.

1. Remove the hands, taking due precautions not to scratch the dial. Then remove the dial, dial washer, hour wheel and canon pinion. If free, remove minute wheel.

    a. Check the fit of the stem bearing in the movement.
    b. Check the meshing of the clutch wheel with the winding pinion, and the winding pinion with the crown wheel. Look out for broken, rusty or worn teeth.
    c. Pull the stem into the setting position and check the meshing of the clutch wheel with the setting wheel. Some setting wheels have a beveled end, which end is placed on the post facing down. Check the setting lever spring. Is it strong enough to hold the clutch wheel and setting wheel together tightly when in the setting position?
    d. Check for freedom of the setting wheel and minute wheel.
    e. Check the clutch lever spring.
    f. Check the setting mechanism as a whole, for any signs of rust. Remove the rust if there is not too much or else replace the part.

2. Let the mainspring down gradually. If it is an automatic, remove as much of the automatic mechanism as is necessary to permit the letting down of the mainspring.
3. Loosen the stud screw. Note the position of the regulator index because it may be located at the correct regulation point.
4. Remove the balance bridge screw and bridge.

    a. If there is no boot on the regulator, the balance assembly will drop free from the bridge.
    b. When there is a boot, the balance bridge and assembly must be turned upside down on the bench and the boot turned away to free the hairspring.
    c. Check for bent, broken or rusty pivots. Check the roller to see if it is loose. Check the roller jewel for looseness and to see if it is broken. Is there excess cement on the active side of the jewel?

5. If the watch has an indirect drive sweep-second mechanism, it should be removed at this time.
6. Check the pallet end shake. Remove the pallet bridge with great care by not tilting it too much as this may break or bend the pallet arbor pivots.
7. Check the end shakes of all train wheels. Spin the train and check the truth of the wheels in the flat. If the fourth wheel has a long pivot for the second hand, check the pivot to see if it spins true. Also check the center wheel post to see if it spins true. A fault in either of these two places could stop the watch and definitely cause the hands to wobble.

    Remove the remaining bridge screws together with the ratchet and crown wheel screws. After removing the ratchet and crown wheel, take care to remove the steel center around which the crown wheel turns. This center is a reinforcement for the brass boss which is part of the barrel bridge.

    Nearly all crown wheel screws have left hand threads and some, but not all, ratchet wheelscrews have left hand threads.

    Take particular note of the length of the barrel bridge screws. Later it will be necessary to replace the shorter ones in their proper place or else damage to some of the winding and setting parts will take place.

    Take special note of the particular characteristics that the screws may have. There are left and right hand screws, screws with shoulders under the screw head, screws of different diameters and lengths. When reassembling the watch it will be necessary to return screws to their proper place or else damage to female threads or to other watch parts will take place.

8. Remove the barrel bridge.
9. Remove the train wheel bridge or bridges.
    When removing the train wheel bridge be sure that it is not tilted too much in any one direction and that no pivots are sticking in their holes. A rusty or bent pivot will stick in a hole jewel.

10. Remove the train wheels.
    a. Check their teeth and pinion leaves for rust, burrs, bends or breakage.
    b. Check the condition of the pivots. Are any broken, bent or rusty?
11. Remove the barrel.
12. Remove the barrel cover. Take note of the direction that the mainspring is wound around the barrel arbor.
13. Remove the mainspring.
14. Remove the cap jewels.
    a. Peg all hole jewels. Check to see if the jewels are cracked and whether or not they are loose or out of upright.
    b. If the watch is provided with shock resisting jewels, they must be disassembled and assembled according to their manufacturer's specifications. Their hole jewels should be pegged. All of the shock-resisting jewel units can be disassembled by picking off the cap jewel by the use of scotch tape in case the cap jewel is stuck on so tight that it will not separate in the cleaning process.
15. Disassemble all winding and setting parts.
16. Repair or replace all parts that have been discovered to be defective.
17. Before removing the hairspring observe the location of the stud as it may be exactly on or near the in-beat point.
18. Check the balance wheel in the truing calipers for truth in the flat and round.
19. Check to see if the balance wheel is poised.

CLEANING THE WATCH

Cleaning always follows disassembly and repair. If the job is done by use of the cleaning machine, then care must be exercised by placing small, delicate parts together and away from heavier parts to avoid damage to the more fragile ones.

There is something to be said for grouping related parts in the same compartment of the basket which will help out in the assembly procedure. Be sure that the parts having long pivots are so placed in the basket that the pivots do not protrude through the wire mesh.

The cleaning time in the machine may be reduced if the plates, bridges, and other parts have been brushed and the jewels pegged. Two or three minutes in each solution should be sufficient.

It is recommended that the balance wheel assembly and pallet be left out of the basket and hand-cleaned.

If the watch is to be cleaned by hand, then brush the parts and peg and jewels first. Follow this by stringing the bridges on a wire, which is then immersed in the cleaning solution and allowed to stand for a few moments and then gently moved back and forth until clean. Next place the wire and parts in the first rinsing solution and proceed as before. A second rinsing solution treatment should follow for the best results. After this, a gentle movement of the parts while still on the wire, through a good grade of heated boxwood sawdust, will dry them. Blow the parts free of dust particles with a hand blower. A thorough examination should be made to be sure that all particles have been removed.

There is on the market a tiny, fine-meshed basket suitable for holding any and all screws found in watches. This provides an efficient and effective means of cleaning screws.

Do not put water-proof crowns in the cleaning solutions, which may have a detrimental chemical effect on the crown gaskets.

Watch bands, excepting leather ones, or those having leather inserts, should be given a preliminary cleaning to remove heavy, loose dirt. Follow this with buffing, then cleaning in a good jewelry cleaning solution. A solution of ammonia, soap and water is very good. This should be followed by using a stiff brush with soap and water. Rinse in alcohol and thoroughly dry.

Follow the same cleaning procedure outlined for the band when cleaning the case. If the case has an unbreakable crystal then it should be removed to avoid possible damage from the cleaning solution. Do not remove the crystal if it is of the permo, seal type and do not use cleaning solutions. Soap and water should be sufficient.

In all cases where the case gaskets are other than metal, they should be kept away from cleaning solutions.

Enameled dials are the only ones that can safely be washed in any cleaning solution. Appropriate precautions must be taken with all other types of dial finishes.

ASSEMBLY

1. Replace all cap jewels.
   All cap jewels, excepting those for the pallet, should be oiled before assembling the train. Check for the oil ring in all cap jewel places.
2. In some watches the winding pinion can be inserted only from the barrel bridge side. In that case put the winding pinion in now. In all other cases it can be inserted from the dial side when assembling the winding and setting mechanism.
3. Oil the mainspring before putting it into the barrel. After the main spring is inserted, put the barrel arbor in place and oil. Be sure it hooks on to the inner terminal of the mainspring and with a good snug grip. Put the cap on.
4. Place the setting lever screw in place, then the setting lever in the pillar plate.
5. Insert the barrel first if it is a direct drive sweep second watch. Put the center wheel in place, then its bridge, followed by its screws and then oil the upper pivot before putting on the center second bearing plate.

6. Insert the train wheels.
7. Put on the train wheel bridge or bridges and their screws.
8. Put on the barrel bridge and its screws, being careful not to place a long screw where a short one should go.
9. The click and its screw, click spring, crown wheel, ratchet wheel and their screws may now be placed.
10. Install the winding and setting mechanism and lubricate.
11. Oil the train wheel pivots. Oil the center wheel post.
12. Try the train reversal test. Place the cannon pinion on the center wheel and again try the train reversal test.
13. Assemble the sweep second mechanism if there is one.
14. Oil the pallet stones on their impulse faces. Place the pallet in the pillar plate followed by its bridge and screws.
15. Check for draw by moving the pallet away from the banking pin, not sufficient to unlock, and see if it pulls back again to the banking pin. Check the pallet end shake, lock and slide.
16. Put the balance wheel assembly and its bridge and screws in place.
17. Check the hairspring for levelness and centeredness. Check the regulator arc to see it it is concentric with the regulator. Does the hairspring vibrate evenly between the regulator pins as the regulator is moved from fast to slow?
18. Put the watch in beat.
19. Wind the watch and check the motion of the balance wheel.
20. Check the rate in the timing machine if one is available.
21. Place the hands in position.

FAULTS

The following list of things to be checked as a matter of routine in disassembly and assembly on general repair work should be learned early and well. The failure to do so could set the stage for poor time keeping by the watch and inevitable comeback work:

Hands touching the dial.
Hands touching the crystal.
Hands touching hands.
Hour hand with sloppy and excessive end shake, indicating need for a dial washer.
Hour hand having no end shake due possibly to its not needing a dial washer or to a bent dial.
Dial feet loose or too long. Dial screws not holding. When cam-locking screws are present be sure they hold firmly.
Case should be checked for dents, burrs, failure to close evenly or whether or not it presses on the watch movement, which might stop it.
Hour wheel binding on the cannon pinion.
Hour wheel tooth bent or missing.
Cannon pinion loose.
Cannon pinion burred and binding in the hour wheel.

Cannon pinion too tight, which will bend or break minute wheel teeth or the minute wheel post when attempting to set the watch.
Minute wheel post loose or missing. Teeth bent or missing.
Setting wheel binding too tight on its post.
Setting wheel not meshing properly with the clutch wheel.
Stem poorly fitted or worn. Loose crown.
Stem bearing does not fit case opening.
Stem pilot loose in its hole or too short or long.
Stem slot too wide or too narrow.
Winding pinion loose or worn and does not wind or set properly.
Clutch wheel slips on stem square when winding or setting.
Clutch spring broken.
Yoke or setting bridge rusty. Setting lever spring broken.
Setting lever broken or rusty.
Crown wheel loose, slipping, teeth binding worn or missing. Not meshing with the ratchet wheel.
Ratchet wheel loose, screw broken, not turning true, missing the click or crown wheel.
Mainspring broken, too strong, too weak or slipping.
Barrel cover on wrong, not turning true. Bent, burred or missing teeth. Binding on the arbor. Excessive end shake of the arbor in the barrel or no end shake. Barrel rubbing pillar plate, barrel bridge or center wheel.
Click or click spring broken or loose. Circular click spring lower leg too long will interfere with barrel and stop the watch.
Center wheel post bent causing hands to bind. Wheel out of flat and rubbing the barrel or bridges. Safety pinion loose will act like a broken mainspring. Teeth or pinion leaves missing. Not enough or too much end shake.
Third wheel, out of flat. On indirect sweep second watches, the long upper pivot is sometimes bent.
Fourth wheel, out of flat. Look for bent pivots or a cracked jewel. Sometimes the second hand is put on carelessly, causing damage. When the fourth wheel is also the center wheel, be sure the post is true and that the hole in the center wheel is clean.
Escape wheel, out of flat, rusty, damaged teeth or magnetized. Lower escape cap jewel screw too long. Broken or bent pivot. Wheel loose.
Pallet loose, broken or tilted stones. Bent fork, guard pin or pivots. No end shake. Out of upright or cracked hole jewels.
Balance Wheel Assembly - Wheel out of round or flat. Touching pallet bridge, center wheel stud or boot. Hairspring out of flat or round, rubbing the bridge, balance arms, center wheel, stud or pins. Collet spread or loose. Loose pin at the stud or collet. Balance staff hub cracked. Damaged or loose rivet. Wheel loose on staff. Roller crushed or loose. Loose or cracked roller jewel. Jewel not long enough to fully engage fork slot. Jewel too long and touching the guard pin. Pivots bent, broken, worn or mushroomed.

THE BULOVA 13AH CHRONOGRAPH

HOW IT FUNCTIONS
DISASSEMBLY
ASSEMBLY

BULOVA 13 AH CHRONOGRAPH

(BASIC CONSTRUCTION)

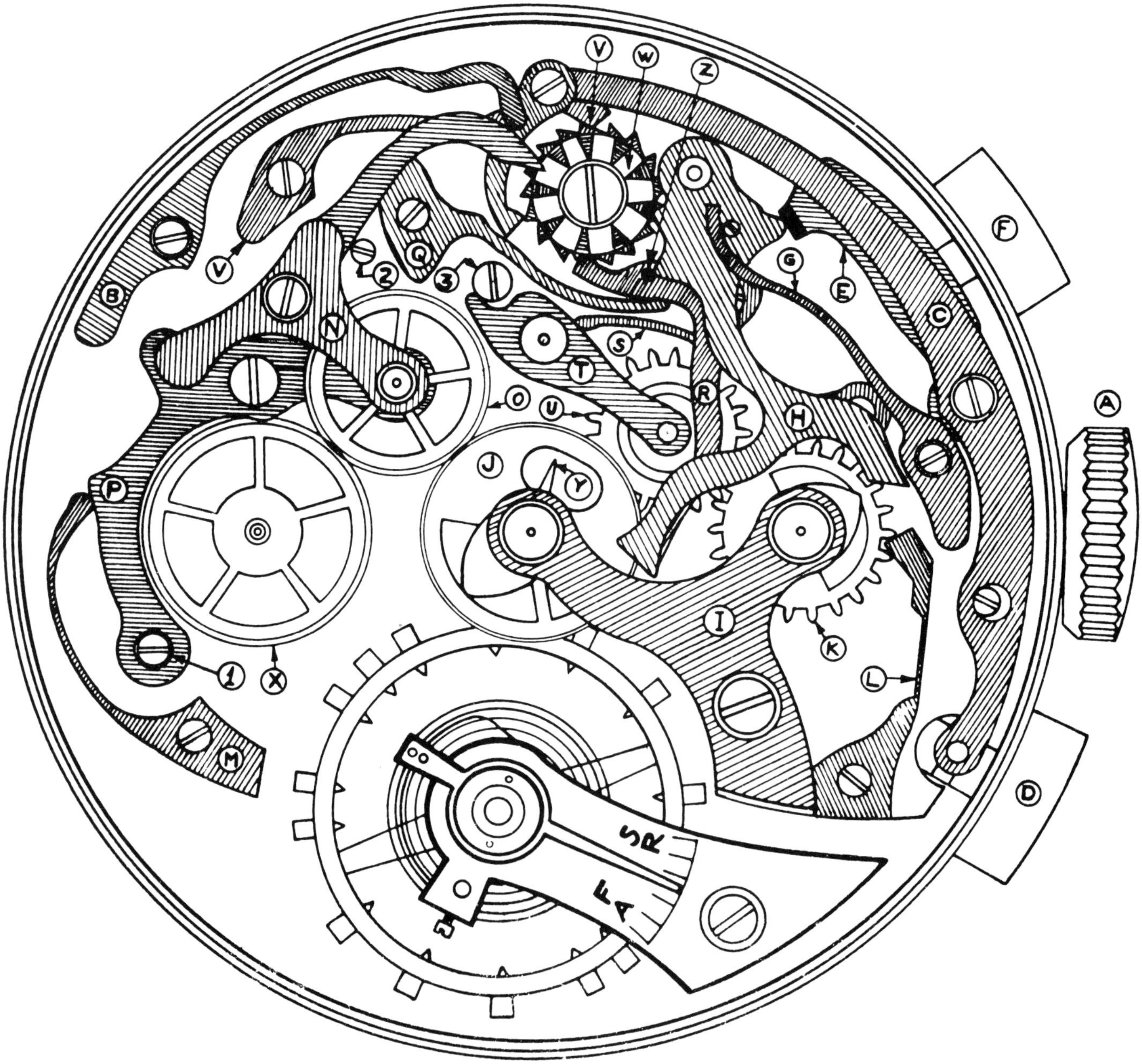

HOW THE BULOVA 13 AH CHRONOGRAPH FUNCTIONS

Before disassembling a chronograph or any complicated watch, it is wiser to study the functioning of the component parts rather than to try to memorize the locations of these parts.

The following explanation should be read very carefully, referring to the very accurate drawing of a Bulova 13 AH. When you understand how every part of the Bulova 13 AH functions, you can then proceed to study other chronographs in the same manner. You will find that basically, all chronographs function in the same way. This will enable you to replace the parts of the chronograph by their relationship to each other. Reliance on memory would only result in confusion and loss of time. A full understanding of the inter-relationship of the parts will also simplify the adjusting of the chronograph mechanism.

To Engage the Sweep Second Heart Wheel

When you press the button, (D) the operating lever (C) pulls the pillar wheel (W) to the next ratchet tooth, causing the coupling lever (P) to drop between the vertical teeth of the pillar wheel. This action makes it possible for the coupling wheel (O) to engage the sweep second heart wheel (J) and the driving wheel (X). Power to turn the sweep second heart wheel is received from the driving wheel (mounted on the upper fourth wheel pivot) through the coupling wheel.

The tension of the coupling lever spring (M) against the coupling lever keeps the coupling wheel engaged with the sweep second heart wheel.

Stud #1 adjusts the depthing between the coupling wheel (O) and the driving wheel (X).

Stud #2 adjusts the depthing between the coupling wheel (O) and the sweep second heart wheel (J).

The pillar wheel spring (V) applies tension against the ratchet teeth of the pillar wheel (W) and so locks the wheel in its particular position.

To Disengage the Sweep Second Heart Wheel

When button (D) is pressed a second time, the operating lever (C) pulls the pillar wheel (W) to the next ratchet tooth, so that now the coupling lever (P) rests against a vertical tooth of the pillar wheel. This action disengages the coupling wheel from the sweep second heart wheel.

Because a continuous spring tension is applied against the operating lever (C) by the operating lever spring (B) the operating lever and the button (D) are always returned to their original positions immediately after the button has been pushed in.

At the time the sweep second heart wheel is disengaged, one end of the brake (R) falls between the vertical teeth of the pillar wheel, causing the other end of the brake to lock against the sweep second heart wheel. In this way, the sweep second heart wheel and its hand on the dial, are stopped instantaneously with the disengaging of the coupling wheel from the sweep second heart wheel.

When the sweep second heart wheel is again engaged, by once more pushing the button (D), the brake (R) will now be against a vertical tooth of the pillar wheel. This action will disengage the brake from contact with the sweep second heart wheel.

Spring tension for the brake is provided by the brake spring (Q).

How the Minute Recording Heart Wheel Functions.

Fixed to the shaft of the sweep second heart wheel (just under the wheel) is a finger (Y). With every complete revolution of the sweep second heart wheel, the finger turns the sliding gear (U) one tooth. The sliding gear which is meshed with the minute recording heart wheel (K) will therefore move the minute recording heart wheel one tooth.

On the dial, a hand which is mounted on the minute recording heart wheel will register a minute for each revolution of the sweep second heart wheel.

The minute recording jumper (L) applies a fine spring tension on the minute recording wheel to prevent it from wavering. However, when the recording heart wheel is rotated one tooth, the minute recording jumper will be lifted gently from its position between two teeth of the minute recording wheel (K) and dropped between the next pair of teeth. In this way, the minutes are accurately registered by the minute recording hand on the dial.

Stud #3 adjusts the depthing between the sliding gear (U) and the finger on the sweep second heart wheel (Y).

Zeroing the Chronograph

To zero the chronograph, push the button (F). This action forces the zero push piece (E) against the hammer (H) at the same time, the hammer displaces the sliding gear bridge (T) disengaging the sliding gear (U) from contact with the finger. Also with this action, both feet of the hammer slide against the flat sides of the hearts, located on the sweep second heart wheel (J) and the minute recording wheel (K), returning the minute recording hand and sweep second hand both to zero.

During the zeroing action, the "chin" of the hammer (H) snaps against an upright pin (Z) attached to the brake (R). This raises the brake from the sweep second heart wheel so that the brake will not injure the fine teeth of the sweep second heart wheel.

A continuous tension is maintained by the hammer spring (G) against the hammer (H) so that the hammer will return to its original position when pressure is released from the button (F).

When the sweep second heart wheel is engaged, the pointed part of the hammer rests on top of a vertical tooth of the pillar wheel. This prevents zeroing while the sweep second heart wheel (J) is engaged with the coupling wheel, at the same time the point of the brake (R) is resting against a vertical tooth of the pillar wheel allowing free rotation of the sweep second heart wheel.

The sliding gear spring (S) applies a continuous tension against the sliding gear bridge (T) making it possible for the bridge and sliding gear (U) to return to original positions when pressure is removed from the button (F).

BULOVA 13 AH

DISASSEMBLING & ASSEMBLING A CHRONOGRAPH

NOTE: All screws should be pressed into a piece of flat pithwood in the order in which they are removed (see fig. 1). String all chronograph parts on wire in order in which they are removed. These parts should correspond with the screws placed in the pithwood. When assembling start with #24 using the last screw inserted in the pithwood.

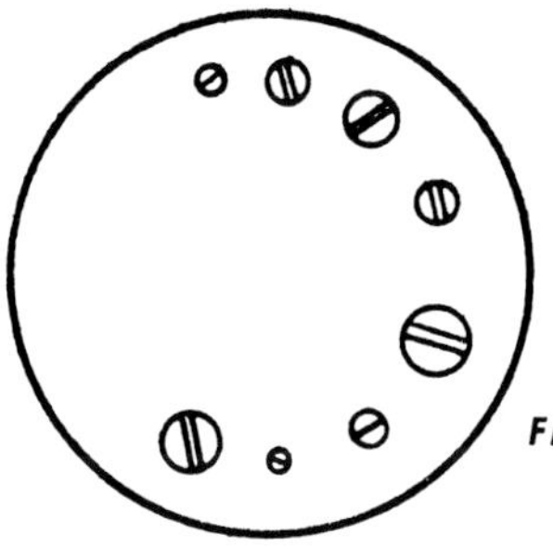

Fig. 1

1. Remove hands

2. Remove stem

3. Remove the operating lever spring (see fig. 2)

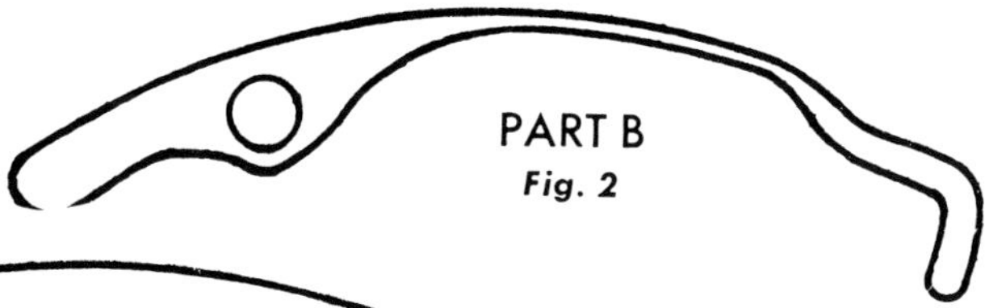

PART B
Fig. 2

4. Remove the operating lever and button (D) (long pin) (see fig. 3). Button (F) zeroes heart and button (D) engages and disengages the coupling wheel and the sweep second heart wheel.

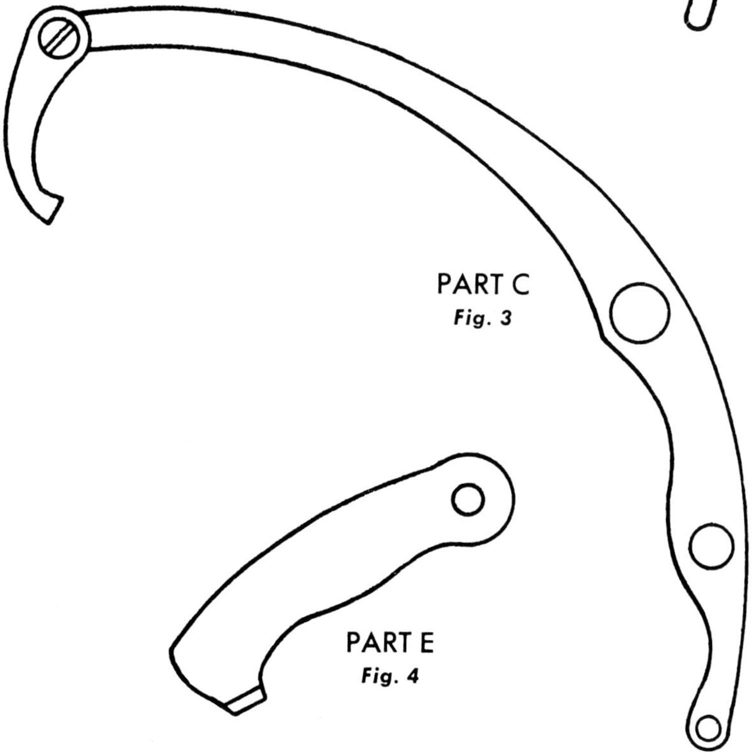

PART C
Fig. 3

5. Pick up zero push piece under preceding lever (see fig. 4).

6. Remove case screws

PART E
Fig. 4

7. Remove movement, dial, hour wheel, and cannon pinion.

8. Remove button (F) (short pin)

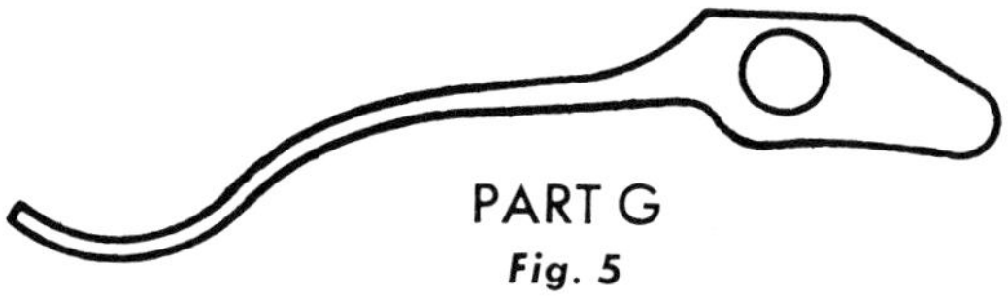

PART G
Fig. 5

9. Replace stem, tighten setting lever screw

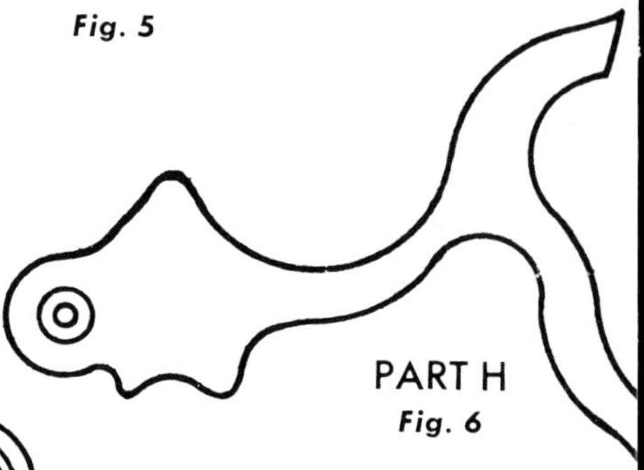

PART H
Fig. 6

10. Remove balance assembly.

11. Remove hammer spring (see fig. 5)

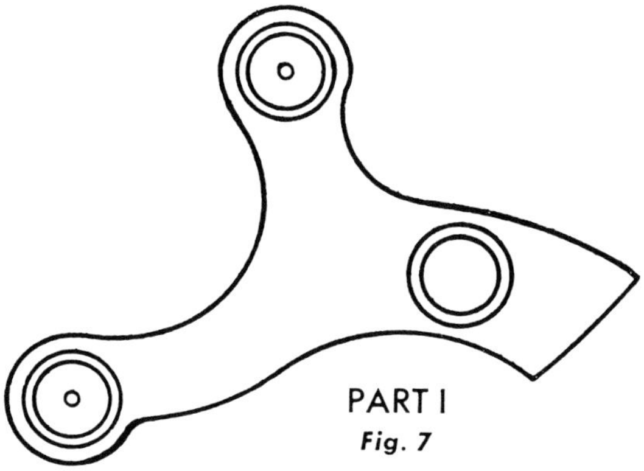

PART I
Fig. 7

12. Remove hammer (see fig. 6)

13. Remove chronograph bridge (see fig. 7)

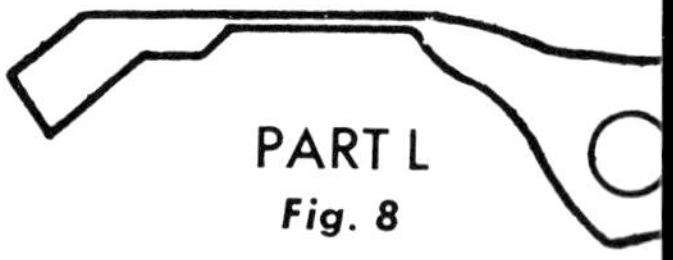

PART L
Fig. 8

14. Pick up sweep second heart wheel and minute wheel recording heart

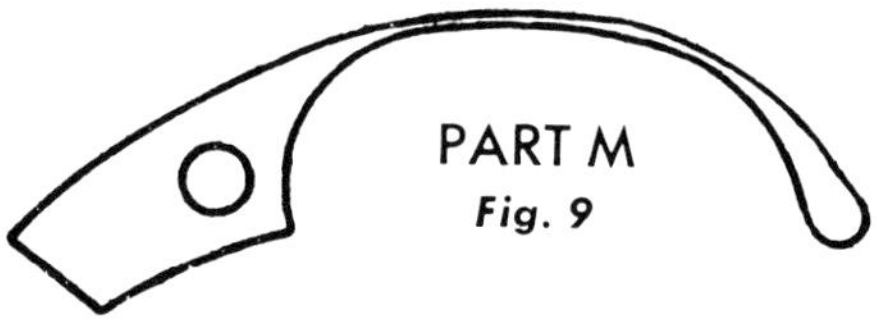

PART M
Fig. 9

15. Remove minute recording jumper against minute recording wheel (see fig. 8)

PART N
Fig. 10

16. Remove coupling lever spring (see fig. 9)

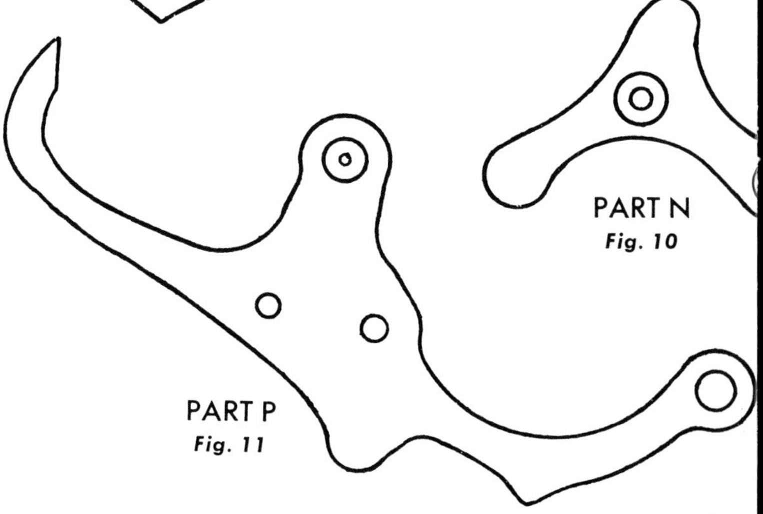

PART P
Fig. 11

17. Remove coupling wheel bridge (see fig. 10)

18. Unscrew the screw on the coupling lever (not the eccentric stud which is eccentric and adjusts the meshing of the teeth between the coupling wheel and the heart wheel) and remove coupling lever (see fig. 11)

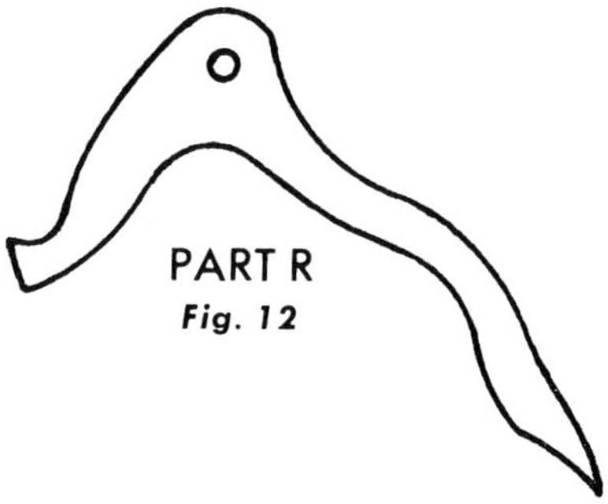

PART R
Fig. 12

PART S
Fig. 13

19. Remove brake lever (see fig. 12)

PART T
Fig. 14

20. Remove sliding gear spring (see fig. 13)

PART Q
Fig. 15

21. Remove sliding gear bridge (see fig. 14)

22. Remove brake spring (see fig. 15)

PART V
Fig. 16

23. Remove pillar wheel spring (see fig. 16)

24. Remove pillar wheel (see fig. 17)

PART W
Fig. 17

NOTE: Disassembling and assembling the rest of the parts is the same as in an ordinary watch. Check the watch carefully before assembling the chronograph parts.

To oil the chronograph parts use a piece of pegwood, moisten with oil and *MOISTEN, NOT OIL, all posts which are* used as bearings for the levers. *MOISTEN, DO NOT OIL,* all parts of chronograph which slide against each other.

Comparison of

HEAT TREATING PROCESSES FOR WATCHMAKERS

Used for Drill Rod (1.00% Carbon)

PROCESS	PURPOSE	TEMPERATURE & ZONE	COLOR	RATE OF COOLING	METHOD OF COOLING
Hardening	To harden in preparation for Tempering	Critical (1450°-1500°) F	Dark Orange	Rapid	Brine, Water, or Oil
Tempering	TO: 1. Relieve Internal Stresses 2. Reduce Brittleness 3. Obtain Specific Properties a. Machinability b. Flexibility c. Toughness 4. Refine grain structure	*Reheat* Below Critical (300° F-950° F) NOTE: Steel must first be hardened before it can be tempered.	Light Straw to Light Blue	Any Convenient rate depending on method of heating	1. AIR—If entire piece was heated uniformly. 2. OIL or WATER—If tempering is done directly over open flame or if job was not heated uniformly because of size or shape.
Annealing (Full)	TO: 1. Completely recondition the steel 2. Soften the steel	High into Critical (1550° F)	Light Orange	Very, very slow	Raise up out of flame very slowly
Annealing (Process)	To soften the steel in preparation for a subsequent machining operation	Any temperature (Usually below critical) (600°-1200° F)	Usually Light Blue	Slow Preferred	Air

HARDENING QUENCHES

RATE OF OLING	QUENCH	HARDNESS	CHARACTERISTICS	SAFETY	USE
FAST	Brine (Salt Water 9% Salt by Weight)		1. Maximum Hardness 2. Best Scale Removal 3. Less Tendency to Leave Soft Spots 4. Wets Faster than plain water		Cutting Tools (gravers, tool bits) Used for low Carbon Steels and whenever water alone would not produce the desired hardness
DECREASES	Water	DECREASES ↓	1. Yields Hardness Values slightly under those of Brine 2. Clean surface 3. Bubbles collecting on surface at large sections may cause soft spots 4. Tends toward hard outer surface with a tough inner core. 5. Cheapest and easiest to obtain.		Used whereever high hardness or a clean surface is required.
	Oil		1. Less Danger of Cracking 2. Less Danger of Warping 3. Deeper Hardness penetration 4. More uniform surface hardness. 5. Increased toughness. 6. Fewer Internal Stresses.	INCREASES ↓	1. Parts having intricate shapes or sudden dimensional changes (stems, winding & setting parts) 2. Springs 3. Tools, Levers and parts requiring maximum strength. 4. Hardening steel which is to be machined after hardening (steel for balance staffs)
OW	Air		Maximum safety with a minimum of distortion. Used only when hardening special Air Hardening Steels. NOT SATISFACTORY FOR DRILL ROD		For tempering where entire piece has been heated uniformly.

EFFECT OF TEMPERING TEMPERATURE UPON THE PHYSICAL PROPERTIES OF HARDENED DRILL ROD (1.00% CARBON)

TEMP. 0° F APPROX.	COLOR	HARDNESS	PHYSICAL PROPERTIES	TOUGHNESS	APPLICATION
300°	Light Straw		Stress Relieved, Very hard, Brittle		Cutting Tools, Gravers
340°	Straw		Hard, Strong		Cutting Tools requiring greater strengths:— (Undercutting tools, Boring Tools, Taps, Drills.) Watch parts requiring hardness and strength (Winding Pinion, Clutch, Click, Clutch Lever.)
375°	Brown	DECREASES	Best combination of Hardness and toughness		Parts and Tools requiring Hardness, Strength and Resistance to shock. (Winding Stem, Ratchet and Crown Wheels, Setting Lever, Chronograph Pillar Wheel Jumper.)
450°	Purple		Brittle		No Practical Use.
600°	Dark Blue		Flexible	INCREASES	Light Springs (Clutch, Click, Yoke or Setting Bridge.) Tools requiring strength and flexibility (Colleting Arbors, Screw Driver Blades.)
650°	Light Blue		Increased Flexibility Machinable		Heavier Springs and springs of lighter tension (Chronograph Hammer Spring, Minute recording jumper.) Steel and parts that are to be machined after hardening and tempering (Steel for balance staffs, stems when slot will be cut after hardening

DECIMAL AND MILLIMETER EQUIVALENCY CHART

NO.	FRACT.	DEC.″	M/M
		.0039	.100
		.0059	.150
		.0078	.200
		0098	.250
		.0118	.300
80		.0135	.343
		.0137	.350
79		.0145	.368
	1/64	.0156	.397
		.0157	.400
78		.0160	.406
		.0177	.450
77		.0180	.457
		.0196	.500
76		.0200	.508
75		.0210	.533
		.0216	.550
74		.0225	.571
		.0236	.600
73		.0240	.609
72		.0250	.635
		.0255	.650
71		.0260	.660
		.0275	.700
70		.0280	.711
69		.0292	.741
		.0295	.750
68		.0310	.787
	1/32	.0312	.794
		.0314	.800
67		.0320	.812
66		.0330	.838
		.0334	.850
65		.0350	.889
		.0354	.900
64		.0360	.914
63		.0370	.939
		.0374	.950
62		.0380	.965
61		.0390	.990
		.0393	1.000
60		.0400	1.016
59		.0410	1.041
58		.0420	1.066
57		.0430	1.092
		.0433	1.100
56		.0465	1.181
	3/64	.0468	1.191
		.0472	1.200
		.0511	1.300

NO.	FRACT.	DEC.″	M/M
55		.0520	1.320
54		.0550	1.397
		.0551	1.400
		.0590	1.500
53		.0595	1.511
	1/16	.0625	1.588
		.0629	1.600
52		.0635	1.612
		.0669	1.700
51		.0670	1.701
50		.0700	1.778
		.0708	1.800
49		.0730	1.854
		.0748	1.900
48		.0760	1.930
	5/64	.0781	1.984
47		.0785	1.993
		.0787	2.000
46		.0810	2.057
45		.0820	2.082
		.0826	2.100
44		.0860	2.184
		.0866	2.200
43		.0890	2.260
		.0905	2.300
42		.0935	2.374
	3/32	.0937	2.381
		.0944	2.400
41		.0960	2.438
40		.0980	2.489
		.0984	2.500
39		.0995	2.527
38		.1015	2.578
		.1023	2.600
37		.1040	2.641
		.1062	2.700
36		.1065	2.705
	7/64	.1093	2.778
35		.1100	2.794
		.1102	2.800
34		.1110	2.819
33		.1130	2.870
		.1145	2.900
32		.1160	2.946
		.1181	3.000
31		.1200	3.048
		.1220	3.100
	1/8	.1250	3.175
		.1259	3.200
30		.1285	3.263

NO.	FRACT.	DEC.″	M/M
		.1299	3.300
		.1338	3.400
29		.1360	3.454
		.1377	3.500
28		.1405	3.568
	9/64	.1406	3.572
		.1417	3.600
27		.1440	3.657
		.1456	3.700
26		.1470	3.733
25		.1495	3.797
		.1496	3.800
24		.1520	3.860
		.1535	3.900
23		.1540	3.911
	5/32	.1562	3.969
22		.1570	3.988
		.1574	4.000
21		.1590	4.039
20		.1610	4.089
		.1614	4.100
		.1653	4.200
19		.1660	4.216
		.1692	4.300
18		.1695	4.305
	11/64	.1718	4.366
17		.1730	4.394
		.1732	4.400
16		.1770	4.496
		.1771	4.500
15		.1800	4.572
		.1811	4.600
14		.1820	4.623
13		.1850	4.699
		.1850	4.700
	3/16	.1875	4.762
		.1889	4.800
12		.1890	4.801
11		.1910	4.851
		.1929	4.900
10		.1935	4.915
9		.1960	4.978
		.1968	5.000
8		.1990	5.055
		.2007	5.100
7		.2010	5.105
	13/64	.2031	5.159
6		.2040	5.182
		.2047	5.200
5		.2055	5.220

DECIMAL AND MILLIMETER EQUIVALENCY CHART

NO.	FRACT.	DEC.″	M/M
		.2086	5.300
4		.2090	5.308
		.2125	5.400
3		.2130	5.410
		.2165	5.500
	7/32	.2187	5.556
		.2204	5.600
2		.2210	5.613
		.2244	5.700
1		.2280	5.791
		.2283	5.800
		.2322	5.900
A		.2340	5.944
	15/64	.2343	5.953
		.2362	6.000
B		.2380	6.045
		.2401	6.100
C		.2420	6.147
		.2440	6.200
D		.2460	6.248
		.2480	6.300
E	1/4	.2500	6.350
		.2519	6.400
		.2559	6.500
F		.2570	6.528
		.2598	6.600
G		.2610	6.629
		.2637	6.700
	17/64	.2656	6.747
H		.2660	6.756
		.2667	6.800
		.2716	6.900
I		.2720	6.909
		.2755	7.000
J		.2770	7.036
		.2795	7.100
K		.2810	7.137
	9/32	.2812	7.144
		.2834	7.200
		.2874	7.300
L		.2900	7.366
		.2913	7.400

NO.	FRACT.	DEC.″	M/M
M		.2950	7.493
		.2952	7.500
	19/64	.2968	7.541
		.2992	7.600
N		.3020	7.671
		.3031	7.700
		.3070	7.800
		.3110	7.900
	5/16	.3125	7.938
		.3149	8.000
O		.3160	8.026
		.3188	8.100
		.3228	8.200
P		.3230	8.204
		.3267	8.300
	21/64	.3281	8.334
		.3307	8.400
Q		.3320	8.433
		.3346	8.500
		.3385	8.600
R		.3390	8.611
		.3425	8.700
	11/32	.3437	8.731
		.3464	8.800
S		.3480	8.839
		.3503	8.900
		.3543	9.000
T		.3580	9.093
		.3582	9.100
	23/64	.3593	9.128
		.3622	9.200
		.3661	9.300
U		.3680	9.347
		.3700	9.400
		.3740	9.500
	3/8	.375	9.525
V		.3770	9.576
		.3779	9.600
		.3818	9.700
		.3858	9.800
W		.3860	9.804
		.3897	9.900

NO.	FRACT.	DEC.″	M/M
	25/64	.3906	9.922
		.3937	10.000
X		.3970	10.083
		.3976	10.100
		.4015	10.200
Y		.4040	10.261
		.4055	10.300
	13/32	.4062	10.319
		.4094	10.400
Z		.413	10.490
		.4133	10.500
		.4173	10.600
		.4212	10.700
	27/64	.4218	10.716
		.4251	10.800
		.4291	10.900
		.4330	11.000
		.4370	11.100
	7/16	.4375	11.112
		.4409	11.200
		.4448	11.300
		.4488	11.400
		.4527	11.500
	29/64	.4531	11.509
		.4566	11.600
		.4606	11.700
		.4645	11.800
		.4685	11.900
	15/32	.4687	11.906
		.4724	12.000
		.4763	12.100
		.4803	12.200
		.4842	12.300
	31/64	.4843	12.303
		.4881	12.400
		.4921	12.500
		.4960	12.600
	1/2	.500	12.700
		.5039	12.800
		.5078	12.900
		.5118	13.000
	33/64	.5156	13.097